Introduction to Construction Operations

Introduction to Construction Operations

RICHARD PATRICK MAHER
Texas A & M University

A Wiley-Interscience Publication

JOHN WILEY & SONS

New York · Chichester · Brisbane · Toronto · Singapore

Library of Congress Cataloging in Publication Data:

Maher, Richard Patrick.
   Introduction to construction operations.

   "A Wiley-Interscience publication."
   Includes index.
   1. Building.   2. Construction industry—Management. I. Title.

TH146.M26   624'.068   82–1884
ISBN 0–471–86136–7   AACR2

Printed in the United States of America

10 9 8 7 6 5 4 3

Preface

University programs directed exclusively to teaching the future contractor are a relatively new experiment in higher education. I do not know of any that existed prior to the turn of this century. Traditionally, the various facets of construction were taught within the disciplines of civil engineering and, to a lesser degree, architecture. Now, however, social and economic demands on the construction process have resulted in programs specifically designed to develop and train future building constructors and contractors to also be professional managers and operators.

These programs are designed primarily to provide the construction process with trained contractors. This has required a considerable change in the educational process: adjunctive construction courses offered in engineering and architecture were traditionally thought to be sufficient to train people who go into construction and its contracting, but now new programs consider construction a specific discipline.

The construction industry also has changed its thinking as to where managers and operators are to be developed. Thirty years ago the industry preferred to acquire most of its management and supervisory personnel directly from the building trades; but advances in technology, coupled with the economic mandate for fast, efficient building at cost, have forced the industry to realize the value of higher education as a means of developing these types of personnel.

One of the unfortunate results of the rapid growth of construction education has been the presumption of academia that sophisticated and specialized course subject matter is needed for the development of the construction professional. I do not agree with this presumption: instruction in the fundamentals of the various aspects of the construction process, with adequate preparation in the arts and sciences, is what is needed to produce competent professionals. The fundamentals of contracts and contracting procedures (the word *contract* after all is contained within the word *contractor*), management principles, and operations are elemental areas of learning in the new discipline. The teaching of sophisticated courses in any of these areas before the fundamentals are learned is not the way to go about training future professionals. While sophisticated techniques are very valuable to the construction process as a whole, they should be taught only after the student has been instructed in more basic fundamentals, and this instruction should start from the very beginning of his learning experience. It is this matter of starting with the "basics" that motivated me to write this book, which examines the basics of just one of the areas of construction—operations.

My desire is to show not only *how* the typical construction operational organization and some of its more important operations function but, just as importantly, *why* they function in the way they do.

I hope that the student of the construction process will benefit from reading this book and that after becoming familiar with the development and procedures of construction operations he will go on to learn and develop advanced techniques in this field of the construction process.

Masculine pronouns are used for convenience throughout this book, although I recognize that women are now entering the construction process in rapidly growing numbers.

RICHARD PATRICK MAHER

College Station, Texas
January 1982

Acknowledgments

To the administration and faculty of the College of Architecture and Environmental Design at Texas A&M University for providing the setting; to my colleagues in the Department of Building Construction of the College for their help and encouragement; to the late Professor Ira E. Montgomery, Jr. for the idea; to Mrs. Marie Prihoda, Miss Kathryn Hansalik, Miss M. I. Helwick, Mrs. Norah Albright, Mrs. Brenda Ryan, Mrs. Jacque Cysewski, Miss Nancy S. Maher, and Mrs. Mary Ellen Reese for their efforts in the physical preparation of the manuscript.

Contents

Introduction to
Construction Operations

Part I

General Principles

1

Comments on the Study of Operations

1.1 INTRODUCTION

The objective of this book is to teach the basic principles of construction operations. It is directed specifically toward the student of construction in the university and in general to all persons interested in the construction process.

This book will endeavor to describe to those interested in the general contracting segment of the construction process how things are done by its operators to get a building constructed within the process. This is not a text on the management of construction: management consists in telling others what or when to do something. Operations consist of actually doing the things required by the process.

Other readers who are in some way participants in the total construction process should be interested in this book to learn not only how buildings are physically constructed within the process but also the thinking processes involved in the development of construction operations. This book should, for example, be of value to the architectural or engineering student who wishes to know why a specific construction operation exists and how it is developed and used. Practitioners in the subcontracting and supply aspects of the industry are a vital segment of the construction process as well. This book may help them gain insight into how other construction operators think and operate. Any insight into the various facets of the construction process through another segment of this process will be valuable to those involved in the total construction process. We hope that all the readers will learn why a particular operation functions in the way it does and that they will realize that construction operations are for the most part the industry's solutions to problems that are caused by the unique nature of the construction process.

The construction industry can use all the help it can get in continuing to solve troublesome and persistent problems of the construction process. This help may indeed come from the various closely related but separate segments of the industry, as a result of a fuller understanding and appreciation of the problems and attempts within each discipline to solve them. This text might also be an aid to the legal and financial professions, helping them to better understand how construction operations are developed and work in the construction process from the point of view of the construction contractor operator.

This book does not attempt to instruct the nongeneral contracting segments or disciplines within the construction process about their business practices or methods of operations; that would be presumptuous. Rather, this book intends to guide and instruct future construction operators about the operations through which they perform their function of constructing things. Hopefully, it will be useful to those about to enter the industry and who aspire to become managers there.

1.2 WHY STUDY CONSTRUCTION OPERATIONS?

In order to justify the expenditure of time in the study of construction operations we must ask the following questions:

1. What makes *construction* operations different from other manufacturing or production operations?
2. Why study construction operations formally as a set course of study?

There are unique differences between the construction process and other manufacturing or production processes, as well as similarities. The agricultural, automobile manufacturing, and construction industries, for example, all make a product. In this way they are similar; but they do not produce their products in the same way.

The construction industry's ways of producing its product are distinctly different from those of other industries. These differences are evident when a comparison is made, and they consist of the following:

1. The uniqueness of the product, evidenced by the individuality of the building or structure.
2. The medium of production, that is, the contract.
3. The importance of time to the contract.
4. The indeterminate physical factors and conditions that govern and control the construction process.

No two construction projects can be identical in either form or construction. The product of construction is always truly unique. This is not to say that one building's design cannot be identical to another, but the construction of any design always calls for its adaptation to the site, climate, seasons, and so forth. The construction process of building a structure can never be the same: the weather, seasons, site conditions, time factors, and so forth are different in each case. This variability distinguishes the construction process from other manufacturing processes, which mass-produce their products in a standard form.

The only way to perform the construction process today is through its medium—the contract. This is a unique feature of the construction process: the contract, besides establishing binding obligations, also forms the basis for its performance. Performance of a contract presupposes the need for operations. It is true that other industries use the contract to regulate performance, but no industry uses the contract to the same extent or in the same way as does the construction industry.

Time is an important factor in most, if not all, other industrial processes; but few industries are governed by time considerations in the same manner as is the construction industry. Whether or not it is specifically stated in the contract, time is of the essence in the construction process and is regarded very differently there than it is in other industrial processes.

There are always indeterminate physical factors that affect the construction process which do not affect other manufacturing processes. Weather, for example, affects farming and construction more than it does automobile manufacturing; in this respect farming and construction are similar. But it is also true that there are great differences between managing farm labor and managing construction labor. Different conditions and physical factors call for different approaches to operations in the various industrial processes.

Now we must answer the second question. Construction operations are now being studied formally in a set course of study because the construction industry requires it. The significant growth in the formal educational programs for the study of construction as a special discipline is a strong indication of this demand. As in other disciplines, industries, and professions, the construction industry simply requires more knowledge from its entering personnel and, particularly, its potential management personnel.

This raises the age-old debate of hands-on experience versus formal training. Very few industries have functioned for as long or as well as the construction industry without having specific formal educational programs to train their personnel. In fact, experience has always been highly regarded by the industry. The industry's thinking has changed, however, because of the following:

1. Rapid advancements in construction technology.
2. Rapid and continuing changes in contract systems, types, and their uses. The vast field of contract law and its application has been affected as well.
3. Ever-changing economic factors that directly or indirectly affect and control the construction industry.

These reasons have led to the requirement of formal education in the construction process, including specific education in construction operations.

1.3   GENERAL COMMENTS ABOUT CONSTRUCTION OPERATIONS

Definitions

The verb *to operate* is defined as "to be in action so as to produce work." The noun form, *operation,* means "the act, process, or method of operating." By simply applying the word *construction* to the noun form, we arrive at the definition of the expression *construction operations* as used in this book. Construction operations are the acts, processes, procedures, and methods used to produce construction work.

Another word that must be defined from the start is *process. Construction process,* for example, is a term that is frequently used throughout this text. The word *process* means "a particular method of doing something, generally involving a number of operations." Thus in this text *construction process* means "the method used to construct structures, generally involving a number of operations done within the parameters established by a contract."

Approaches to the Study

In this book we use two approaches to study construction operations:

1. One approach is to study the traditional divisions of the construction organization, which is the study of operational organization.
2. The other approach is to study some of the basic processes and procedures that have been traditionally developed from the need to implement the

construction contract, including those that have proven to be valuable tools in producing construction work. This is the study of operational procedures and processes.

1.4 THE CONTRACT AND CONSTRUCTION OPERATIONS

The construction contract is the only medium through which the industry builds today. Therefore it is necessary for the construction operator to know about contracts. For the purposes of this book, however, the study of contracts will be kept separate from the study of construction operations. While the contract can never be completely separated from operations, it forms the basis for such an extensive area of study that space in this book does not allow for much more than its mention in these few lines. It must be understood that the contract is assumed to be the starting point in the study of operations: it is the common denominator of the whole construction process; it starts the construction process and its effects continue through its completion.

Certain things must be assumed about the contract when studying construction operations, after which we will leave the contract alone as much as possible:

1. The contract is the only medium through which construction is accomplished.
2. The contract is the starting point for construction operations and remains the basis for the development and use of all construction operations.
3. All of the procedures and processes in construction operations result in one way or another from the requirements of the contract.
4. Most, if not all, of the separate divisions of responsibility in the construction operational organization result from practical demands made necessary by the contract's performance functions.

Construction operations are assumed here to be the children of the construction contract. No further conscious attempt will be made in this text to explain or justify the relation between the contract and operations.

1.5 TRADITION AND CONSTRUCTION OPERATIONS

The role of tradition in the field of construction operations is another concept the reader should understand. This book is not about a new field in the construction process; rather, it examines proven organizational operational structures and construction operational processes and procedures. We use these existing developments to explain the "hows and whys" of construction operations. No attempt is made to develop or advance new ideas about organizations and procedures. Understanding the role of tradition in the construction process is therefore essential to an understanding of the contents of this book.

Example. The operational organization division known as job supervision is not a new one; no doubt the pyramids had a job superintendent. Tradition was

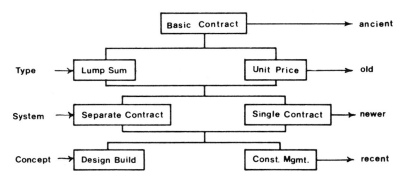

Figure 1. A time chart showing the development of contract types, systems, and concepts from the basic contract.

important in developing the role of the job superintendent. Present means and methods used in job supervision have developed through tradition and have evolved into established construction operational procedures and processes. Since we are studying construction operations, the importance of tradition in their development must be emphasized from the beginning.

Tradition and experience have refined the construction process and continue to do so. We observe that the demands and requirements of the current era are forcing new developments in construction operations as well as in other parts of the construction process; but these new developments are still based on traditional concepts.

Example. The contract process is an old one. Ancient contract processes used in the beginning of contract construction have evolved into different types of contracts, used for special purposes. Fitting types of contracts into systems of contracts is a more recent development but is still fairly old. Fitting contract systems into new concepts in the construction process is a very new development of the contract subprocess. Figure 1 shows the development of the contract subprocess.

If there is to be a full understanding of the most recent introduction of new contract concepts into the construction process, the history of the development of contract systems, methods, and types must be carefully studied. This study will enable the operator to have a better understanding of the development of operations within the contract subprocess.

1.6 "THINKING CONSTRUCTION"

This book has another major objective—to put the reader into a state of "thinking construction." *Thinking* means "to center one's thoughts on, to have one's mind full of." This "thinking" can then be adapted to the development and application of construction operations to fit any situation or set of circumstances encountered. The understanding of "basic principles" is necessary for this adaptation.

This book examines specific areas of the operational organization, as well as certain specific operational processes and procedures. It does not, however, attempt to further subdivide the organization, nor does it cover all of the operational processes and procedures.

Specific examples of operational organization and procedures are used to show the reader how they work and to give him the opportunity of learning the basic general principles of construction operations. The reader may then adapt his knowledge to any construction operational problem. He will develop a way of thinking that will be useful in devising and developing his own methods in construction operations. Construction operations are easily observed and readily remembered; they can be learned, stored away, and then adapted and applied to particular construction situations. This is the process of "thinking" construction.

Conceptions and Misconceptions

It is necessary to identify and examine some common conceptions about construction operations and to then establish their validity and try to correct any errors in them that might adversely affect good construction thinking:

1. The mastery of construction operation skills depends to a large extent on personal experience in operations.
2. Construction operations can be formalized.
3. Rapid advances in the construction process are currently taking place and they require learning specifics rather than acquiring a basic and general knowledge of construction operations.

All of these statements are correct to some extent, but careful examination will also reveal certain inconsistencies.

Some personal experience is necessary and valuable in learning any skill, and it may be extremely important in acquiring skills in construction operations. But personal experience is by no means the only way to gain expertise: construction operations can and should be learned through the development of thinking processes based on basic principles, and basic principles are most efficiently learned through formal, directed study. It is also a fact that personal experience often creates false conceptions. Learning basic principles through formal study generally eliminates the possibility of misconceptions arising from emotional or psychological interferences, which occur many times in the trial-and-error type of learning.

The statement that construction operations can be formalized is almost entirely false. One may formalize construction methods and procedures to a limited extent, but it is unlikely that construction operations as previously defined can be completely formalized. This is true because of the nature of the product of the operations. It is illogical to assume that the end product of the construction operations—the work—will always be entirely the same. Mastering the subject of construction operations involves formalizing thinking processes rather than mastering specific or formalized methods and procedures. There is no one correct way to perform any specific construction operation; one way may be better than another under a certain set of circumstances, but it would be wrong to believe that the same method of operating is the best in all situations.

For example, even when an established procedure has been found to work under one set of circumstances, specific thinking is still required to see whether it works as well in another situation. Specific thinking is always necessary. To rely on a long-established payment procedure, for example, without considering its application to the particular project at hand would be impractical, as well as inappropriate.

The final common conception to be discussed is the most difficult to examine. There is no doubt that there are rapid advancements taking place in construction technology, financing, law, contracting, and other fields that relate to the construction process, all of which directly affect construction operations. Much time is required just to keep up with these advances, time the construction operator cannot afford to waste; but neither can he afford to ignore these advances. The time he spends investigating the ways in which these advancements apply to his own operations must be rationed: he must not examine new questions about law and financing in relation to each other at great length, for enough of his time will be spent in examining how each relates separately to construction operations. Therefore the constructor must at some point in his thinking start from certain assumptions about law and contracts. Resulting questions must be answered through construction operational thinking rather than legal, contract, or financial thinking. There is no time to interrelate all these parts of the process to each other in developing and using construction operations.

1.7 FINANCIAL OPERATIONS

Financial operations in construction are usually categorized as construction operations. Financial operations, with the exception of payment procedures, are deliberately excluded from this book.

Payment procedures are such an integral part of the production of the product that to exclude them would interrupt the sequence of the study of construction operations. Another reason for including payment operations and procedures is that they are not specifically departmentalized in a construction organization; they are not, for example, solely within the exclusive jurisdiction of the financial operators in the organization. Most of the other divisions of construction operational organization are involved in payments, and a discussion of payments cannot be avoided when these divisions are studied. Other financial operations are excluded, however, because they are too involved to include their examination in this book.

This text presents the general and basic principles of the other divisions of the construction organization and their operations and procedures. The study of financial operations does not lend itself to the same approach, because it requires a detailed study of too many specifics, and this would not be adhering to this text's approach to learning the basics of construction operations in general.

1.8 PARAMETERS FOR THE STUDY

To help the reader relate the contents of this text to something concrete we have established parameters within which construction operations are to be studied.

This book is not written for construction operators in the largest or smallest construction firms, because these firms have probably established their own operational organization, using their own titles for their personnel. However, we do make the following claim as a basic tenet:

Each and every construction operation examined and discussed in this book is performed by someone in every construction company doing multiple work under contract.

Certain specifically named divisions are used and certain terminology applied to personnel in the construction organization for the purpose of presenting some recognizable and convenient format to the reader. This format also gives specific names to people relative to the construction operations they perform and for which they are responsible.

This book has been written for those who do not know or fully understand who performs the construction operations or how construction operations work. These readers need something to visualize so that they can relate the contents of this book to practical application.

The Model Construction Company

The following is a model construction organization that establishes parameters within which we examine construction operational organization, processes, and procedures:

1. It is primarily a general contracting company.
2. It is a growing company.
3. Its increasing sales reflect the growth of the company.
4. Its billed completions are consistent with its volume of sales and financial and organizational structures and capacities.
5. The number of its jobs within its sales volume are spread between small and large jobs, relative to capacity. There are approximately 6 to 12 major jobs in various stages of construction, and 12 to 16 small jobs to be finished within a year.
6. It has maintained its bonding capacity with steady increases thereto.
7. It has the ability to finance the work by its own forces and to financially handle subcontractors in the proportion in which they are used within the volume.
8. It has adequate and competent personnel in the field and office, operating within a planned organizational structure.
9. It engages in the construction of commercial, industrial, institutional, and public works projects.
10. It occasionally acts as a subcontractor in the concrete and masonry trades.
11. It bids primarily with the competitive bid in both the public and private sectors.

Ideally, completions must be accomplished on time and in order. They must be projected and planned by the construction operators responsible for those operations. Projections of completions are necessary for the following business reasons:

1. To allow for new business and volume.
2. To allow for the proper flow of money and the maintenance or increase of bonding capacity.
3. To allow for the full use of production, personnel, equipment, and overhead organization.
4. To allow for profit and thus growth.

Planning Completions

Completions must be planned: they do not just happen by themselves; they require limits and directions. Once planned, they must be accomplished by the construction operators in every division of the construction operational organization. The projection and planning of completions are themselves construction operations. Each operation, in turn, calls for the development and use of other construction operations.

Example. Construction project scheduling is a construction operation that is directly related to the planning and projection of completions. The primary purpose of the schedule is to give visible direction to the constructors as to the building sequence of the project. The schedule can serve other operational purposes as well, one of which is to illustrate a projection and plan for the completions on the project. Construction project scheduling operations serve this purpose, even if one regards the schedule's production sequencing features as the main reason for its formation.

Consistency and Order

The construction operator must also view completions from another standpoint: he must see that they are in order and relatively consistent in money amounts forthcoming for them. Order and consistency are beneficial to the financial, business, and organizational management of the construction company. Ideally, the most orderly and consistent completions are those in which equal divisions of the total contract time are assigned an equal proportion of required completions in construction and money amounts. Thus a $1,200,000 project to be contractually completed in 12 months would have 11 payments of $100,000 each (less retention), plus the final payment, which includes retention. If each project in the company work volume is handled in roughly the same manner, then it becomes easier to manage the total work volume of the organization, particularly in its financial and organizational capacities.

Interferences With Completions

These ideal completion goals are very often unattainable because of certain unpredictable interferences. These interferences are almost always indeterminate in nature: weather and conditions of site would be examples of interferences that often delay completions.

Operational Practices and Completions

There are operational practices that can seriously affect the objectives of completion. It is not good operational procedure, for example, to have significant imbalances in completions at the front or rear end of a project's time allowance, even if these imbalances result from honest assessments of time and work; this often causes imbalance in the management of the construction organization's financial resources. Substantial inconsistencies in the amount of billings within the divided time frame would be an indication of poor operational control over completions.

Illegitimate operational practices can also seriously affect completion objectives.

Example. Illegitimate overbilling and front end loading are the most common deliberate operational practices that adversely affect the objective of completion and result in many problems. Front end loading is the practice of overbilling for costs during the early partial payment phases on the project. This overbilling is not based on actual completions. There can be legitimate reasons for front end loading. One reason would be to minimize the financial burden on the construction company in the beginning stages of the job. The early payment of bond premiums can be an enormous help in cash flow in financial areas if it is paid in full by the owner at the beginning of the job, and it is a good example of a legitimate front end load. However, in most cases billing for money for front end loading taken from subcontract prices is an example of an illegitimate front end load. Overbilling is simply collecting more money than has been earned; it is an artificial and arbitrary overcharging for completed work. Front end loading can simply be overbilling in many cases. Overbilling, except for legitimate front end loading, is very risky and poor operational practice. The main danger in overbilling is that at the end of the project the operator will find that he does not have enough money to pay his materialmen and subcontractors. Overbilling creates false impressions of completions to everyone concerned, even to the operator who has made the overbill.

These kinds of ill-advised operational practices fall mainly within the scope of financial operations in the construction process. They are examined here only as they involve other divisions of the construction operational organization. Of the construction operators discussed in this text, the expeditor is the most involved in overbilling and front end loading practices. Even though the expeditor is directly and extensively involved in the payment process, directions in the matter of overbilling usually come from the financial operators or organization executives.

Conclusions

It is important for the construction operator to realize that he is the instigator and director of operations governing completions. He is responsible for working around and adjusting to indeterminate conditions that restrict the meeting of this operational objective. Operators must develop and use means, methods, processes, and procedures so that their completions are in fact actual completions. Projected and planned completions in all their phases that turn out to be actual completions furnish good proof of the operational competency of the construction organization.

Quality

Building a product is the second essential objective of all construction operations. Meeting this objective also facilitates reaching the objectives of completions and costs. It is foolish for the construction operator to direct his own operations and those of others toward anything less than the best product that he and his organization can furnish. Those entering the construction industry must realize the importance of this objective of quality. The following are the four major reasons for a construction operator to do first-class work.

1. Quality work throughout the construction project gives integrity to the entire building or structure, and it results in a safe and sound building in all its component parts, both while it is being built and when it is used in the future.
2. Quality work makes the administration of the contract easier and faster. The time needed to implement the performance of the contract is minimized. All of the performance provisions of the contract become easier to define and accomplish. Contradictions and adverse interpretations of the contract are less likely to arise, thus making the contract easier to complete.
3. Quality work results in fine buildings and structures and makes the construction operator's task more satisfying to himself, his company, his customer, and the public; it gives him visible purpose and pride in his job. It also builds a good reputation for himself and for his company and creates new business opportunities.
4. Quality work saves time and increases profit. Slovenly work wastes time, talent, and money.

In most cases the judgment of quality work is subjective. Nevertheless, objective standards of quality work are mandatory because the health and safety of the general public must always be assured. The application of objective standards is also mandatory as far as the safety of the workers on the construction site is concerned. There can be no compromise in maintaining these standards in the construction of a building or structure. The construction operator cannot shift the responsibility for an unsafe or unsafely constructed building onto others if he has not done his duty in seeing that its construction has been done properly. This is true not only because of the strict obligations arising out of the contract to build with quality, but because it is simply the economical and professional way to build. No construction organization will continue to function for long if it fails to recognize that building in accordance to codes and technical specifications is required in the construction process. Those organizations or their operators who would cut costs by sacrificing quality will fail, and the sooner the better.

The Contract and Quality

Contractually, the quality of the product is guaranteed in only an indirect way by our present-day construction process. It almost seems as though quality work is assumed to be a result of the contract. There is actually very little in the typical construction contract that refers specifically to the quality of work; usually quality work is forced indirectly through conditions that implement the performance of the contract. Some professionals claim that quality supposedly results from forcing strict compliance with the technical specifications, but, in fact, this sometimes only partially assures quality work.

Example. A technical contract specification on concrete is an example of a specification that does not necessarily insure quality work, even when strictly complied with. It gives requirements for testing and even certain procedures in the methods of placing, curing, and so forth. But if the concrete is to be placed in forms, what does the specification say about the formwork? Usually it says in a very general way that the formwork is to be safe and sound, and the constructor is then assumed to be responsible for this safety and soundness. This responsibility, even though it is not detailed in the contract specifications, is as much involved with quality work as with the physical quality of the concrete itself in the finished product. Another example of this feature of quality construction can be illustrated by examining the construction methods used in building a masonry wall. The masonry structure itself might be of first-class construction, but the scaffolds that were used for building it may have been of the poorest quality. A construction operator must ask himself, therefore, whether the scaffolding, although it is a temporary product of construction, should be of as good a quality as the wall of the permanent product. The answer is that it must.

Let us examine another way in which the contract indirectly forces quality work. Payment is dependent on completions. But how is the quality of the completed work determined? Usually, the construction contract says that it must be acceptable to the owner or his agent; therefore the determination of quality is based on the subjective judgment of such people.

There is a traditional method for determining the acceptable quality of completed work for substantial or final payment; it is called the punch list. The punch list is a listing of uncompleted or unacceptable items of work to be finished or corrected by the contractor before final payment or even payment for substantial completions. The punch list itemizes poor-quality work, and poor-quality work is considered as being contractually uncompleted work. Poor-quality work will not be regarded as completed work by any reasonable person. The punch list is the basis for determining acceptable quality and it thus determines actual completions.

When the construction operator really analyzes what a punch list is and how it affects the three main objectives of construction operations, he will realize that his thinking must be directed towards minimizing its causes and effects. The listed items concern quality work or workmanship, and the operator must remedy both the situations and circumstances that have caused the poor work and, in addition, have the work corrected. This affects completions. The punch list can be the single most expensive part of the construction of a project relative to the unit costs of its component parts and, indeed, its total cost. Operations directed to the punch list's performance in an efficient, orderly, and conscientious manner are essential; otherwise it adversely affects the cost objective.

A Cause of Controversy

There is another reason for the construction operator to give his closest attention to the objective of quality. Poor-quality work causes great controversy and adverse situations within the total construction process of a project, as the following shows.

Example.   When a wall is constructed, its base structure is built first; then comes the finishes on the wall itself, and then the work (both rough and finish) that is connected or abuts to it (ceilings and floors). If the first operation in building this wall is not done correctly and the following work is applied, there will be three serious and troublesome problems to face. Firstly, the initial cause of the problem must be fixed. In many situations this is a very complex task, particularly when concurrent and consecutive construction operations were involved in building the structure, and this becomes even more complex when the building has been done under separate contract arrangements. Secondly, the poor construction must be corrected. The corrections usually create operational problems in both the demolition and rebuilding of the wall. Thirdly, the entire remedial operation must be paid for on an equitable basis. But liabilities cannot necessarily be determined on a contractual basis, therefore many problems arise. The consequences of carelessness cause situations that are ripe for controversy and dispute.

Conclusions

We wish to repeat again for emphasis that of the three main operational objectives, quality is really the only one that cannot be compromised. While completion and cost objectives can almost always be compromised, at least as they relate to each other, quality objectives must be firmly established and then met in their entirety.

Costs

It is reasonable to say that almost anything can be built if there is enough money to build it. In one way or another, society and the economy impose restrictions on the cost of construction today. Various factors govern, control, and limit the costs of the construction product; even the least cost restrictive methods of contracting, that is, the "cost plus" method, involves probable cost restrictions because of budgetary limitations.

Generally, three main types of costs involved in the production subprocess are directly related to a general contractor's construction operations:

1.   The direct labor costs of the construction company's employed forces, including their fringe costs.

2. The direct material and equipment costs necessary for the work of the firm's own forces or directly incorporated by the firm into the project by its own forces, including sales and use taxes and property insurance.
3. The direct costs of subcontracts and material supplies.

There are other direct costs involved in the construction process, such as bond premiums, public liability insurance, and so forth. We recognize that these and other costs can be of a significant amount, but the construction operator cannot control these on a continuing basis as he can the three listed direct costs.

The construction operator attempts to operate at or below the allowed costs in the three main cost categories. His objective of at least meeting estimated cost is attained through the operational controls, procedures, and processes and the team effort of all of the divisions of the operational organization. Of the three costs, the first and second are under the direct control of the general contracting operational organization. The cost of subcontracts and material supplies are controlled directly by the general contractor's operations until they are subcontracted, after which time there remains only an indirect control. This indirect control of subcontractors and suppliers, however, is of extreme importance in most cases in maintaining control of the labor, material, and equipment costs. Its importance should not be minimized when viewing the total cost objective. The construction operator must develop procedures for directing the operations of subcontractors, because their operations are very critical to the completion and cost objectives. The task of directing and managing subcontractors is of such magnitude that it constitutes an entire separate phase of construction operations. Construction operations that govern and direct subcontractor operations are usually of vital importance in attaining all three of the main objectives of completions, quality, and costs.

2.3 CONSTRUCTION OPERATIONS AND THE CONSTRUCTION FIRM

The Type of Firm and the Nature of Its Contracting Services

The size and makeup of the operational organizational structure of a construction firm depend to a great extent on the type of firm and the nature of its work and services. Three types of construction firms will be examined here, all of them falling within the model company parameters established in Chapter 1:

1. The traditional general contracting firm (which can include the firm working as a separate contractor).
2. The broker contracting firm.
3. The design and build firm (which can include the firm working as a construction manager).

The traditional general contracting firm performs certain significant categories of work with its own forces and subcontracts specific categories of work to others. Traditional industrial practice has established which of the trade categories are usually subcontracted in today's construction process.

The broker contracting firm performs its work with little other than field and office overhead forces and subcontracts to other firms all, or nearly all, of the categories of work in the project.

The design and build firm can be either in the mode of a traditional general contracting firm or a broker-type firm, but, in addition, it furnishes and performs additional services. It most commonly offers services in the design subprocess of the construction process, but it may include furnishing financial and business services as well. This type of firm will only be briefly discussed in this book. Since the design/build and new construction management concepts used for building construction today demand greatly expanded operational resources that cannot be fully covered in this text, we will direct our attention primarily to studying only the general contracting- and broker-type firms' requirements for operational organization.

The Nature of Its Work

Categories of Work

When the size and makeup of the operational organization's structure is examined, there is another factor that is important for the construction company to consider; it involves thinking about the nature of the work the firm specializes in performing, so that a structure can be determined which will fit the needs of doing the particular type or types of work. Building construction work falls into the following categories:

1. Institutional work.
2. Commercial work.
3. Industrial work.
4. Utility work and so forth.

A construction firm may perform one, several, or all of these kinds of work. The category of work undertaken also effects the extent and kinds of operational organizational structure the firm requires.

Public Versus Private Work

Another basic characteristic of the firm's work that can affect operational structures and needs is whether it engages in public or private work and to what extent. It can engage in one or both areas. It can engage itself in one area while working the other continuously or only at intervals—or it can constantly alternate between the two.

Specialized Work Within a Category

Another characteristic affecting operational structures would be whether the firm specializes in one particular area within the categories listed above or more than one. For example, a firm engaged in utility work could specialize in the construction of sewage treatment, electrical, telephone, and natural gas facilities within this category of work, or it could specialize exclusively in sewage treatment work.

The nature of the firm's business always strongly influences its operational organizational philosophy, structure, needs, extent, and development.

Basic Criteria for Developing Operational Organization

A contracting firm usually uses two separate criteria in determining basic operational organization needs; both are of equal importance in developing operational capabilities:

1. The amount of work performed by the firm's directly employed forces compared to the actual and projected amounts of work subcontracted within the firm's total volume.
2. The need for expert and competent personnel in the operational organization based on the nature of the work making up the volume of the firm.

Applying the first criteria, it should be obvious that a firm that subcontracts 80% of its volume of work does not demand the same operational structure as the firm which performs the same percentage of its volume of work with its own forces. The broker-type firm needs more office personnel and fewer field forces, whereas a traditional general contracting firm may require the transposed proportions of operational personnel in its structure.

Example. A firm that performs a major portion of its volume of work by its own forces (and this major portion is projected to remain so) needs significant numbers of people in its field operations, namely, superintendents, engineers, foremen, yardmen, cost personnel, and general supervision. Expeditors, estimators, and financial operators are also needed for the office operational functions. A firm that brokers most of its work needs fewer superintendents, expeditors, and estimators, and has even less need for engineers, foremen, and personnel to handle labor costs and payment functions. The broker firm seldom needs yard and equipment personnel. Usually a broker-type company can operate with a much smaller operational organization than can the traditional general contracting firm.

It does not follow, however, that the broker-type firm, because of its lesser need for numbers, can operate with personnel of less quality, expertise, and talent. This must be kept firmly in mind when organizing the operational structure of the broker-type firm.

When making determinations based on the first criteria, further examination of the quality and characteristics of the individual operator who will best suit the requirements of each of the two types of firms must be made. The following questions illustrate the importance of this point in developing operational organizations.

Is the type of construction superintendent needed for running broker work, with its emphasis on subcontracts, the same as the one needed to run work for the general contracting type of construction firm that places heavy emphasis on working its own forces?

Is the work load for the expeditor in the general contracting-type firm likely to be the same as it is for the expeditor in the broker-type firm, even though the expeditor's operational tasks and responsibilities in both types of firms might be quite similar?

Are the operations to be performed in one type of firm more complex and of a nature different than those in the other type?

In any case, the ideal approach in selecting one or another person for the job is to proceed on the basis that the individual selected should be of such ability, personality, and character that he or she will be able to adapt to either type of firm. This would be the ideal approach. The practical approach, however, must consider that individual people (especially talented people) are different in many respects: they differ in their goals, motivations, viewpoints, and orientation to their work. It is best for the individual construction operator to like what he is doing and to be enthusiastic about it. It is better for the firm to have the operator doing what he does best. There are also times in a firm's life when business considerations require the interchanging of personnel into different operations within the operational organization. This can occur because of up- and downturns in the volume of work in the company, or, more importantly, when a firm alternates between traditional general contracting and broker contracting. This calls for determining during the selection process the individual that will best fit into all of the operational divisions of the organization.

When applying the second criteria—the nature of the work in the volume—there are other things to consider about the makeup of the operational organization. Even within the listed categories the nature of the work can be varied: institutional work suggests, for example, schools, hospitals, and museums; industrial work includes manufacturing or process facilities, in-plant or renovation work, warehousing, and other work; and commercial work might consist of hotels, retail facilities, or multiple-housing developments.

This wide variety in the nature of work also affects the development of the construction operational organization and, in particular, the individual traits and characteristics of its personnel. This is illustrated by the following questions about the natures of the various types of construction work:

1. Would the building of an in-plant industrial production facility within an owner's production line process be the same as building an unencumbered school building project relative to the considerations of time and other pressures?

2. Do the same time and pressure problems exist in estimating a sewage treatment plant as in estimating a commercial building, irrespective of whether or not the firm is a broker contractor or a general contractor?

3. Which of the two individual job superintendents would be the best to use in building a substructure job—the pusher (cost-effective) type or the expert mechanic (quality-effective) type? Which one would be more adaptable to building a museum? To building a factory?

Again, when we attempt to find answers to these and a whole range of similar questions, we realize that they could be based on selecting and developing ideal people for an ideal organization. But to answer just these three questions, some criteria must be used to select the best personnel that is based on the real appreciation of people as they are. Not all people, for example, react to time pressures in the same way; not all of them give the same consideration to time or even consider it in the same way. One operator may be strongly motivated by cost considerations but less concerned with quality and appearance; another may be strongly influenced by the appearance and quality of the buildings he constructs, giving much less thought to costs.

The type and nature of the work of the construction firm therefore seriously affect or control the selection of the best personnel for construction operations. They affect the composition and structure of the construction operational organization and determine the required personality and character of the individual construction operator.

2.4   CONSTRUCTION OPERATIONAL CONTROL

Acquiring Control Skills

Many construction operations involve the direction of work forces. Direction presupposes some kind of control over the forces, which can be either of two types—direct or indirect.

The *type* of forces control in the construction process is important to construction firms and their operators, because the type of control exercised usually determines the operational organizational structure required and affects the functions of the construction operator as well. Whether or not the control is direct or indirect is determined by where the responsibility for the direction of the various work forces is placed within either of the two subprocesses of the construction process—the contracting subprocess and the production subprocess. Indirect control over various subcontracting forces, for example, is given to the construction operator by the contracting subprocess; direct control over his own forces is given to him by the production subprocess.

Acquiring skills in exercising the direct control of forces is usually a much simpler task for the construction operator than it is in exercising indirect control. The diagram of the job organization structure in Figure 2, Chapter 3, illustrates direct control. The main lesson to be learned in controlling forces is how to establish indirect control over all the forces involved in the total construction process that the operator must use to meet his objectives.

Example.   Contractually, there is no direct control given to the general contractor's expeditor or superintendent over a material supplier's operations. Even so, the supplier's operations affecting the project must somehow be controlled. Therefore these operators control the supplier's operations indirectly by and through the skillful use of the procedures available. For example, through contractual procedures of payment the expeditor can exercise meaningful and forceful control over sub-subcontractors and suppliers. Similarly, within the production subprocess the job superintendent, while having no direct control over the supplier, may influence him through control of storage areas on the job. Both these operators, the expeditor and the superintendent, must plan their operations carefully so as to obtain this indirect control over these work forces. Both operators must control all the other forces on the job in one way or another in order to efficiently carry out their respective responsibilities for meeting the objectives of completions, quality, and cost.

Developing Control Skills to the Type of Work Force

To conduct construction operations successfully the operator must develop innovative thinking processes whereby he can control all the forces involved with the project. The operator must, however, be very careful in establishing the kinds of control he exercises over his and other work forces on the project. Each type of work force requires a different procedure for controlling it. Ordinarily, the general contractor's operator controls the following work forces:

1. Work forces that are in the direct employ of the general contractor.
2. Subcontractors and their work forces.
3. The owner (client) and his representative's forces.
4. Materialmen and suppliers and their work forces.
5. Supplementary forces such as labor business agents, inspectors, and any other party involved with the construction of the project.

What are some of the things the construction operator must consider concerning the direction and control of these forces on his project?

Firstly, he must consider their specific positions in relation to the construction contract. He cannot control and direct a subcontractor's work forces in the same way he does his own work forces; he does not have the contractual right to do so. A job superintendent's direct intervention with a subcontractor's personnel, for example, is a very different thing from his perfectly legitimate intervention with his own directly employed work forces. If he undertakes with a subcontractor actions permitted to him in dealing with his own forces, there could be serious consequences when viewed in the light of the principles of contract law.

Secondly, a construction operator must consider the relative importance of each component of the total work force in accomplishing the work and meeting the objectives of construction operations. Thus an operator cannot consider the labor costs of a subcontractor's work force in the same light in which he considers the labor costs of his own work forces. Although he may be concerned in a general way about a subcontractor's labor costs, his own costs are far more important to him, since they are a main concern of his own employer.

Thirdly, a job superintendent cannot consider an elevator subcontractor, for example, to be of the same importance in his operational thinking as an electrical subcontractor, simply because he needs the electrical contractor's services for a longer period of time than he does those of the elevator subcontractor.

Fourthly, an operator must treat all of these different forces as individuals, groups, and organizational entities, and all in relation to their importance in gaining his objectives. An operator must look at his own work forces (as individuals, crews of a trade, or as a job organization) in one light, and subcontractors, design firms, labor organizations, or other entities in another. When establishing control over work forces, the operator knows that the individual operators in the total job organizational structure of his own company probably have a common interest in gaining the objectives, but the work force of another entity does not necessarily have the same interests at heart.

An example that illustrates the need for developing special methods of control over another type of special entity is when the job superintendent learns to use the labor organization and its representatives to work for him. In developing his operational skills in dealing with unions, he must look at the labor organization in different ways in order to bring it under his control. He can, for example, conclude that a union is being a help or a hindrance to his operations on a project, depending on the circumstances. Depending on his conclusion, he can maximize or minimize the influence of the union, but he must in some way gain some control over it.

Acquiring Some General Skills

The construction operator must consider delegating authority and responsibility to the various work forces he controls. In operations involving his own work a superintendent may delegate either part or all of the responsibility for a building operation to a specific foreman or he may spread it out to other foremen. He can give responsibility for cleanup and job safety to his individual subcontractors or he may determine that a central control would be more efficient and maintain control over all of these operations himself.

The construction operator must consider his own personality and disposition when establishing his authority in the direction and control over the project work forces. For example, in operations involving subcontractors a construction superintendent can be stern and uncompromising, or he can be just the opposite; to establish control he can deal with the subcontractors individually or as a group. The choice depends on the situation at hand and the superintendent's personal relationships with the subcontractors involved.

The operator must consider the same basic qualities and characteristics of each type of force he controls and directs. This calls for operators to give a wide range of different considerations as to the size and capacities of the work forces they direct and control. An expeditor must consider, for example, whether a subcontractor is large or small, financially capable or not, and local or nonlocal; the specific way in which he operates with each subcontractor depends on such differences.

The job superintendent, on the other hand, must consider the differences between individuals when he looks, for example, at the physical work his laborers do in a concrete pour operation as opposed to the hoisting engineer's physical work in hoisting operations. Each of these operations requires different physical demands and skills from the worker. Both workers must be controlled directly. The inherent differences in the physical operations required of each must be clearly understood by the superintendent before he can gain the necessary control over the individuals or entities that make up the work force.

2.5   SPECIFICS IN DEVELOPING CONTROL SKILLS

Control of Own Forces

It is now necessary to get a closer look at the differences between controlling the various work forces on the project. Here the expression *own work forces* means those work forces that are directly employed and controlled by the construction organization; these forces, in turn, govern the whole unit of work. The whole unit of work is split into three separate units:

1. The labor unit.
2. The equipment unit.
3. The material unit.

The primary concern of the construction operator in controlling his own work forces is the direction and control of the labor unit from both an operational and cost standpoint. The operator is also concerned about control over material and equipment usage and costs as well, but usually not to the same extent as with the labor unit. There is good reason for this greater concern: a labor unit of time, once used, is gone forever. It does not return. It is not reusable like material and equipment. As time passes, money is owed to the labor on the job. This debt must be paid and the labor unit will have to have already produced something of at least equal value in return. This fact makes it absolutely essential that the construction operator do certain things in relation to controlling his own work forces.

1.   The operator must make sure that all the operations he uses or directs in the construction of a building are directed primarily toward the economical and productive use of the labor unit. This is particularly important for a traditional general contracting construction firm. In this type of firm the amount of money assigned to the labor of an operator's own work forces within the contract price is a substantial portion of this price and almost always the prime area of the contractor's risk.

2.   The construction operator must thoroughly understand the following characteristics of the labor unit itself:

a.   How it was arrived at in the estimate.
b.   How it fits into the total price of the job.
c.   How its proportion to the total price affects and governs operations and vice versa.
d.   How it can be affected by job conditions and weather.
e.   How it is affected by subcontractors and their operations.
f.   How it is affected by union, non-union, or jurisdictional labor problems.
g.   How it is affected by the way work forces are assembled and used within the labor unit.
h.   How it is to be adjusted for loss or gain in the production subprocess.
i.   How its performance is recorded, reported, and used in other construction operations.

j. How it is controlled.

k. How the unit of production differs from the job overhead units.

3. The operator must understand that the labor unit is primarily controlled through individuals in the operator's own work force. Therefore all operators must learn how the other divisions in the firm's construction operational organization contribute to the maintenance of the unit. This is not merely a matter of fitting people into slots in an organizational diagram, hoping that somehow the organization will produce good results; it is a matter of understanding individuals and their thoughts and actions. Thus it is important for the operator, when controlling his forces, to know such things as the following:

    a. What each individual thinks about himself and his relationships with others. Pride in doing a good job or in seeing that it is done correctly, for example, are positive individual traits. Carelessness and indifference are negative traits.

    b. The attitudes an individual has about his company, his superiors, his subordinates, and his fellow workers. An individual whose primary goal is personal gain presents a different control problem than does an individual who is company oriented.

    c. The means and methods of disciplining, rewarding, compensating, and delegating responsibility that suit each individual in the work force.

    d. The personal habits, peculiarities, dispositions, and characteristics each individual has. For example, controlling a drinking man is generally not as easy as controlling a sober man.

The foregoing has been a discussion about controlling work forces in relation to the labor unit. It is also necessary to control work forces with regard to the equipment and material units, because these units relate directly to the labor unit and cannot, therefore, be ignored.

How do they relate? To answer this question we will examine an example of how a good operator should think about labor, equipment, and material units as a whole.

Example 1. A general superintendent will usually allow only the necessary hoisting equipment onto a job site. What equipment is necessary is generally determined by consideration of only two situations:

    1. When it is physically impossible to move the materials any other way.

    2. When the cost of the equipment is less than the cost of the labor and savings in money can be shown by using equipment.

In either or both of these situations equipment is necessary. Consideration of equipment costs are then directly related to the labor unit.

Example 2. A job superintendent knows that if brick materials are not available to the mechanic, the labor unit will suffer. This operator knows, therefore, that controlling the delivery of the brick and properly storing and moving it are operations that require his closest attention. But even when he can not control the material delivery, storage, or its movement, he still has control over his labor unit.

How? Through the control and direction of his own forces he has options to minimize the effects of his lack of control over material and equipment. He can shift labor, minimize it, or get rid of it, depending on the availability of materials and equipment. Even though other forces may have control over the material and equipment units, he can still adjust the whole unit through the control of his own labor.

The construction operator must constantly be aware of the relationships between labor, material, and equipment units in conducting his construction operations.

Control of Other Forces

As stated earlier, there are several other work forces the operator must control besides his own. We will examine here in more detail some of the more important ones to help the reader develop thinking processes that can be applied to controlling all of them. Some of these forces are the following:

1. Subcontractors.
2. Material suppliers.
3. Labor organizations.
4. Owners and their representatives.

(We discuss and examine in some detail subcontractors and subcontracting procedures in Chapter 8. Here our discussion is about controlling subcontractors in operations in general.)

Each force in the construction process is controlled differently, according to its role in the process. A labor organization requires controls that are different from those of an architect's firm. They cannot be treated or handled in the same way. Subcontractors, for example, play a more important role in the chain of events in construction operations than does the labor organization, because the subcontractor organization is needed for a long period of time in the production of a building, whereas a union organization is involved only intermittently.

Subcontractors

Subcontractors are a vital component of the construction process and its contracting subprocess; therefore the study of subcontracting is an important part of the study of construction operations. The contractual relationship between the general contractor and subcontractors presents the major control problem for the construction operator. We have said before that construction operators must always conduct their operations with subcontractors with the understanding that the subcontractor is an independent contractor. The construction operator cannot avoid the issue of subcontractor independence in aspects of control. This fact is especially critical in controlling subcontractor performance on the job.

Because they are independent contractors, subcontractors, at their own risk, can breach their contract at any time. Although they can be held responsible for this breach, remedies for breach of contract do not usually benefit the construction process or the objectives of its operations. The required operational skills that give the operator sufficient control, avoiding at the same time the action of forcing breach, must be carefully developed by the successful construction operator.

In any case, subcontractors must be controlled by the operator. They can be controlled in several ways without jeopardizing their independent status.

We suggest that subcontractors be controlled by the following procedures:

1.   Using legitimate contractual payment practices to "encourage" the subcontractor to follow directions and perform as requested.
2.   Making the subcontractor aware of the business consequences that might result if he does not perform or is uncooperative.
3.   Being reasonable and fair in dealings with the subcontractor. When skills are developed in this method, it is by far the easiest and most practical way to exercise control.

Materialmen and Suppliers

In the case of materialmen and suppliers one can see the practical necessity for the construction operators having operational control over them (see Example 2 in the previous section). The construction operator gives much the same thought and consideration to this group of forces as he does to the subcontractors, with one significant difference: a subcontractor usually supplies labor in addition to materials, whereas suppliers rarely do so except in certain special cases and then only when labor is supplementary to their main function of supply. This means that construction operators do not have to concern themselves with controlling suppliers to the same extent as they do with subcontractors. But there are at least two things the operator must carefully consider when contracting with suppliers:

1.   Their location relative to the project.
2.   Their capacity to produce and deliver on time.

In addition to the supplier or materialmen directly in contract with the general contractor, the reader must also understand that subcontractors too can have their own suppliers. These suppliers are contractually and operationally remote from the general contractor's construction operator and thus his control over them is liable to be considerably diminished. Nevertheless, in many instances the operator must give the same considerations to these suppliers as he gives to his own.

The Labor Organization

Let us now examine an operator's control of another outside force, the organized labor organization. How does an operator control this force when the need arises?

1.   The operator exercises control by knowing and being reasonably able to interpret the union contract thoroughly and correctly. He knows what the contract says and means.
2.   The operator exercises control through his direct and indirect dealings with

the organization's representatives. His dealings with them should be based on (1) above, but, his manner, attitudes, and techniques in dealing with them become personal. The operator's tactics and strategies must be deliberate and directed towards solving the problem in the fastest way.

3. The operator should have a basic understanding of the law of labor relations and its applications. He needs this knowledge to counter representations and claims made by the union representative in favor of the union's position.

4. The operator exercises control as much as possible by using union members in his own work force to counter the union's position, particularly at the beginning of a labor problem. If he can, he should use the union tradesman, the trade foreman, or the job superintendent to control a problem situation as much as possible. It is not advisable at this level of a dispute to enlist the aid of those people who are antagonistic or unrealistic about labor, the labor movement, or its unions.

The Owner and His Representatives

The last force that must be controlled by the construction operator is the owner and his representatives—namely, the architect/engineer. Here the situation that creates the major control problem is that the owner is the other party to the construction contract, and along with the owner comes his agent, the architect/engineer. The owner and architect/engineer are clearly defined within the general contract itself. Each of them have clearly defined contractual duties and responsibilities; but these clear definitions in no way assure that they will remain so in the contract's actual performance.

When we discuss developing the method for controlling the owner and the architect, we emphasize learning to control the architect/engineer rather than the owner. The construction operator controls the owner both within and without the contract, although most of the control is indirect. Clear examples of control occur when, for example, the contractor chooses to give the owner optional alternate prices for reducing costs in a project after award of the contract or when offering optional prices in his quotations for additional or extra work. Developing skills in controlling the architect/engineer is our main concern here. There is no contractual relationship at all between the architect and the operator's firm in the traditional construction process. Nonetheless, it is essential, for efficient operations, to have some control over the architect in specific instances.

Example.   The operator must have some control over the time it takes for the architect to approve submittals. He must control this, because his other operations and the other party's operations depend on it. The approvals are critical for schedules and deliveries. The production of a building depends on the operator's ability to obtain the information from the architect/engineer so that it may be distributed for fabrication and installation purposes.

We cannot easily categorize the necessary numbers, kinds, or extents of control over these professionals as we can with other involved forces, nor can we list reliable means and methods for controlling them. Nevertheless, the construction operator must be constantly aware of the effects of the architect/engineer's actions, judgments, and decisions in order to develop means and methods of exercising some degree of control over them, because they are so essential to the whole construction process and its operations. This control must be established with discretion, good sense, and respect for professional sensibilities. The establishment of control must be based on the premise that the architect/engineer's functions are an essential part of the contracting and production subprocesses. An operator can best control the architect/engineer by cooperating with rather than fighting them. Mutuality of interest makes this approach sensible, thus making it practical and profitable for all.

Reminders

This section cannot be concluded without reminding the reader of the following.

   1.   In the construction process the sum total of the construction operator's functions are only a part of the total operations required by the construction process to successfully build the product. This requires the operator to direct and control his and all others' work forces in order to make the process produce properly. At the same time he must realize that his operations are not the only ones that are important.

   2.   In directing and controlling operations and work forces, the construction operator must be aware that he too is being manipulated and controlled by the same forces he is seeking to control himself. This constant contest between different participants in the construction process is (if reasonably and legitimately conducted) a benefit to the process. In addition, it makes the construction business more exciting and challenging.

Part II

Operational Organization

3

General Supervision

Record it for the grandson of your son—
A city is not builded in a day.

Vachel Lindsay

3.1 INTRODUCTION

Definition

The operation of general supervision consists of the control and direction of all field operations by a single central authority within the organization of a general contracting company.

The general superintendent is the person in the construction operational organization who is the head of all field operations of the construction firm. Not only is he the head of the field force, but he is its heart as well. The quality of the construction field production force is a direct reflection of this individual and his operational abilities.

Whether or not an official title is given to the general superintendent within the construction organization is immaterial: someone in the organization must perform general supervisory operations and be responsible for meeting the main operational objectives of field operations. A construction company cannot perform its field work efficiently without someone handling the operations described in the following pages. Most construction companies require a central control over field operations, and they give this control to the general superintendent.

The old, experienced field professional is generally found in the position of general superintendent; experience is his trademark. Experience forms the foundation for the talents required for operating in this position and the confidence his people have in him rests on it. In governing and guiding the total field operation of the organization, the general superintendent relies heavily on his past experiences in dealing with people. The reader may ask why it is necessary to discuss the general superintendent in this text; most likely it will be a long time before a young construction operator will assume this position. It is necessary for field operational personnel, because what the general superintendent does, and how he does it, specifically affects them and their future directly: he hires the various construction operators for the field operations and determines the abilities and capacities of his personnel and the compensations, promotions, and dismissals of all of the field personnel.

No doubt his subordinates will sometimes consider the general superintendent's actions to be unreasonable or even illogical. A good general superintendent exercises sound judgement based on a solid analysis of construction problems. Many times certain of his actions may seem to his subordinates arbitrary or based on poor logic, but they are in fact usually based on factors or information that are exclusive to his knowledge and position. His duties and responsibilities are explained here in order to promote a better understanding by his subordinates of his organizational position in construction operations.

38

Authority

Of all of the other various construction operational positions the general superintendent has the most clearly identified position of authority. A good general superintendent rarely leaves any doubt as to his authority over the field forces employed by the company organization. He is really and truly the "boss" to those who work for him. Generally, there is a clear separation of office and field functions and forces in the typical construction organization. Responsibility for all field functions belong to the general superintendent. It is probable that this person has the last word (and sometimes the only word) to the field forces. Since field forces can make up the major portion of a construction organization's forces, it is easy to understand the great weight and responsibility this operational position carries. The field work forces' part in the construction is the actual manufacturing of the product—the end to which all other functions of the total operational organizations are directed and to which they give support.

The exclusiveness of the general superintendent's position and authority in the organization is another facet that must be considered and understood by the reader. He is indeed a unique character in the organization. Because of his unique position of responsibility and authority, the general superintendent often becomes the natural bridge between the office and field forces. Therefore, in addition to having the last word to the field forces, he is often quite influential in the direction of the other parts of the operational organization as well.

3.2 THE OBJECTIVES AND THE GENERAL SUPERINTENDENT

The examination of the general superintendent's position begins with an analysis of his role in accomplishing the objectives of the operational organization in the construction process. (The reader will recall that these objectives are completions, quality, and costs.) Since all of his subordinate field forces are responsible to him for these objectives in their respective capacities, he has a special role in accomplishing these objectives for his superiors.

The general superintendent's first responsibility is to be certain that his subordinates clearly understand what these objectives mean and how important they are to their employer. His second responsibility is to see that the objectives are attained by competent field construction operations and operators under his control.

There must be a clear understanding of the scope of the general superintendent's responsibility for meeting the objectives. He is responsible for the overall company field operations. This means he is responsible for overall completions, good-quality construction on all jobs, and the complete and overall job labor, material, and equipment costs. As stated above, his responsibilities run downward to his subordinates; but they also run upward to his superiors as far as meeting the objectives are concerned.

It is his exclusive responsibility, for example, to give accurate projections of completions to his superiors for the long and short terms within both the company's fiscal and work years. He must realize that these projections are vitally important to the firm: they determine its capacity to take on new work, to efficiently utilize its resources, and to effectively operate for a profit.

Completions

The general superintendent must *demand* completions from his immediate subordinates—the job superintendents. This demand must be based on a realistic and accurate appraisal of what they are capable of. The general superintendent cannot allow too much leeway or too many excuses for nonperformance from these subordinates. Certain unknowns such as weather will at times furnish excuses for failure to complete work and this will affect his projections, but since he has the authority to adapt the total company's field operations to the completion objective, he must utilize it to mitigate the effects of the unknowns on the overall completion picture of the company.

A successful general superintendent will insist on a realistic quantity of completions from his subordinates. He must extend his best efforts within his authority and responsibilities to see that these completions are accomplished. He will determine the reasons and suggest solutions for unsatisfactory completions. He will correct any deficiencies, using unbiased and objective means. Completions of the product are the prime objective of the construction general superintendent.

Quality

It is not practical to build without quality. This objective of the construction process directly affects the other two objectives of completions and cost. A construction operator is simply fooling himself if he assumes that fast completions (where the quality of the construction is poor) are a valid goal of construction operations, for two reasons.

First, in almost all cases poor quality results in longer completion times and greater costs. Second, poor-quality construction work in a building is there for the whole world to see. While it is true that deficiencies causing poor-quality construction work can be hidden, over a period of time the effects of hidden defects can become an advertisement of the construction operator's and his firm's failure to build properly.

The general superintendent, who is the superior of all production personnel, *must* insist on quality work and workmanship. If he is to be credited for properly constructed buildings, he must bear the ultimate responsibility for the poor-quality structures as well. This responsibility requires direct action with those who built carelessly, generally taking the form of a face-to-face no-compromise confrontation with the superintendents, foremen, and even tradesmen who do not build with care. This becomes a continuing responsibility: it is often necessary to remind those who have performed satisfactorily in the past, but who have become careless, to mend their ways—and immediately. There is a simple rule that the general superintendent can cite to his subordinates when insisting on good-quality work,

"If something is worth doing at all, it is worth doing right."

This is an extremely important rule for the general superintendent to firmly establish in his organization. His insistence on quality construction is based on good, solid reasoning. This same reasoning should be evident to the competent subordinate construction operator. One should not, for example, have to explain to a good subordinate who has been assigned the corrections on an extensive punch list the value of quality construction. The punch list clearly illustrates the value of insisting on quality construction.

Costs

The cost objective is the general superintendent's third and probably the most time-consuming responsibility. The general superintendent is responsible for the labor cost on all jobs, which is considered the major field expenditure in this text. He is also responsible for the equipment and job-used material costs on all jobs as well, but the labor cost is usually the company's major risk.

There are substantial differences in the way the general superintendent looks at overall company labor costs as opposed to the way the job superintendent views his individual projects' labor costs. One main difference is in their respective appreciations of overhead costs. The general superintendent is responsible for the control of the total overhead labor costs of all superintendents, engineers, and timekeeping forces and whatever supplementary personnel they require in their operations. An individual job superintendent is not concerned with personnel not actually engaged on his job.

Another difference results from the general superintendent's control over the placement of key foremen and mechanics within the field organization. It is his responsibility to place or move all of these people around as the total field production requires. In making these assignments, however, he often takes over the primary responsibility for overhead cost from the individual job superintendent; it is his responsibility to assign personnel as he sees fit, but he must accept the consequences of his actions. The job superintendent does not have this same authority or responsibility.

The general superintendent's ability to see the overall company job labor costs enables him to adjust the labor costs of individual jobs. He may choose to move certain key men to certain jobs that are not making their labor costs in order to correct the problem. He may authorize premium time on a particular job in order to keep the schedule if he knows that the overall company job labor costs justify such an action. The general superintendent is much less restricted in controlling the labor costs for the whole field operation than is the individual job superintendent on his individual job.

The general superintendent controls labor costs in different ways, including the following:

1. Moving key men as they are needed to control production and costs.
2. Authorizing overtime other than spot overtime.
3. Determining the type of equipment needed for an individual job after an analysis of its ability to remedy unacceptable labor costs or to simply improve costs.

   4. Authorizing weather work in a general way.
   5. Determining the overhead personnel needs of a particular job according
           to its size or particular type of construction.
   6. Assigning available materials and equipment to a particular job merely by
           judging that job's priority in the entire company operation.

All of these ways of controlling the cost of production are directed primarily toward labor cost savings and, secondarily, toward improving completions. The general superintendent has substantial control over both of these directions.

The General Superintendent and the Subcontractor—The Objectives

There is one feature of the main objectives that is not directly under the general superintendent's control: The general superintendent does not have direct control over the subcontractors on the job; nevertheless, it is still his responsibility to control. Subcontractors can be a very significant factor in controlling the progress and sequencing of a construction job; they somewhat control the completions and, indirectly, the labor, equipment, and overhead costs on the project. This is particularly true in relation to the job overhead costs.

In spite of this fact, the general superintendent still contributes substantially to the objective of cost in this phase of the construction process. He can best accomplish this objective by carefully selecting job personnel that can get subcontractors to work the most efficiently. These job personnel must control the subcontractors and not allow them to control the job. Judgmental skill and a thorough knowledge of his personnel are necessary for the general superintendent to select those job superintendents who can best handle subcontractors. The higher the percentage of subcontracted work on a partciular job, the greater the need for field construction operators who can push subcontractors to their full capacity.

Some job superintendents, on the other hand, have personal preferences for pushing the work of their own forces, with little enthusiasm for pushing subcontractors; others prefer pushing subcontractors over pushing their own work. Either preference does not necessarily decrease these peoples' value in a total construction operational organization. The general superintendent's rsponsibility is to determine these preferences and utilize them to the best advantage in construction field operations.

These remarks about personnel selection are all directed towards illustrating that one person, the general superintendent, has the responsibility of meeting the objectives of completions, cost, and quality to the advantage and profit of the entire firm. Subordinate operators sometimes have a difficult time appreciating the actions of the general superintendent in this matter of personnel selection and assignment. This is particularly true when these actions affect them directly. Subordinate operators should understand, however, that the general superintendent must quite often make purely objective decisions for the sake of the whole field operation. A good general superintendent learns how to soften the effects of any of his unpopular selections, assignments, and decisions on his subordinates; but the young construction operator must also understand that the general superintendent must make these decisions in order for him to do his job properly. He has no other choice at times.

3.3 THE TYPICAL CONSTRUCTION FIELD OPERATIONAL ORGANIZATION

Figure 2 illustrates a typical construction field operational organization. The general superintendent's position on the chart clearly shows that he has responsibilities running upward and authority running downward within the operational structure. His position relative to his authority over and responsibility to his subordinates should be of particular interest to the reader. Figure 2 shows that his immediate subordinate is the job superintendent, then, in descending order, the engineers, timekeepers, and trade foremen. Figure 2 also shows that the number of major projects assigned to his operation can vary with the number of new contracts obtained. Small jobs are diagramed as they are, because many times several of them can be manned by one set of supervisory personnel. This is particularly true with industrial or commercial in-plant work of different sorts.

Aside from being the superior operator of all jobs, the general superintendent is in charge of two important supplementary operations: the pay operation and the yard operation. These two subsidiary operations are adaptable to central operational control. By being centrally controlled, they aid the general superinten-

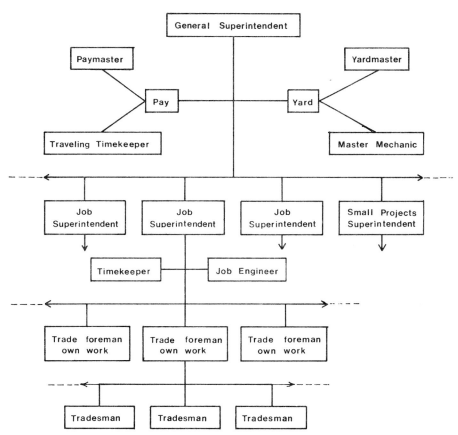

Figure 2. A typical operational organization for the total field forces of a general contractor.

dent in meeting his responsibilities of running the total field operation. The reader should be absolutely clear about the function and "reason for being" of these two subsidiary construction operations: both function solely for the purpose of serving the production of work. While they are centrally controlled by the general superintendent, their function is entirely subservient to the needs of the various individual projects. It is the general superintendent's duty to have those in charge of the pay and yard operations completely understand their relative positions in field operations.

The Pay Operation

The pay operation is a vitally important function of field operations. By law, mechanics and field personnel must be paid the proper due amount of wages and they must be paid on time. In union areas the union contract makes this abundantly clear. It is the absolute responsibility of the general superintendent to see that the operations of timekeeping, time reporting, and labor cost reporting and the actual payment of mechanics are carried out properly and on time. There are many reasons why the pay operation should be under field operational control (even though the bulk of its operation may be office oriented).

Since the time clock is not traditionally used in the construction process, the hours of the working mechanic must be recorded either from personal observations by the timekeeper or from a transcript of time reported in the foreman's time books. The timekeeper observes, records, and reports the time and this information is translated into the paycheck, which is traditionally paid at the end of the workweek. This means that there is much close and attentive work that must be done continually by and between the timekeeper and the paymaster in order to pay properly. Another supplemental reason for having field control over the pay operations is the need for prompt payoffs.

Operations that must meet the legal requirements of payment (and in union areas this includes the proper distribution and payment of funded fringe benefits) can be handled better under field operational control than anywhere else. The various procedures for the payment of fringes, the withholding of taxes, and other labor-related pay items are so closely interrelated with other field operations that they should be coordinated by the total field organization.

The Yard Operation

The yard operation is the other subsidiary operation that falls under the direct control of the general superintendent. The yard is a physical place. It is a central distribution point for equipment and field-used materials (that is, materials not incorporated into the project itself). When a job needs these things, it orders them from the yard; when the job is finished using them, they are returned to the yard. The yard is the place where equipment and materials are stored and maintained between uses.

The yard is generally under the control of a yardmaster, whose direct superior is the general superintendent. The master mechanic is another construction operator in the yard who is responsible to the yardmaster, but in some matters he or she is directly responsible to the general superintendent. The master mechanic's main responsibility is rather simple: to maintain the equipment in a serviceable state at all times and at any place it may be located. The general superintendent's operational responsibilities include furnishing sound and maintained equipment that operates properly to the company jobs that need it. He insures this by personally directing the master mechanic. The master mechanic sees that the equipment operates when it goes out of the yard and also repairs breakdowns when they occur on a job site.

The yard operation must be set up and managed as a service organization. The bulk buying of form materials by the yard operators for example (because of its storage and maintenance needs), is often more economical and efficient than individual job buying and storage. In addition, the yard's control of reusable materials is often the best way of insuring their reuse when compared to the alternative of leaving unused but reusable materials on the individual job site. The yard exists to serve. If the yardmaster keeps a careful inventory of material on hand, this service supplies reference data that is extremely helpful to the general superintendent in the running of his operations. The prefabrication of formwork and the fabrication of reinforcing steel are other types of service operations that sometimes take place in the yard because it is better to perform them there.

The general superintendent pays close attention to the yard operation and the behavior of its key personnel. He must make clear to the yard personnel what their position is in the total production of work. Selection of the yardmaster is of great importance, since he deals constantly with job personnel. Cooperation between the field and yard is essential to good field operations and therefore the yardmaster's position should be filled by someone who understands the need for this cooperation. The yardmaster holds a key position in construction operations.

3.4 THE THEORY OF CENTRAL CONTROL

The need for the operational position of general superintendent (like any other operational position) must be justified. It is justified simply because of the need for the central control of field operations, which arises from the complex nature of the problems of the many concurrent field operations involved in a construction firm's field production. Control must be exercised over three main areas of construction operations: the utilization of people, materials, and equipment. The central position of general superintendent allows for the most efficient utilization of these resources.

Resources are almost always relatively limited. No two men or two pieces of equipment or material can be in the same place at the same time. Business economy generally does not allow for duplication of these resources. Qualified men and expensive machinery and tools are not things that are always available.

The general superintendent's position allows him to control and direct operations through a central point. His position of authority allows him to utilize the resources made available to him by the company to the best advantages. Let us examine how this control works within the model construction organization.

Example. The company has a certain number of contracts to construct varying types of buildings at a specific point in time. Each job is in varying stages of completion; thus there are varying but simultaneous demands on the company's resources of men, equipment, and materials. Because the stages of completions vary, so do the essential needs of each of the individual jobs for these resources. Let us say that there are six jobs concurrently involved with reinforced concrete framing of one sort or another. These jobs require men, equipment, and materials of a certain nature. In addition to the form/concrete work, four jobs (two of which are included in the six above) are involved in masonry operations and three jobs are in the late subcontract, punch list, or close-out stages. There are six small in-plant jobs also being done at the same time. In addition to this work already on hand, bidding of new work for the company is heavy. One project might be in the stages of negotiations to contract. This is the company's work and contract status at this point in time.

Consider the present and future field operational problems raised for the construction firm in this one example. The company has base field operational crews consisting of superintendents, engineers, and trade foremen. All of these base people are competent, but some are more adaptable to particular types or sizes of projects than others. The company has only so much equipment and materials available at any one time. Limited capital resources do not allow for an increase (to any great degree at one time) in equipment, materials, or overhead personnel. The company operates on the theory that equipment and materials usage must be as efficient as possible and field operations conducted with what is on hand. This efficiency always results in a strain on resources.

How is the company's field operation managed under the situation in the example? It is managed and controlled by and through the general superintendent. He is the person who solves the problem of making resources available that are necessary to complete work, thus allowing for completions and the contracting of new work. Central control prevents the duplication of efforts and resources. It allows for the orderly, efficient production of work through the proper utilization of resources. In any multijob construction operation there must be some central control of field resources. Through the proper planning and analysis of present and future demands on available resources recommendations can be made to the company for adding new resources.

3.5   THE DIRECT RESPONSIBILITIES OF THE GENERAL SUPERINTENDENT

The direct responsibilities of the general superintendent flow from the theory of central control of field operations. Because of these responsibilities, the general superintendent's time is at a premium. The following list of some of his responsibilities is based on the general principle that he must interrelate and coordinate all jobs with respect to labor, materials, and equipment.

Hiring

The general superintendent personally hires all key operational field personnel. This responsibility includes hiring superintendents, trade foremen, engineers, and sometimes even key mechanics in special categories, such as riggers or operators of heavy or special equipment.

We must examine closely why the general superintendent hires key personnel himself, instead of allowing his job superintendents to do so. There may be solid reasons at times for allowing job superintendents to hire their own key personnel, thus fostering good cooperation and effort on the job owing to the personnel's familiarity with each other's way of operating. Many construction crews like to follow each other as closely as possible from job to job; but the main disadvantage of this hiring practice is that it conflicts with the necessary mobility, adaptability, or associability of key men within the total company field operation. It is desirable that these key people realize they are working primarily for the general superintendent (and thus the company) rather than for an individual job superintendent.

When specialized personnel such as riggers or other tradesmen are employed in the company's operations as individuals or in very small numbers, it is wise for the general superintendent to exercise direct control over them. Needless to say, when expensive equipment requires a certain tradesman to operate it, it is necessary that the general superintendent control that operator's movement with the machine.

One of the general superintendent's main problems in carrying out his hiring responsibilties consists of matching the personalities of key personnel with each other; sometimes there are such personality conflicts between the job superintendent and the subordinate key personnel assigned to him that they cannot work together. The general superintendent must exercise caution when this situation arises. He may have to compromise now and then, but he must still maintain his authority to man the job with key personnel as he sees fit. When extremely competent people are involved in this type of situation, the general superintendent's appeal to their good sense will usually solve the problem. He may convince them of the benefits derived for the company and themselves from their cooperation in spite of their differences. When this fails, the general superintendent must adapt to the situation and change the key personnel makeup of the job.

Equipment Acquisition

The general superintendent must determine and recommend to his superiors the company-wide needs for plant and equipment. There is really no other individual in operations who can take on this responsibility better than the general superintendent. These needs require determining the proper numbers, types, costs, and feasibility of all different kinds of plant and equipment. It should be clear, however, that while the general superintendent must give some reasonable consideration to the cost of these needs, generally his responsibility does not extend beyond an objective recommendation. The final acquisition decision is the responsibility of his superiors.

In his determination of the needs the general superintendent must exercise discretion and good judgement. Just because his field forces ask for plant and equipment does not justify his recommendation. There is a good general rule to follow in justifying and making determinations on acquisitions:

The acquisition of each piece of plant and equipment must be justified through its ability to save manual labor.

The general superintendent must be very cautious about recommending to his superior the acquisition of equipment that cannot be used at maximum efficiency. Many construction companies put themselves in the position of being overloaded with unwisely purchased equipment. Too often it is acquired on the spur of the moment to satisfy a present need. When there has been no projection of future maximum efficient use of the equipment, acquiring too much equipment can cause financial problems for a construction firm.

It is the general superintendent's option to recommend rental, lease/rental purchase, or outright purchase to the executive. While rental and lease arrangements make the equipment more immediately available for use, the general superintendent should realize its cost compared to owned equipment, and the same general rule stated should apply just as strongly to these arrangements.

Equipment Usage

The general superintendent is responsible for the proper distribution, use, and care of the equipment. A good general superintendent will personally handle the distribution of all heavy equipment. He will determine which piece will be used on which job for what period of time. This determination is made on the basis of his solid analysis of the current and future needs of all the company's projects. He usually pays particular attention to hoisting equipment. In assigning hoisting equipment to the project, he must be aware of the length of time the equipment will be tied to one job. It is not only a matter of the general superintendent being able to adjust immediately to different needs when they occur, where heavy equipment is involved; he must also carefully consider that the costs of the breakdown and movement of this equipment are very high. It is important that he make the right decision about the real need for heavy equipment in the first place to avoid unnecessary costs in its shifting and reshifting from job to job.

There are certain attitudes that reflect the competency of subordinate personnel that a general superintendent relies on when judging their basic value to company operations. A striking indication of competence and worth is a subordinate's care of the equipment in his charge. Good care and maintenance of equipment shows competence and personal interest in the company's welfare. Proper maintenance of equipment is essential to good construction operations; it indicates a true realization of the value of the equipment. Poorly maintained equipment results in extra costs, particularly when one or a series of construction operations depend on it. A bad vibrator, for example, that breaks down during a concrete pour causes serious problems for the entire operation. The capable general superintendent will not only insist on the proper care of tools and equipment but will also instruct his entire organization about the cost and safety hazards of working with poorly maintained or unattended equipment.

When discussing this responsibility of the general superintendent, we must also include in it the distribution, use, and care of reusable materials. The main reusable materials used in the construction process are such things as form lumber, scaffold materials, and so forth. Take, for example, form material. These are materials that are reusable and priced as such in the company's estimates, and they must actually be reused as estimated; otherwise there is a direct job cost not covered in the cost of the work. While these materials are reusable, they are also expendable—the number and kinds of reuse diminishes their usability. A cut on a piece of form lumber, for example, generally restricts its reuse. The general superintendent is responsible for seeing that his personnel are strongly aware of the reuse capacity of these materials. Cutting a scaffold plank to fit some special condition diminishes its value as a scaffold plank. If it is cut, it usually cannot be used again in a standard scaffolding installation. The cut, in effect, makes the plank inadequate for the use it was originally intended and purchased for. The general superintendent, through his job superintendents and yardmaster, must control such wasteful use of these kinds of reusable materials.

As these materials are used and reused, their value diminishes until they become scrap. New materials must be constantly refed into the inventory to replace these materials. Buying in bulk usually has cost advantages, but bulk buying means large expenditures of money. Careful consideration then has to be given as to when the bulk or piece purchase is to be made in relation to the actual or immediate need for these materials. It is the general superintendent's responsibility to say when the bulk buying of material is necessary and to what extent and location it is to be initially delivered.

Reusable materials must also be maintained. Cleaning and maintaining the material is the major part of the cost in labor when reusing them; but cleaning is also a labor operation in itself. A general superintendent must then decide on balancing the cost of cleaning old reusable material to the current market cost of new material; he decides when they are expendable because of cost considerations. Making decisions of this sort are the regular and continuing responsibility of a general superintendent.

Costs

The general superintendent is responsible for the maintenance of good costs on all jobs in the total field operation. This requires that he establish a workable and efficient cost-keeping and recording system for the field which must be uniform and adaptable to the capabilities of his subordinates. The general superintendent must insist on good cost records, and their proper upkeep, on a set time basis in order to meet his responsibilities in this area of operations. This is extremely important: he should allow no excuses for the nonperformance of this work by his field forces once the system has been set up and learned by them. Cost reporting is an essential operation of the field forces that cannot be delayed. Procrastination or carelessness on the part of his subordinates in these operations is never acceptable.

The general superintendent is the first to know the true figures of actual costs, good and bad, on all of the individual jobs. Good costs call for his examination as to why they are good. Excessive costs call for an examination of their causes and, in addition, for immediate remedial action. Usually these actions consist of moving men or changing the means and methods of construction.

The general superintendent is in a unique position to know the meaning and consequences of total job payroll figures. He can easily project these cost figures to the end of each job, compare actual cost to allowed cost, and quickly react to to remedy what the projection shows. He can order an increase in production forces when necessary or the abrupt or gradual decrease in the jobs' work forces. The general superintendent knows the true relationship of maintaining acceptable completions to bad actual costs on a particular job. He can force a reduction in payroll, and without authorization from a superior.

Unsuccessful means and methods of construction can be changed by the general superintendent, but only after a complete analysis of such actions' effects on the total field operations. Shifting away from predetermined methods or sequencings of operations is always risky, because it usually involves a major expense, but sometimes this must be done to adjust for unacceptable projected cost overruns. The reader must not think that such overruns can always be blamed on poor initial planning. Many of the unknown conditions encountered in the construction of buildings can cause bad costs through no fault of any of the construction operators. Making the decision to change operational methods in midstream is always a difficult and risky task for the general superintendent, primarily because of the uncertainties of the causes of bad costs and the potential harmful effects of changes.

Whether to relieve or shift existing forces or individual personnel because of cost considerations is a much more difficult decision. The problems encountered in this situation are not so clear-cut, nor are they as easily solved as those where means and methods are the cause. A general superintendent will exercise extreme care and discretion in solving problems involving bad costs that are based primarily on personnel considerations. This is true because the causes of this type of problem are usually indeterminate and additional problems may be created. The following is an example of such a problem.

Example.   The general superintendent initially assigns the job organization that he wants to run the project and its authority on the job. Once set, it becomes exceedingly difficult for the general superintendent to countermand the authority he established. He should only do so when this authority is clearly not producing good completions and costs.

As key personnel come onto the job in its various stages, they individually begin to pick up and store vital information about the project. Experience and knowledge about the work is gained by each of these people as they spend more and more time on the job. At certain stages of a job it would seem inconceivable to change or shift the organization for this reason alone, but some very real problems can develop because of personal relationships, behaviors or temperaments that result in conflicts, which can become the direct cause of poor costs. The situation must be remedied. But who should be changed or shifted? Who is to blame? What if the real cause of the problem is the highest authority on the job— the superintendent? What is to be done about the situation when the best and most extensive knowledge about the job is already in the heads of personnel that are parties to the conflict? Which replacements in the organization can pick up this knowledge the most efficiently? Can they be made available without damage being done to other jobs or to the whole field operation? These questions and others have to be answered and acted upon by the general superintendent, based on the premise that bad costs resulting from a conflict situation must be remedied as quickly as possible.

This example offers another solid reason for giving the general superintendent central control over the entire field operation.

Supplemental Operations

The general superintendent is responsible for the direction and control of all supplemental operations that affect the production of work. These supplemental operations include pay and yard operations and the purchase of expendable items of equipment and materials. The general superintendent is responsible for these operations because they support and supplement field production operations. Usually his operational base is the yard facility. This locates the field authority between the support operations and individual projects they serve.

The general superintendent must clearly and firmly establish the positions of these operations relative to the production of the work. Clear-cut areas of responsibility and lines of authority must be set up between them and the individual projects. Proper and efficient channels of communications must be established between the field and support operations. Simple rules should suffice to guide and set up an efficient system. Relevant statements of policy, like "men must be paid and on time" or "equipment must function and be available" and so on are criteria for establishing the systems that produce the desired functions of these supplemental operations.

Safety

The general superintendent is responsible for setting up and implementing the safety programs and policies for all field operations. This responsibility entails more than just lip service on the part of the general superintendent. It means that rules and guidelines for safety are to be firmly established in the minds of his subordinate construction operators and then followed.

Safety is a prime responsibility of every field construction operator even though job safety is the direct responsibility of the job superintendent. The general superintendent sets the whole tone of safety consciousness for the entire organization. He must be alert, firm, and forceful on the matter of safety; subordinates who ignore or refuse to consider safety must be dealt with severely. The general superintendent must carefully weigh and balance the value of men who will not or cannot be safety minded. Pure economic considerations should significantly tilt the balance in favor of their dismissal if they will not or cannot operate their job safely. A general superintendent must convince all key personnel, by his strict attitudes about safe operations, of the importance of this matter to the construction firm.

Labor Relations

The general superintendent is responsible for handling field labor relations and their problems at the level where union business agents enter the scene in labor disputes. Many labor problems start among the tradesmen at the job site, most of which can be resolved on the spot with proper handling by job supervisory personnel. This usually involves appealing to the good common sense of all the parties concerned. The job superintendent and trade foreman in particular can exercise great influence at these early stages in resolving labor disputes and grievances. The general superintendent should encourage their handling of these types of labor problems. With minimum instruction in the simplest elements of labor relations, labor law, and the labor contract, these people are often more capable of taking care of these minor situations than anyone else.

When the labor business agent does become involved, however, the general superintendent should become concerned about his job personnel handling the situation. The reason for not encouraging participation by job construction operators with business agents is simple: the business agent is likely to be a professional in labor relations, and job personnel generally are not. The main purpose of the business agent's job is to protect and foster the interests of his union's membership. Labor relations can only constitute a very small part of the construction operator's expertise, simply because he does not have the time to become an expert in this field. The highly technical aspects of labor relations are not, as a rule, a significant part of his responsibilities when they are compared with his more direct production responsibilities.

Does this mean that the general superintendent will have the expertise and skill to match the business agent's skills? Not necessarily; but a good general superintendent should make it a specific goal to learn about labor relations and the labor contract to a degree where he can competently protect the company's interest against the labor union's. Experience is also important in this area of operations. Even though the general superintendent does not have the time to learn about labor relations that the business agent does, he certainly should have learned more about them than the individual job superintendent. It is a matter of putting the best expertise available in the firm against the business agent in an adversary situation at this level of dispute.

At the business agent level the dispute is generally of a more serious and involved nature than it is at the job grievance level. The business agent represents the highest local union authority and the general superintendent represents the highest authority in the field construction organization; it is therefore wise to match the two against each other. This is the best way to keep the dispute from becoming even more involved or uncontrolled. The main advantage of the general superintendent's participation in a serious labor problem lies in the fact that he is the only person who knows its overall potential effect on company operations. It is not likely that he will take a stand on a problem that will have overall effects on the whole field construction operation without carefully analyzing the consequences. The job superintendent, on the other hand, might very well take unwise actions in the same situation, because he only sees them in relation to his job. The general superintendent must look at the entire picture and is therefore more likely to be cautious, which is necessary in a bad labor situation.

There is one area of labor relations where the general superintendent should not be made solely responsible for the resolution of problems. This is in the area of the jurisdictional dispute.

Example.   In many instances the operational authority to deal with a jurisdictional problem must be shared with the expeditor, who is the operator in whom the subcontracting authority rests. He makes the assignment. He must also be included when subcontractors have jurisdictional problems. The initial assignment of work is not the responsibility of the general superintendent, and he should therefore exercise great care in his actions in jurisdictional disputes. He can generally only insist on fast action by responsible parties who have particular expertise in labor relations in this complicated and touchy labor relations problem.

Field–Office Liaison

The unique position of the general superintendent in the construction organization forms the basis for his responsibility of setting up and maintaining liaison between office and field operational forces. Both office and field operators have their own individual responsibilities. Generally, the whole operational organization will work together to accomplish the work. Since the entire operational organization must always work towards the main objectives of completions, quality, and costs, a close association that fosters good cooperation between office and field operations is essential to successful operations.

Many times, however, there are occasions where operators from both areas are unable to resolve differences in their viewpoints on a particular approach to a problem. These occasions perhaps arise out of the most sincere but conflicting appreciations of a certain situation. Sometimes personal or philosophical attitudes can become overly involved in approaching similiar goals. At other times the responsible operators from both the office and field just do not get along with each other.

The general superintendent can get these people together and, using the authority of his position, insist upon a resolution of these problems. He can also act as a judge or arbitrator in such a dispute. He can transmit accurate information and facts in order to clear up misunderstandings. In a word, he can settle these problems because of his unique position in the total construction organization. While his prime responsibility is to direct field operations, he must not allow this to influence the way in which he approaches this type of problem; he must be fair.

Example.　The timely furnishing of doors and frames is initially the responsibility of the expeditor. He starts the process of furnishing the materials to the job site. If they do not arrive on time and job progress is affected, then the resulting delay becomes a production problem. The general superintendent can expedite the delivery by pushing the expeditor and hold him responsible for the delay if he was at fault. But if the expeditor was informed by the job superintendent at the start of the job that only certain doors and frames were needed or that all of them could not be handled at a certain time because of lack of storage space and the superintendent suddenly changes his mind and demands more frames on the spur of the moment, then the fault shifts to the superintendent. The general superintendent should at this point require the job superintendent to adjust his operations to suit the lack of doors and frames. It would not be fair to blame or harass the expeditor for causing the problem.

There are all kinds of similar examples of the usefulness of the general superintendent in solving interorganizational problems. He must, however, supply himself with accurate background information on the situation and act accordingly. The new construction operator should clearly understand the position and function of the general superintendent in meeting this responsibility. Understanding will lessen the pain when the general superintendent calls on him to account.

In another interorganizational area the general superintendent becomes an especially valuable medium for passing on accurate information that is vitally important to the estimating department—information about actual unit costs. The labor cost reports show the actual costs but do not say why they are the way they are. The centrally involved operator who can best interpret and then transmit the reasons for these costs is the general superintendent. This does not simply mean that he just complains about the estimator's quantity take-off mistakes or his mistakes in judgment in pricing; it means that he takes a constructive approach to the communication of constructive information to the estimator, based on careful observations of the unit of work itself, giving true reasons why there is a deviation from estimated cost. Estimators are usually grateful for receiving such information.

Morale

Another responsibility of the general superintendent is the establishment and maintenance of the field forces' morale. This goes beyond merely giving compensation and reward: it extends to the orientation of field personnel toward attitudes that are beneficial to the company and its operations, as well as instilling in these forces good attitudes towards safety, intrafield cooperation, and general attitudes of helpfulness and cooperation to the entire construction operational organization. Positive attitudes toward preventing waste and keeping equipment clean and serviceable are other attitudes to be encouraged that will benefit the company.

Making the company policies and positions clear to key supervisory personnel is one area in which the general superintendent should provide open, clear explanations. He must relay honest and accurate information to his subordinates about these matters.

The good general superintendent needs to promote affirmative positive thinking on the part of his subordinates. There is no better way to accomplish this than establishing free and clear communication between all members of the construction organization. Few people can act affirmatively under clouds of uncertainty about their own individual position within an organization. Subordinates who are undercut or not reasonably supported in their acts and decisions are likely to produce a serious organizational morale problem. Morale should not be a difficult problem for the general superintendent provided that he is fair, honest, and loyal to his subordinates.

Subcontractor Relations

The last responsibility of the general superintendent listed here is fairly restricted in scope; it has to do with the general superintendent's role within the subcontracting process. He must be restricted in this responsibility; otherwise he interfers with other operators who are primarily responsible for these operations. Each and every problem, for example, the job superintendent has with his subcontractors is not the general superintendent's responsibility to solve. In fact, only under the most special circumstances should the general superintendent interfer.

Not every insignificant problem or dissatisfaction a subordinate operator has with a subcontractor should take up the valuable time of the general superintendent. The general superintendent can and should give assistance in situations where his subordinates have particular or peculiar problems with subcontractors. Certainly when there is a deliberate vital holdup of delivery on a job caused by a subcontractor or material supplier, the general superintendent should assist in solving the problem. There is every possibility that the attitudes of a habitually uncooperative subcontractor can be corrected by the general superintendent; this is because the subcontractor recognizes the general superintendent's authority, position, and influence. There are even some instances when this type of subcontractor, on the general superintendent's recommendation, can be excluded from future consideration for work by those who have the responsibility for subcontracting in the operational organization. When subcontractors realize that the general superintendent can affect their future business by his recommendations, they tend to reform their uncooperative attitudes.

The effectiveness of the general superintendent in these types of problems is based directly on his position of authority and influence in his own firm. Therefore overuse of the general superintendent in insignificant problems will reduce his effectiveness. The general superintendent must exercise the greatest discretion before stepping into a situation of this kind.

Demands on Time

The foregoing list is a description of only some of the main responsibilities of the general superintendent. Even if he had only these responsibilities to meet, the reader should recognize that he is a very busy individual exercising great influence and authority in meeting the main operational objectives of completions, quality, and costs.

Because of the demands on his time, difficulties sometimes arise between him and his subordinates. Perhaps abruptness with subordinates on his part or the unnecessary insistence by his subordinates on detailed explanations of his decisions or actions may cause trouble in the personal relationship that must exist between them. An understanding of his time problem is essential by those who deal regularly with him. The general superintendent must regulate his time carefully; he must ask himself how he can accomplish all of the things he is charged with doing.

3.6 THE DUTIES OF THE GENERAL SUPERINTENDENT

Personal Tasks

In meeting his responsibilities, the general superintendent has specific operational duties to perform. Generally, the responsibility for the performance of these duties is exclusive to his position. They are based on practical considerations of efficiency and authority. Some are of a technical nature and some of a nontechnical nature; some are specific and some are general. The following list of a few of these duties (with brief explanations) should illustrate the range of the performance functions of this construction operator.

1. The general superintendent has the duty of seeing that the original survey and job layout is checked for accuracy. This is his duty because only he has the authority to either order other company operators to do the checking or to hire outside surveying and layout crews to do the work. It is hard to imagine a greater embarrassment for field personnel than when they find that the building is located in the wrong place. Embarrassment should be reason enough to require control over the operation functions, but the cost consequences of mistakes in layout can be much more serious.

2. The general superintendent must keep his subordinates informed on the long-term and immediate availability of the company's material and equipment. This duty extends beyond his responsibility for the original assignment and distribution of these resources to the job. The general superintendent must know, or be able to find out quickly, what is available within the company inventory when his subordinates request material and equipment. He must give accurate and full information to those who need it.

3. The general superintendent must personally examine and take note of equipment use and misuse, its maintenance, and its cost of operation. This duty includes the close observation of both the labor and equipment costs involved in overequipping a job.

4. The general superintendent must assign key personnel to jobs and move them as required by the needs of the total field operation. This duty requires observing these people as individuals, noting their skills and personal strengths and weaknesses. It is the rare trade foreman or mechanic who has both mechanical and managerial skills of the same degree. On the one hand, a foreman who is overly particular about workmanship will probably not show the best results in meeting costs; on the other hand, a hard labor pusher is not the most likely man to be the strictest about the workmanship of the mechanics of his trade. It is the general superintendent's duty to select the best man for the particular work to be done, balancing carefully the objectives of quality and cost.

5. It is the general superintendent's duty to confer with the job superintendent in the selection of means and methods of construction and the scheduling of the job. Selection and scheduling are both directly tied to the factors of men and material availability. Means and methods are always directly related to reaching the main objectives of completions, quality of production, and costs. The general superintendent must also fit standard means and methods, developed for the total field operational capacity of the company, to the individual projects in the company's volume.

6. The general superintendent has the duty to determine or at least project the short- and long-term needs for key personnel. This is at times very difficult. Keeping key people on the payroll who are not being productive always presents a difficult operational problem. Selecting key people in relation to the projected needs of an organization is difficult, because this involves a reasonably accurate

projection of future work possibilities. One important point the general superintendent must remember here is the following:

It is easy to say that key people who are not productive are expendable, but the good general superintendent cannot expect good people to be available only when he needs them. Good people do not stay idle.

When the need comes for the special skills these people possess, the general superintendent is faced with real problems when he cannot reach into available and proven manpower resources within his own organization and find them there.

The general superintendent must weigh very carefully the value of keeping key people (whatever their attendent cost may be) employed when he realistically projects a future need for their services. Sometimes it is possible to shift or replace other less competent people with his key employees. Sometimes key employees will be of such great value that their availability can be insured by using them only in a partially productive way, as, for example, working a foreman with his tools until he can be utilized to his full potential.

7.   The general superintendent must evaluate key personnel. It is best for him to make his evaluations as objectively as possible. The performance, cost effectiveness, quality, and cooperation of individual key people are some basic guidelines for the general superintendent to be aware of. In any case, he should realize the strong and weak points of each subordinate. His analysis of the relative values of his personnel can be accomplished either by his personal observations or through references from their co-workers.

Compensation

The general superintendent has the pleasant but sometimes difficult task of setting compensation for his subordinates. This includes the determination of compensation over and above any recognized (usually union) scales of compensation. It includes determining extra compensation and rewards and giving promotions. Usually, tradesmen and foremen in construction industries represent themselves simply for what they are, that is, journeymen or foremen. In union areas minimum rates of compensation for these people are established by the labor contract. This facilitates the general superintendent's task, because the ordinary journeyman or foreman will be compensated according to the contract. When exceptional people are involved, however, good sense requires giving them extra compensation in some way. The union contract does not restrict extra compensation.

The general superintendent has the same determinations to make about nonunion segments of the field operational organization. There is probably no industry group that passes on personal information among themselves faster than the construction industry personnel. Keeping a secret about differentials in pay is almost impossible in construction companies. Construction company organizations are usually compact and their employees correspondingly work very closely with each other. Thus intercompany members become rivals not only in actual pay scale differentials but in their own analysis of how they are regarded by the company. The general superintendent obviously cannot keep everybody satisfied at all times, but he can mitigate the results of these rivalries by establishing fair and equitable standards in determining differential compensations and promotions. Seniority is one standard; the degree of responsibility is another. The size of a crew and the type and complexity of a project are other criteria by which to establish the amount of responsibility a subordinate is capable of handling. Even though such criteria are open to questions of application and have their inequitable features at times, at least they are recognizable standards his subordinates can understand. Once subordinates understand that standards have been established and they are closely adhered to, potential problems in this area become less numerous.

One of the difficult matters in the area of compensation for the general superintendent to act on fairly and objectively is when more compensation is being paid to people who actually have less responsibility than do other co-workers. This is particularly apparent in the lower management positions in the operational organization. Union contracts can actually force payment of more compensation to people that have less overall responsibilities than others. Carpenter foremen, for example, are often paid more (on an hourly basis) under the terms of a union contract than the job superintendent or job engineers. Even in the non-union situation it is still possible for this apparent inequality of treatment to exist, particularly when the demand for highly skilled trade foremen is greater than the available supply. This presents an extremely difficult problem for the general superintendent, because he fully realizes the necessity and value of these lower management people; yet he cannot pay them equally. He can only go so far in solving this problem by furnishing an acceptable explanation. He can give some explanation to justify the situation by comparing the total yearly earnings of the management people with the high weekly compensation of the journeymen. Statistics tend to indicate that the average journeyman mechanic only works so many hours a week averaged over a year's time. This is a rather weak explanation, because his young operators are fully aware that good people work all of the time. By making his lower management salaried employees aware of the salary pay system's advantages over the hourly system the general superintendent will often decrease the differential considerably in the minds of his non-union subordinates; at least it should make these people feel better. But the most influential of all the arguments he can use, however, is to show that by developing their managerial and operational skills, these lower management operators will more quickly move into positions of greater responsibility, resulting in higher pay over the long term.

Another set of difficult problems encountered in the area of compensation are bonus systems, which are used extensively in the construction industry. Bonuses usually result from a pool of money that is available for distribution as a result of good business and operations over the past year. In most cases this money will be taxed if not distributed. Company executives feel that it is better to gain some advantages in goodwill with their employees rather than to pay taxes on the money. Distributing this money would seem to be good personal relations policy, but the system creates other problems for the general superintendent.

Usually some standards for the distribution of bonus monies to employees must be established by the construction firm. The bonus is intended to reward good effort. Standards identify both the participants who are to receive the bonus and the amounts of money to be paid to each of them. Another standard that should be established is the performance criterion by which the bonus is paid. This criterion is set by answering the following questions. What effort is the most deserving of this extra compensation? Is the size of the job to be the determinant? Are savings on labor allowances to be considered? Is seniority to be a factor?

It is almost always the case that the general superintendent has a major involvement in this bonus distribution process, at least to the field employees. When he is involved, he should try to avoid creating unnecessary problems for himself, which might happen because he is the one who decides which person goes to which job, large or small, simple or complex. He is the one who decides on putting his most talented subordinates on the really tough labor/cost ratio jobs. It is likely that he will assign his best people on the most difficult work, irrespective of the capacity of that work to make good costs. If good cost furnishes a major criterion for bonus distribution, then obvious injustice occurs and his problems are multiplied. Then, too, if seniority should be another criterion the general superintendent is faced with the problem of encouraging talent irrespective of age or service. All of these problems must be carefully considered before establishing bonus standards.

No greater skill is required in handling personnel relations than in this area of extra compensation. The best thing to do here is to keep the bonuses as even in amount as possible, with as wide a distribution as is practical.

Adjunctive Help

It is the general superintendent's duty to determine the need for adjunctive help in construction operations. This generally involves employing consultants in areas that are obviously beyond the capacity and ability of his own personnel. This is his duty, because he exercises central control over field operations and is the one person who is best able to determine the inability of his own forces to cope with problems. Outside consulting requirements are most common in situations involving structural and soil mechanics problems. Hiring legal consultants specifically is not handled by the general superintendent, because this is the responsibility of his superiors in all cases.

Developing Cost Systems

The general superintendent must demand absolutely proper and complete cost recording and reporting from the field forces. This is essential to his responsibility of attaining the objectives of completions, quality, and costs. The general superintendent has great use for cost reports and records and should therefore take great pains in insuring the validity and accuracy of their contents; he must thus insist that they be kept exact and complete. He must also realize that those subordinates who will not do the cost recording and reporting correctly or who are sloppy about doing so are in fact detriments to the total construction organization and to the company. Such people are not only a hindrance to these essential elemental operations, but they become a positive menace to the business and prosperity of the construction firm.

There can be simply no exceptions to the replacement of these irresponsible subordinates, and the sooner they are replaced the better. This is so, irregardless of whatever other real talents they may have. No construction firm can honestly call itself efficient or professional if it does not have a valid cost-reporting and recording system. But it must be admitted here, with great reluctance, that no system at all is better than a system which is misused, mishandled, or carelessly maintained, simply because of the resulting confusion and damage such actions will create. The general superintendent must correct such behavior or promptly discharge those who will not perform these operations properly.

Correcting Bad Costs

When costs are bad, it is the general superintendent's task to identify and correct them as soon as possible. The skill required in identifying the causes of good and bad costs results from experience. The remedy for bad costs consists of using forceful managerial and operational skills, as fast as possible. Realization of the causes of good costs is a great advantage to operators, and to the firm as well.

Safety and Cleanup

Two other major areas of concern in the operations of the general superintendent arise from two separate aspects of operational discipline that are similiar in nature and in some ways interrelated—safety and cleanup.

The general superintendent may exercise more patience with subordinates who will not run a clean and safe job than he can with those who will not keep good costs—but not much more. Again, it is a "do it or else" attitude that the general superintendent must clearly demonstrate to his subordinates. This attitude and insistence is fully justified by the consequences of not being clean and safe on a job site. The justification for this arbitrary type of attitude is based on competitive, economic, business, and public relations grounds; there is also the truism that says that a sloppy and unsafe job is a direct reflection on the building competence of those who are in charge. Conditions of safety and cleanliness on the job tell stories, both good and bad, about the people running it. The general superintendent cannot allow sloppy, careless housekeeping on his projects and he certainly cannot permit adverse testimonials to the company's reputation.

There are many more duties of the general superintendent that are supplemental to his major responsibilities. They are general in nature but important enough to mention here.

1.   The general superintendent should attend job meetings when his authority and presence will have some significant meaning or impact.

2.   He has supplementary duties relative to his labor, and labor relations responsibilities, particularly in keeping his subordinates informed of new developments in these areas. Transmitting information about current labor contract condition problems is also an important task.

3.   There are supplemental duties that the general superintendent has in the operational set up of the yard and its relation to the field. Determinations of yard layouts, plant layouts, and the distribution of space fall into this category.

4. Another of his duties is building confidence and pride in his subordinates. This includes seeking and insisting upon their mutual cooperation, and discouraging internal stress and buck passing within the total operational organization.

5. The general superintendent also has the continuous task of fully preparing himself with facts, figures, and pertinent information for his superiors. His superiors set the requirements here and he adapts his operation to these requirements. Clear, concise reporting on his phase of operations will build confidence in his ability in the minds of his employer.

3.7 DEVELOPMENT

We have said before that experience is the trademark of the general superintendent operator. But experience is not the only background requirement a general superintendent should have. He must develop personal characteristics that are essential to becoming expert in this operational position and then use them wisely. There are several personal qualities a person can develop, along with experience, that will be of great assistance in carrying out operations:

1. A sense of fairness.
2. A sense of firmness.
3. Foresight.
4. Thorough thinking and analyses.
5. Open-mindedness to a reasonable point.
6. A sense of toughness.

The value of the first four characteristics should be fairly apparent to the reader; the other two may not be.

A general superintendent must learn to listen. Even though his job entails mostly ordering, directing, or explaining, it is very important for him to listen carefully to subordinates. A great deal of valuable information can be stored by an open mind. Listening carefully, with an open mind, to ideas, whether they appear at the time to be valuable or not, is an extremely useful talent for a superior to develop. Just one good idea, listened to with patience, can result in real benefits in either its immediate or future application in construction operations.

But a hazard any general superintendent must guard against is listening too long to ideas, discussions, and arguments of his subordinates that may adversely effect his decision making. The general superintendent must realistically establish for himself when to close off discussion and arrive at an appropriate and timely decision.

When the word *toughness* is used here it does not mean that the general superintendent is to be rough, rude, insolent, or inconsiderate; it means that he must be able to be objective in his relations with the subordinates he controls and directs, at all levels. Toughness means the ability to discharge a subordinate on his performance, regardless of any personal considerations to the contrary, and dealing with familiar subcontractors who do not perform properly as if they were strangers—if this is what it takes to make them perform properly and on time. It means honest, forthright statements of facts, in spite of whom these facts may disturb or hurt.

The general superintendent in the construction operational organization is truly a unique operator among many others in the organization. He plays a major part because he controls and directs the operational organization that produces the essential product of construction, which is a demanding and time-consuming job. Both the success of the business and operational organization, and thus the profit to the firm, depend greatly on the performance of the general superintendent.

4

Job Supervision

*Experience is the child of thought and thought
is the child of action. We cannot learn men
from books.*

Benjamin Disraeli

4.1 DEFINITIONS

The main purpose of the construction operation of job supervision is the on-site direction and management of the individual construction project. It requires knowledge and application of all project-related operations and procedures. Job supervision is performed by an operator designated as the job superintendent by construction tradition and practice.

The function of the job superintendent is to direct and control *all* operations of work on the site of the project. This direction and control includes the direction of the labor of his own forces and the materials they use, handle, or install. This operation also includes the direction and control of subcontractors, but only in relation to their being independent contractors and not as direct subordinates.

The job superintendent may from time to time be responsible for the direction of off-site work when it is directly related to his project.

The project the superintendent controls and directs is usually located at one site. Sometimes the project may consist of multiple places of work on the one site. His project, for example, may consist of one or several buildings on the site. Usually, the general superintendent decides whether or not individual building projects on one site each require their own superintendent; he may decide on a single job superintendent per building or a single job superintendent for the entire site, assigning assisting supervisory personnel to the job superintendent as required.

The work performed by the job superintendent is that of overseeing all of the work necessary to construct the building. This work is not only confined to the actual physical building operations but includes all supplementary and complementary work as well.

Example. When a project has a concrete substructure and frame, the job superintendent is ultimately responsible for the proper placement of forms for concrete. These operations are generally delegated by him to the carpenter foreman, who directs the physical erection of the forms. Whatever else is necessary to accomplish this erection constitutes the work for which the job superintendent is ultimately responsible. Thus his responsibility includes proper form design, proper detailing of the framework, the safety of the form operations, and so forth. His work, therefore, consists of more than the mere direction of men and materials to a place on the project in order to accomplish the form erection. Since much of his work is delegated to subordinates, part of the superintendent's task lies in the proper delegation of functions and authority to people who are competent to perform the erection of the forms. Here his responsibility becomes part of his work as a construction job superintendent.

4.2   A CONDITION OF CONTRACT

There is one feature that distinguishes the job superintendent from the other operators of the organization: his is the only position that is a directly stated requirement of contract. Usually, the conditions of contract specify the job superintendent. They specify this person by the name *superintendent,* and they describe him and his main responsibilities contractually. Let us examine such a condition.

In the American Institute of Architects (AIA) General Conditions of the Contract for Construction (AIA Document A201) condition 4.9 obligates the contractor as follows*:

> *The Contractor shall employ a competent superintendent* and necessary assistants who shall be in attendance at the project site during the progress of the work. The superintendent shall be satisfactory to the Architect, and shall not be changed except with the consent of the Architect unless the superintendent proves to be unsatisfactory to the contractor and ceases to be in his employ. *The superintendent shall represent the Contractor and all communications given to the superintendent shall be as binding as if given to the Contractor.* Important communications will be confirmed in writing. Other communications will be so confirmed on the written request in each case.*

All the statements contained in this provision are important, but the statements in italics are of greater importance in knowing the job superintendent's position within the construction contract and his relation to his own construction operational organization.

The condition states quite clearly that the contractor shall employ a competent person to be superintendent. This creates a clear and definite contractual obligation for the contractor to fulfill. The condition also states that the superintendent "represents" the contractor. Here a question arises. Is there an implied admission by the contractor that the job superintendent speaks and acts as if he were the contractor in person? (*Person,* as used in this context, means the company, or entity, for whom he works.) The answer to this depends on many factors and reveals some unusual features of this operation. There is no doubt, for instance, that the consequences of the job superintendent's actions in representing the contractor make it necessary for the job superintendent to be fully competent in conducting his construction operations. His operational abilities are somewhat assumed by the contract. Because of this common contract condition, the job superintendent's position is unique in the operational organization of the cost of the contract form. None of the other divisions of the operational organization are conditioned by contract, directly or by specific name. This is not to say that the other operational positions are not required by other provisions of the conditions of the contract. The superintendent's position is, however, made more clear in the contract than the position of any other operator, including the chief executive of the contracting company.

* Italics are author's.
* Reprinted by permission of the American Institute of Architects.

4.3   THE JOB SUPERINTENDENT AND THE MAIN OBJECTIVES

The objectives of construction operations discussed in Chapter 2 are the same for the job superintendent as for the other divisions in the construction organization—completions, quality, and costs—but for his project only.

Completions

The completion objective for the job superintendent differs from that of the general superintendent. The job superintendent must make it his responsibility to have the work completed on his job in a continuous, steady, and orderly manner. His job, however, is only one of several in the company's volume. In many instances it is necessary for him to have direction and assistance from the general superintendent and the expeditor in meeting his goals. There are reasons for requiring the direction of the general superintendent as to the completions on the single job and the total production of the firm, including considerations of foreman availability, the delivery of essential but unavailable materials, and weather conditions requiring the acceleration or decleration of phases of the single job.

Usually, job completions are somewhat predetermined by the single job's own progress schedule. If the single job's schedule is followed and maintained, its completions should be satisfactory to the company's total completions situation.

It is the responsibility of the job superintendent to regard the completions on his job as separate and distinct from other company projects, unless he is specifically directed otherwise. This may seem to be a self-serving and shortsighted attitude to some extent, but it is practical. If each job superintendent concentrates on completing his job, the sum of the total completions in the whole organization should be acceptable.

These are certain things the job superintendent must consider about completions that are unique to his position:

1. The availability of competent and sufficient labor and its use within the job schedule.
2. The availability of materials and equipment in relation to their use within the job schedule as well as the consequences of delays in deliveries.
3. Weather and all of its effects when it is liable to disrupt completions.
4. The availability of subcontractors and the consequences of their failure to perform on the total completions situation.
5. The interrelationship of completions, quality, and costs, and especially keeping a reasonable balance between the completion and costs objectives. There can be no compromise on the quality objective.

Problems With Completions

The job superintendent must be keenly aware of common problems in completions that arise for the following reasons:

1. Interruptions and delays.
2. Partial overcompletions.
3. Incontinuity of operations.

It is improbable that a construction project will not experience some interruptions or delays; there are numerous ways in which these can happen on a job. It is therefore necessary that the superintendent use sound judgment and reasoning in adjusting to these conditions. Keeping the objective of completions in mind, he must be able to adapt his operations to any delay situation as immediately as is reasonable. It may be necessary that he even readapt operations over and over again until completions are back in phase—both in direction and in amount.

There are also some serious consequences caused by partially overcompleting the work. *Overcompleting* means completing work out of phase and balance in relation to the job's schedule of operations. It is not likely that the superintendent's superiors are going to object initially to these overcompletions; but overcompletions often make it extremely difficult for the superintendent to maintain efficient use of labor and overhead on his job. In succeeding operations on the job, problems with labor and overhead relate directly to the objectives of cost as well as completions. Unless partial overcompletions are planned for, they can seriously hamper total completion objectives and adversely affect costs.

The job superintendent's problems with continuity are somewhat different. He knows that the building of a project takes on a rhythm; almost any production operation must have a rhythm. Adjustments must be made when this rhythm is broken. Some of these adjustments, unless they are very carefully thought out, are liable to cause disruption in production.

Example.   In the concrete framing of a high-rise building completing the frame depends on the proper sequencing of labor and materials. This sequence is usually a form operation followed by a pour operation, a curing operation, and a strip operation. All of these operations are planned into a time frame. Once they start they are repeated until the frame is complete. There is a definite rhythm to the framing operation. If one floor is planned for completion in two weeks and the superintendent has it done in one, some problems arise that can become very complex. The costs and completion objectives seem to be satisfied. Completions are twice those anticipated and costs would normally be better because of this. But what has happened here? The rhythm has been broken. Consider the curing operation: the curing requirements depend on time and thus this operation will restrict the availability of form materials. This, in turn, will affect succeeding operations, thereby causing a total disruption of the rhythm of the framing operation. This break in rhythm will eventually affect the completion objective if the interruption in continuity has not been planned for. It will affect labor and material costs as well.

This example shows that it is wise for the construction superintendent to consider very carefully the problem of maintaining continuity of operations in relation to completions.

Quality

No one in authority in the operational organization is closer to the quality of the total product than the job superintendent. The mason foreman is closer to the brickwork and the carpenter foreman is closer to the form and concrete work, but the superintendent is the one person in the entire construction organization who is directly responsible for maintaining the quality of the total work in production operations. He should insist on quality construction of the total product.

Since the job superintendent is responsible for quality, he must give special consideration to this objective. He must be determined and must develop some special skills based on good judgment and objectivity. He usually has leeway to weigh the objectives of completions and costs in relation to each other. He should never compromise on quality.

When examining the objective of quality construction in Chapter 2, we looked at the general guidelines that all members of the operational organization should consider about quality construction.

The job superintendent must give special consideration in meeting this objective of quality to the following:

1. He is the initial authority for the demand of quality construction and is unique in the operational organization in this respect.
2. His own professional reputation depends on his production of quality work as reflected in both his past and present jobs.
3. The way he approaches his demand for quality, depends on whom he is demanding it from.
4. He must assess the kind of quality required for the whole project or the simplest individual building operation at hand.

Construction mechanics are usually proud of their abilities to do quality work. Too often, however, mechanics will take note of the degree of workmanship expected of them by the job superintendent. This applies to subcontractors as well. The superintendent is the initiator of the objective of quality: in one way or another he sets the requirements for what is to be expected in quality workmanship; he establishes what is acceptable and what is not; he corrects false impressions of what is considered acceptable quality workmanship. He sets up the tone of the job in this respect. Indifference and carelessness are not to be tolerated in anyone who works for him. While somewhat different approaches to quality work are used in regard to subcontractors than with his own forces, the objectives are the same. Quality, as the job superintendent determines it to be, is the only acceptable standard.

Construction superintendents become well known in their local geographic areas for their dispositions, peculiarities, and competence, as well as for their attitudes on quality workmanship on their jobs. Among construction people there is a common tendency to directly relate a job with its job superintendent. These people in particular relate a quality building to its job superintendent. There is no doubt that the superintendent's employer also relates the superintendent's production of quality buildings to his reputation, value, and position within the company organization.

The superintendent must carefully consider how he approaches his demands for quality. He cannot, for example, approach his own people with demands that are not consistent with his approach to subcontractors. Good quality is good quality no matter who is to give it. The subcontractor is simply not in the same position relative to the superintendent as are the superintendent's own work forces. Even so, the superintendent must show consistency of attitude and determination in his demands for quality between these two work forces.

The job superintendent cannot approach his demands or vary his corrections differently between his own trade foremen in the different trades. This is sometimes a delicate problem, because all of the foremen work directly for him. While it is true that there exists a direct employer–employee relationship in this situation, it does not necessarily follow that the same quality of work demanded of one trade should be demanded from another. For example, the quality of work demanded from his concrete labor foreman might be quite different from that demanded of his mason foreman. It may be true that the prime quality to be demanded from the concrete labor foreman is maintenance of the structural integrity of the concrete, whereas the prime quality to be demanded for masonry work is appearance. The job superintendent must recognize these differences and approach his demands for quality with discretion and good judgment.

How does he approach quality as it applies to work? He uses subjective criteria based on good common sense. For example, does a sewage treatment facility have to be built with the same kinds of quality as an art museum? Do the underground effluent structures in the sewage treatment facilities have to have the same kind of quality construction as the aboveground construction? Are the below-grade nonexposed walls of the art museum to be built with the same kind of quality as its exposed exterior walls?

The answer to all of the foregoing questions is always yes, if applied to the basic integrity of the structure. This kind of quality work is always required, no matter what type of structure it is. But what constitutes integrity? The integrity required, for example, in building the below-grade concrete wall might be in building a structurally sound and waterproof wall, with little consideration for its appearance. The integrity required in building the above-grade wall might be, in addition to structural and tightness requirements, the appearance of the work. The superintendent must know how to distinguish the quality demanded for the type of project, structure, or building operation he is working on. This is where his determination and special skills in judgment and objectivity serve him best.

Costs

The third main objective of construction operations is cost. This has been examined in Chapter 2 as it applies to the entire operational organization. The job superintendent on his own job is equally responsible for this, as well as for the other two objectives. Many of his responsibilities relative to this objective closely parallel those of the general superintendent: both operators must place more emphasis on labor costs than all of the other costs; both are responsible for knowing the concept of the labor unit and its application in operations; both have the responsibilities of the assignment of labor; and there are many other similarities in the operational approaches that both must take in meeting the objective.

The difference between the two responsibilities results from the closer relationship the job superintendent has to the actual building operations. He directly observes the building of the job and all its component parts. He has a daily personal relationship with his foremen and mechanics and is on hand to personally observe and note cost. He is expected to be labor and cost conscious and cannot delegate this responsibility to anyone else. He is on the scene.

He is the first to note bad costs and the first to know their causes; he is in the best position to order their correction. He has the authority to order changes in the means and methods of construction to adjust costs. He determines where fault lies—in his foreman, mechanics, or both. He is the first person most likely to realize the effects of the nonperformance of his subcontractors on his own labor costs.

He is the one who is on the job when it rains; he is also there the next day to observe the effects of the rain on site access and thus its effects on costs.

He is the man who examines and directs equipment usage—a significant factor of cost. It is his job that is stopped when a labor dispute arises and cannot be settled. The labor dispute is on his job only, and he can see the dispute's effects on its cost and completions.

These and other factors cause cost problems that the superintendent faces and solves: they become his exclusive problems because he is the highest authority on the site; he witnesses the situations that create problems of costs. His main thrusts in meeting the main objective of cost in operations must be directed to carefully observing the causes of problems and doing something to solve them.

4.4 JOB ORGANIZATIONAL STRUCTURE

Figure 3 shows the typical operational control structure on a construction project, whereas the diagrams shown in Figure 2 (Chapter 3) and 4 represent only the

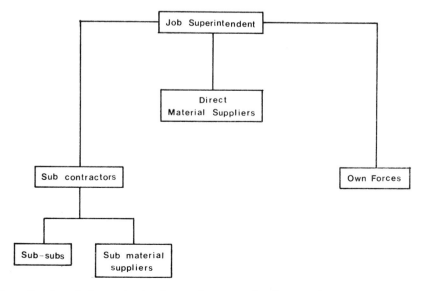

Figure 3. A typical organizational control structure for the typical construction project.

basic field organizational structures used for construction operations. There are many possible variations of these structures, but these illustrate job organizations that have been proven by their traditional use to be efficient and effective for conducting field production operations.

None of these structures are to be regarded as rigid. Flexibility in their application in practice is necessary for the field organization to meet the main objectives of operations. Certain features of these structures should be noted by the reader. The following is shown in Figure 3:

1. There is a direct line running between the superintendent and his own (general contractor) forces and also between the superintendent and his subcontractors, but no lines run between the subcontractors and the general contractor's forces.

2. There is an indirect line running between the job superintendent and the subcontractor's material suppliers. This is necessary because while the superintendent has no direct control over these people, he has a direct interest, for example, in their movements and storage spaces on the job.

3. There is a direct line running between the superintendent and the material suppliers, but no direct line between the material suppliers and the superintendent's own forces. Even though his own forces use these materials, it is necessary to make this separation in control to prevent duplicating aquisitions of materials and wasting effort.

In Figure 4 we need to explain an apparent contradiction to what has been stated elsewhere in this text: Figure 4 shows no direct line running between the superintendent and his mechanics; he operates through his foremen. But the reader should realize, when interpreting this diagram, that the job superintendent is indeed personally responsible for trade operations when the objectives of cost or quality considerations are being compromised. The job superintendent does not, for example, take his foreman's word that the men are working efficiently or producing quality work; he must observe these things for himself. He must develop ways of observing production while at the same time preserving the authority of his supervisory subordinates.

We should now bring to the reader's attention and further explain some other features of Figure 4.

1. There is a direct line between the superintendent and his foremen, but there are none between the foremen and the job support positions. This part of the diagram requires a bit of explanation.

In fact, there is a good deal of direct cooperation between these people. In practice, it is not efficient to rigidly adhere to this structure. Certainly there is a direct relationship between the job engineer laying out the work and a carpenter or foreman helping him, and the timekeeper is often in direct contact with the foremen and their time responsibilities. The operating engineers usually work for and under the direct control of a trade foreman when their machines are working for that trade.

These instances of direct cooperation versus the diagram's lines of control show that this formal field organizational structure must be used with great flexibility.

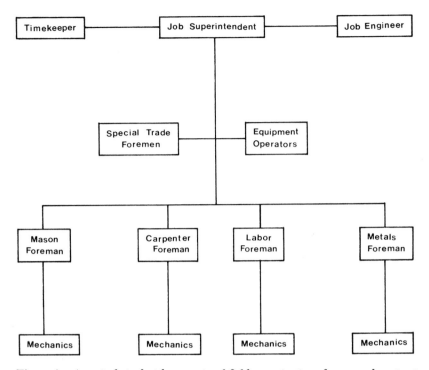

Figure 4.   A typical single job operational field organization of a general contractor.

This kind of organizational structure is necessary, however, because it accomplishes the purpose of giving the superintendent primary control over all of the support personnel and their operational functions. It allows him, for example, to say which trade will have a machine and when they will have it, thus giving control of the machine in the total operation; at the same time it allows the direct control of the equipment to pass to the trade foreman when the superintendent's decision has been made as to which trade is to use it.

2.   There are other variations of the formal structure that are required in practice which might not be immediately apparent from an examination of Figure 4.

The following example illustrates a variation.

Example.   The concrete form operation usually requires many trades working concurrently or consecutively on the forming: carpenters, laborers, finishers, rodmen, and operators are usually involved. Their objective is to prepare for the final operation—the concrete pour. Which of the trade foremen is to be in charge of the various trade operations? When does the responsibility pass from the forming operation to the pouring operation and then to the stripping operation? Which foreman runs each operation and to what point?

Only the superintendent can answer these questions. He may decide, for example, that when the rods are being placed, the rod foreman will have direct control of the entire operation until the rod installation is completed; then control will be passed to the carpenter foreman, or maybe the carpenter foreman, since

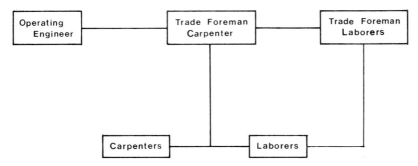

Figure 5. A typical specific trade operational organization on a general contractor's construction job.

he initiated the operation, will have general control over the entire operation, including both forming and rod placement. The superintendent may then put the labor foreman in charge of the pour and the carpenter foreman back in charge of the stripping operation. There are any number of variations in the operational control of his job organization that the superintendent can use.

This last example of a particular operation's direction should reinforce in the reader's mind the need for the flexibility of the organizational structure in practice. The superintendent is concerned with all of the operations—the forming, the rod placement, the mechanical/electrical installation, the pouring, and the stripping. It is up to him to see that all of these things are accomplished on time and in the proper sequence so that the main objectives of completions, costs, and quality can be met. His authority establishes who is responsible for these job operations at any and all times in the construction.

Expansion of the Structure

Figure 5 is a diagram of the typical individual trade organizational setup, which is also varied and expanded when the superintendent sees fit.

Variations in formal organization structure are necessary for several reasons. When variations are made, they cause a split between the job superintendent's and the general superintendent's responsibilities. The former is responsible to adapt the structure to his job, the latter to set up a workable and adaptable structure for the company. Neither should demand variations without the approval of the other. It is likely that the general superintendent has established the company structure of trade organization in some detail. He has done this to establish a job organization pattern for the company as a whole. Variations in the structure must often take the form of expansions of areas of responsibilities when the job superintendent sees the real need for it. Usually the necessity for these expansions is caused by the size of the particular trade operations on any job. An example of such an expansion is shown in Figure 6, which expands Figure 5 but does not basically change it. Figure 6 represents the job organization for a project where there is a very substantial amount of concrete forming on a large site.

The expanded trade organization illustrated in Figure 6 must obviously be jus-

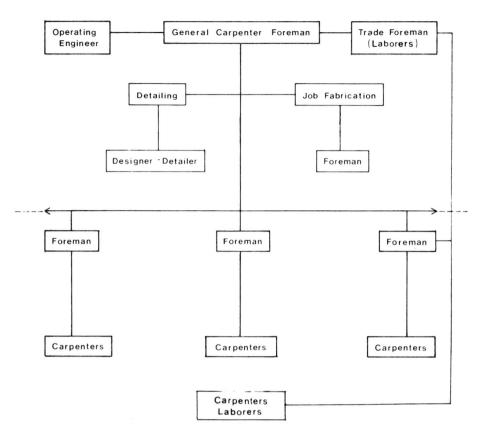

Figure 6.   An expanded specific trade operational organization for a concrete framing operation.

tified by the needs of the job. It must be shown that the organization shown in Figure 5 would not be sufficient (and thus, efficient) to get proper completions, quality, and cost. There is no question that the organization in Figure 6 increases cost simply because of its addition of nonproductive support trade personnel. But is it more costly than trying to operate under the organization shown in Figure 5? Does it decrease the possibility of poor workmanship because supervision and direction are too remote from operations? Does it increase completions? The general superintendent and job superintendent must answer these questions.

Here the job superintendent has a special responsibility. While the general superintendent is his superior, the job superintendent has the responsibility to impose on the general superintendent his views based on his personal observations of job operations. He must be able to show to his superior, in any case at hand, the relative efficiency of using a larger organization. Generally, he has to do this on the basis of cost effectiveness, and this is difficult because the expanded organization, on its face, increases the costs. Nevertheless, the job superintendent is responsible for the performance of his organization and should be forceful in his demands for adequate and efficient forces when he knows they are needed.

The diagrams contained in this chapter are included to show the basic design of field organization structure. Field operations require that these or similar organizational structures be set up for practical operational purposes. Once job organizations are firmly established, they minimize the efforts and time of the superintendent in directing and controlling the project.

Using the Structure

There are some additional considerations the superintendent must give to the matter of job organizational structure:

1. He must use it for the purpose intended.
2. He must fit persons to it.
3. He must make it adaptive to ever-changing job conditions.
4. He must try to maintain it.
5. He must be fair and firm in running it.

Often the superintendent will find that his organization is not accomplishing the main objectives of construction operations. This usually happens because he is not using it as it was intended to function. What if, for example, there is one dominant foreman on the job? This foreman, well intentioned or not, forces the operations of the job in a certain direction, which is often pointed at accomplishing the operations conducted solely by his trade. Sometimes this is a good thing for the job, particularly when this foreman, because he pushes his work, pushes all the rest of the operations. But what if this forcefulness destroys the initiative of the other foremen to the extent that they cannot perform the functions of their own trade? This is obviously undesirable.

The organization is set up to carefully define areas of responsibilities. As we have said before, there must be flexibility in the application of organizational structure to the job. We did not say, however, that the flexibility should extend to destroying the purpose of the organizational structure altogether. Each trade depends upon the other and is vital to the other's operation, but each is still separate and distinct in such things as the type and skill of the manual labor requirements, the materials used and handled, and the time required for the performance of the respective operations. It is presumed that the purpose of the separate trade foremen is to see that the peculiarities and distinctions of their trades are made to fit into the total operation. The job superintendent must see that each trade is properly utilized in the construction of the project. The easiest way to do this is to keep the work in conformance with the separations called for in the organizational structure as much as possible.

One of the most difficult things for the superintendent to do in practice is to fit what people he has been assigned or who are available to him into the organizational structure. People, being individuals, are different, and in many aspects. A job organizational structure chart just says *carpenter foreman*. The job superintendent must fit a person to this designation, considering not only differences in people but also how each potential subordinate will fit in with the people in the other trade designations. These people must all fit together into the structure in spite of any differences.

The job superintendent has an ideal situation as his goal. He knows that the job where everyone gets along and cooperates is a good job. He knows that when he has people who are relatively competent, they will most likely contribute to a smooth-running job. He knows that cooperation, mutual reliance, and common goals are necessary for successful operations.

While the assignment of key personnel is the responsibility and prerogative of the general superintendent, the job superintendent is still responsible for using the personnel to their best advantage on the job. He does this by carefully analyzing their strengths and weaknesses relative to his ability to meet the main objectives. For example, on the one hand, he knows that a pusher foreman might be only a pusher, who can be fairly well relied upon for good completions, but not so much for the objectives of cost and quality. On the other hand, he assumes that when he gets the mechanic foreman who takes pride in workmanship, this type of foreman will produce quality work. This person, however, might cause completion and cost problems. The job superintendent must give careful thought to seeing that the objectives are met with the personnel he is given to build the job.

Usually in a construction company most of the personal traits of its individual key foremen are recognized and are well known to the company's superintendents. How many men can a certain foreman push effectively? Does he insist on good workmanship? Does he detail for his men? Does he watch his supplementary operations closely enough? All of these and other questions are raised when the job superintendent is fitting a person to the organizational structure of his job.

The job superintendent's only alternative (if allowed to him) would be to hire his own key personnel. This is risky business. At least the job superintendent knows something about people assigned to him from within the existing operational organization from prior experience with them or through their "reputations" within the company. The job superintendent has at least some indications of their strengths and weaknesses. This is a much better situation than having to hire people of unknown quality or capacity. There are advantages to filling these positions by asking for available company-employed personnel or in being directed to use them by the general superintendent. The job superintendent should not have a negative attitude when key personnel are assigned to his job. There are positive features in this situation that can make his job easier and which will allow him to devote his restricted time to other important matters.

The job superintendent must frequently adapt the organizational structure to changing job conditions. Problems occur when a project is substantially increased in size during its construction or when it is necessary to change personnel for the overall benefit of total company operations. If, for example, there should be a substantial increase in the size of the job, the job superintendent is immediately faced with adapting the existing organization to operating a larger job. This becomes a difficult task when the original job organization was based on the needs of the smaller job. A carpenter foreman, capable of handling his responsibilities on the smaller project, might need considerably more assistance from, say, support personnel on the larger work. The superintendent must then adjust the duties and responsibilities of the support personnel to meet this need. Indeed, his present foreman may be wholly incapable of operationally handling the larger work.

Closely following the problem the superintendent has in adapting organization to changing job conditions is the problem of unavoidable or sudden changes in key personnel. The problem is serious when the changes are sudden. It is not at all uncommon for the general superintendent to require the movement of a foreman from one job to another. This reassignment can be made on very short notice and is usually based on the general superintendent's sound judgment that it is the best move for the benefit of the overall company operations. This creates some very serious problems for the job superintendent.

The foreman who is moved may be right in the thick of things on the job when the reassignment is made: he knows the drawings, the job, where everything is that his men need, and he knows his men. If he is to be moved, someone must take his place. Until the new foreman learns all that the old foreman knows about the job there can be a serious disruption in operations.

Needless to say, the above situation is easier for the superintendent to handle than if a foreman quits, is discharged, or is incapable. The superintendent must do something to remedy the situation he is faced with; the talent he loses must be replaced. In fact, the replacement might have to be more talented than the foreman who has left in order to save time and trouble in getting disrupted operations back on course.

There are a few things the superintendent can do to minimize the consequence of changing personnel:

1. He can look for and develop backup people already on the job to be his key personnel in the organizational structure, at least partially filling the gap when a change comes about.
2. He can immediately spread some of the load of the responsibilities to the other trade foremen of different trades where it is practical, thus allowing time for the replacement to give priority to the essentials and learn the job when he arrives.
3. He can, through proper analysis of the situation, direct total job operations away from the affected trade as much as possible until the replacement becomes familiar with the job.

This problem has more serious consequences on smaller jobs than on larger ones. It is more likely that the slack can be taken up in an expanded trade organizational structure on a larger job than in the typical structure used on a small job. In most instances the expanded structure can withstand such a change without much disruption in operations.

An actual experience with a discharged foreman will illustrate quite clearly to a job superintendent (particularly one who is new in his position) a simple rule in operational thinking: before discharging a key member of a job organizational structure for other than the most serious reasons, a job superintendent should think twice about the consequences, because the person usually must be replaced under conditions of severe pressure and uncertainty. Experience aids judgment on these matters. A job superintendent can live to regret hasty or thoughtless actions when he discharges members of his key operational organization. The responsibilities and work load these people carry are not discharged along with them.

The last thing to consider here about handling the job organization is in its actual application to operations. It is wise for the job superintendent to be fair and firm with all of his subordinates. The demonstration of undue favoritism, for example, while being a trait of human nature, is highly disruptive to any type of operation, because this (and similar traits) minimizes the effectiveness of an organization. Undue harshness, harping, failure to compromise, continual harassment, the consistent assignment of blame, and other similar tendencies will not serve the job superintendent well in running his operational organization.

Observable vacillation or uncertainty of a job superintendent can be just as disruptive to the good operational control of the job. When a subordinate observes these traits in a job superintendent, he loses confidence in the superintendent's abilities to judge and decide. When subordinates know that the superintendent is not firm in his decisions and judgments or will not stand behind them, they also lose respect for him. In many cases, advantage is taken by subordinates when they find that the superintendent will not or cannot be firm in enforcing his determinations and decisions on the job.

4.5   THE JOB TRADE FOREMAN

The job superintendent must be particularly aware of the position the job foreman holds in the construction organizational structure. The job foreman has a unique and, in some ways, contradictory position in the construction industry. Although he is usually from the ranks of labor, he performs primarily management functions. The foreman is usually considered to be a manager by the industry, the company, his men, and the law.

He is oriented towards traditional or known means and methods of construction operations. The foreman is genuinely proud of his knowledge of his trade and, in addition, is likely to be rather protective of its jurisdiction as well. He is usually interested only in the production of his own trade and is generally cost conscious only about his portion of the overall work. He relies heavily on himself and is motivated principally by the efficient handling of his men, material, and equipment in his trade.

The superintendent will find great advantage in developing a close relationship with his foremen that gives them full credit and responsibility for the positions they hold in the field construction organization. They should be made to feel that they are vital parts of the job organization. Ideally, no antagonism should exist between the job superintendent and his foremen; the superintendent is the head man with certain responsibilities to carry out, and the foremen are his subordinates with their own responsibilities. Mutual respect and the understanding of respective responsibilities between the foremen and their superintendent are absolutely essential ingredients in the proper conduct of construction operations.

4.6   THE DIRECT RESPONSIBILITIES

The job superintendent has some direct responsibilities in conducting his direction and control of the project. These responsibilities are unique and special to the position of the job superintendent. His main operational responsibilities are as follows:

1. Crew sizes and the number of crews required for individual operations and for the total construction of the building.
2. The assignment of specific construction operations to specific trades and trade foremen.
3. The assignment of areas of work, including assignments to his own and his subcontractors' forces.
4. The entire sequence of total operations.
5. The methods and means of construction after initial consultation with the general superintendent.
6. The proper distribution of labor costs.
7. Directions on the usage of equipment on the job and its proper maintenance.
8. Adaptations to conditions of weather, site, and access.
9. Job safety and cleanliness.

While these responsibilities may be exclusive to the personal jurisdiction of the job superintendent, their execution is often delegated by him to his foremen. Some of these responsibilities are acted upon jointly by the job superintendent, the general superintendent, and the expeditor. Let us briefly examine each of these responsibilities.

Crew Sizes and Numbers

The superintendent, with his respective trade foremen, determines the size of the crew to be used on any construction operation, as well as the numbers of crews to be used on a particular operation and on the total operation. Several influences guide these determinations:

1. The materials available for the operation.
2. The space allowed by the physical dimensions and conditions of the project, the site, and the place of installation.
3. The individual foreman's ability to properly handle the size and number of the crews.
4. The adaptability and capability of the crews in performing a range of work operations in their respective trades.
5. The relative costs of different crew sizes and the numbers of crews doing the work.
6. Any factor that affects the continuity of operations.

Assignment of Specific Operations to Specific Trades

Because of jurisdictional mandates in the labor agreement, the assignment of work items to a trade in a job superintendent's own forces is decided for him in union areas of the country; but even in these union areas there is some need for the job superintendent to assign specific trades to specific work. Usually, the assignments of his own forces result from his consideration of which trade will perform the operation most efficiently. Thus, for example, he might assign form stripping below grade to laborers and form stripping above grade to carpenters, even though form stripping raises jurisdictional questions. Frequently the job superintendent can be faced with these and other work assignment problems on the job. The reader should understand that at times this direction even goes beyond the effective assignment of trades in the subcontracts. When the subcontract effectively assigns trades, his problems in this area become more complex.

Assignment of Areas of Work

Construction operations are conducted both concurrently and consecutively in specific areas of the project as a whole. Both the superintendent's own work forces and those of the various subcontractors work together. Most of the time there are several operations going on simultaneously. Operations for one trade must not interfere with another. The superintendent must carefully assign areas of work to both his own forces and subcontractors. He must direct his attention closely to the following considerations in order to most efficiently complete the whole project and meet the main operational objectives:

1. The direction in which the project is to proceed, area by area, and the numbers of forces and different trades that are required to work in each area.
2. The feasibility of either the consecutive or current use of forces on the project in relation to the actual space limitations of specific areas and the whole project.
3. The number of men and different trades that can work efficiently in the same place at one time, with an emphasis on the nature of the work of each trade.
4. The objective of completions in selecting areas to proceed with or speed up operations and, when necessary, selecting areas to stop or slow operations.

The assignment of work areas is quite clearly connected with the job superintendent's next responsibility—the sequencing of operations.

The Entire Sequence of Total Operations

Laterally, a construction job may start at one end and proceed to the other or it may start from opposite directions and proceed to the center, or it may even proceed in a seemingly haphazard order. Vertically, succeeding operations can proceed from bottom to top or from top to bottom. The order of construction operations is usually somewhat predetermined by the progress schedule processed at the beginning of the job. The more detailed the method used to schedule the work, the more detailed the analysis and direction of forces must be to do the work, area by area, on the project within the scheduled time.

The job superintendent is actually in a position to observe the work as it progresses, and he has total control over its direction. He can observe, for instance, inefficiency caused by the crowding of forces into spaces that even the sophisticated schedule will not foresee. He is consistently faced with assigning and reassigning work areas, even when his assignments are at variance with the schedule. The difficult question then arises as to whether the schedule runs the superintendent or whether the superintendent runs the schedule.

The job superintendent must be aware of the fact that when certain forces are not able to perform within an area where they are scheduled, the situation calls for an adjustment of both the number and sequence of these forces and all succeeding forces. The reader, therefore, should realize that the superintendent's responsibility in assigning areas of work involves more than just simply maintaining the progress schedule as it is written. He must assign and assess areas of work and consistently reassess and reassign forces to work in them, irrespective of the schedule, if this is necessary to meet the main operational objectives. This does not mean that he can ignore the schedule entirely, but it is not to be regarded as a rigid direction of the sequencing of the job that allows for no deviations. He must digress from the schedule when it is practical to do so. Indeed, the superintendent, while giving every consideration to the schedule as written, must not allow it to run his project; he must be careful.

Probably no other responsibility of the job superintendent more clearly reflects his talents and is of more concern or affects subcontractor's forces more than this one. Subcontractors quickly question a superintendent's abilities when, for example, they are constantly shifted around on the job by a superintendent who has not carefully considered the consequences before assigning and reassigning areas of work. When a subcontractor's forces are unable to work efficiently because of constant shifts in job sequence, they become very displeased.

For this responsibility the job superintendent must learn to adapt, again and again, to changing circumstances. He must learn to adapt quickly and responsibly. Changing sequences cannot be a hit or miss operation. Improper adaptations of area assignments and job sequencing will raise even more complications, which, in turn, will cause more. Sequencing to meet the needs of the job can create some high-pressure problems for the job superintendent. It is unwise for the job superintendent to jump in to change a sequence too quickly.

The job superintendent must also give special consideration to the effect on expediting operations that involve men, materials, and subcontractors when he adapts and changes the sequence of operations on his job.

Means and Methods of Construction

The job superintendent is responsible for the means and methods of construction after initial consultation with his superior. This responsibility generally means detailing the general means and methods of production agreed to at the beginning of the job by the superintendent and general superintendent. But it also means that adaptations and expansions of these general means and methods, when the objectives of completion and costs are not being met, are also the responsibility of the job superintendent.

The following is an example of how the initial determinations of means and methods are to be expanded by the job superintendent. When forming a structural floor frame, the following general means and methods will be determined by the general and job superintendents:

1. Site versus off-site fabrication.
2. The hoisting equipment required and available.
3. The form materials required and available.
4. The storage and availability of storage on and off the job site.
5. The time requirements for the erection, pouring, and stripping of each component part of the frame.

These are all initial determinations that set the means and methods to be used on the job. The job superintendent decides the following details from these initial determinations:

1. The placement of the site of fabrication facilities and their operations.
2. The equipment usage for each operation in concrete forming operations.
3. The storage areas for form materials and resteel and the means and methods for placing, protecting, cleaning, and reusing (when reusable) these materials.
4. The type of packaging for storage. The type of packaging of stored materials is usually determined by the storage space available.
5. The types of specific building operations. For example, are columns to be formed and poured before deck erection (where scaffolds are required for pouring), or are columns to be formed underneath the deck forming (where scaffolding is not required)? Methods of resteel placement, working scaffolds, and choosing types of bulkheads are other examples of building operations decided upon individually by the job superintendent and his foremen.
6. The time required for each operation in relation to the progress schedule, that is, the time for the forming, pouring, curing, and stripping operations.

Each and every major construction operation on the project requires similar planning and detailing of the proper means and methods required by job conditions.

Proper Distribution of Costs

It is the responsibility of the job superintendent to rigidly insist (and without compromise) on the proper distribution of costs by his subordinates. Usually, construction firms will have their own cost-reporting procedures. Whatever the procedure, it is mandatory that the job superintendent see that his foremen report their costs correctly within the system. This means properly assigning the labor time units spent to the work actually done on a day-to-day basis.

Very little patience is to be used with foremen who misreport or misrepresent the expenditures of their men's time. Such practices not only harm the project at hand but have a serious effect on the entire company's operations and business. If reported costs are used as historical data for estimating purposes, for example, the use of misrepresented costs has serious consequences on the construction company's business.

Sometimes it is difficult for the foreman to understand the critical importance of this matter. This is one area in which the superintendent must be firm in his insistence, even to the point of discharging the refusing or reluctant foreman.

Usage of Equipment

The job superintendent is responsible for determining when, where, and how the equipment he has been assigned by the general superintendent is to be used. Take, for example, a hoisting crane on a multiple trade operation project. This machine can be used for the following purposes:

1. To hoist form materials.
2. To hoist resteel.
3. To hoist concrete.
4. To hoist masonry materials such as brick or mortar.
5. To hoist subcontractor materials.

What happens if there are multiple concurrent operations of the job—where there is simultaneous forming, pouring, resteel, masonry, and subcontractor work? Obviously, someone must establish how, when, and where the available equipment is to be used on a day-to-day or even an hour-to-hour basis. Simply getting more equipment is not usually the answer to the problems that arise here; the objectives of cost will not be met by adding more equipment. The superintendent faces many problems with this responsibility. Let us examine one of them as an example.

Example. The reader will note that in the purposes listed above, item (4) is that of serving masonry operations. A problem occurs here: while the use of the machine, as far as hoisting masonry pieces is concerned, can be a phased operation, the fact that the hoisting of mortar is an almost continuous operation presents a hoisting problem for the entire masonry operation. Other problems with equipment usage must be worked out by the job superintendent for all operations requiring hoisting. A different problem which might arise is that of hoisting resteel, caused by the length and weight of the material versus the space available at its landing point.

Insisting on proper maintenance is another responsibility of the job superintendent. Aside from it simply being good practice to maintain equipment for its useful life, equipment on the job that will not work is worse than useless. It is a real detriment to operations for equipment not to function when its use is planned.

Adaptation of Operations to Weather and Access

The job superintendent cannot be held responsible for the weather, but he is responsible for adapting his operations to it and the job conditions it creates.

There is a specific purpose for emphasizing this direct responsibility of the job superintendent. It becomes too easy for a superintendent to excuse himself and his job organization from responsibilities of weather, site, and job access conditions. Excuses produce nothing of value and are, in fact, counterproductive to sound operational thinking, and their continued use will not long be acceptable to superiors. Instead of offering excuses, the job superintendent should adapt his operational thinking to realities as soon as possible.

Example. A rainstorm affects the progress of the work depending on where the job stands at a point in time. It can affect the work in many ways. At the start of the job rain causes a disruption of operations on the day it falls. Production ceases on that day, but the resulting conditions affect production the next day, and so on. One can even visualize a rain every third day completely shutting down the job site for an extended time at this stage of the job.

To be ready to proceed as quickly as possible after a rain, the superintendent has to have preplanned procedures to mitigate delays. For example, he has to preplan for the removal of water from the site. Leaving standing water will only aggravate the site condition. If the superintendent has not preplanned the removal of water and the maintenance of the site, weather will not be an acceptable excuse for resulting bad costs and poor completions.

Example. Adapting to rain conditions later on in the job calls for different thinking on the part of the superintendent. He must plan, for example, to get the building waterproofed as soon as possible. In particular, he must consider such things as the season of the year and the relative costs of additional and possibly even out-of-phase work to accomplish the waterproofing of the building.

How does the job superintendent adapt operations to these and similar conditions of weather and site? Let us examine some of the ideas the superintendent can think of relative to the adverse effects of bad weather and site conditions.

1. He can plan to utilize permanent drainage facilities for temporary purposes and have these systems installed as soon as possible, even if they are out of schedule phase.

2. At site preparation time he can look at areas immediately adjacent to the building to make sure they drain away from instead of into the work areas. He must examine the necessity and practicality of quick backfilling to avoid the collection of water. The fast backfilling of trench work in the building would be an example of planning for or adapting to an adverse condition.

3. He can waterproof and have below-grade depressions filled and the roofing installed as quickly as possible, even though none of these operations can be done in their entirety or in the scheduled sequence.

4. He can consider the overall cost effectiveness of doing major operations out of sequence. Running backup masonry separately instead of concurrently with the veneer (the ordinary sequence of operations) in order to get a faster close-in of the building, even when the cost appears to be extensive is an example of this consideration, particularly when the separate operation furnishes cold weather protection.

5. He can consider how materials (concrete and masonry) protection is to be built, maintained, and heated, and think about what is required to get the facility ready for space heat as quickly as possible.

All of these ideas and many others are to be carefully considered by the superintendent in meeting this responsibility. Each objective of the construction operation—completions, quality, and particularly costs—must be weighed and balanced with one another.

Job Safety and Cleanliness

Job safety is the direct responsibility of the job superintendent. Safe operations in the production subprocess are directly related to cost and competitiveness in the construction business. All other considerations set aside, it is simply profitable to be safe on the job.

Since the job superintendent is on the job, he is directly responsible for seeing that the general safety procedures and policies, as established by the general superintendent, are carried out to the letter. The general superintendent can make rules and regulations, but he cannot see whether they are being followed. The job superintendent can.

The job superintendent is also able to detect unsafe conditions and practices on the job. He must correct and stop them. His responsibility for safety extends to subcontractor and supplier forces, his own forces, and to the public.

The job superintendent must develop safety consciousness in conducting construction operations. He would be wise to establish an "atmosphere" of safety on his job. Safe conditions and practices should become a matter of course in the conduct of operations; they should be the rule and not the exception. The attitudes of all the people on the job should be directed towards safety. That safety is cumbersome, inefficient, or undesirable is an attitude that should never be allowed in the conduct of operations on the jobs. When it becomes known that a job is safely operated, all of the people connected with it will be more willing to work and cooperate. When it is not safe, quite the opposite attitudes are prevalent.

The old adage of "a clean job is a safe job" is true; cleanliness and safety go together. There are some attitudes that have to be developed by the job superintendent about keeping his job clean. It is difficult to assign the responsibility for the trash and rubbish that accumulate on a job to any particular section of the work force. It is easy enough to attribute a piece of scrap conduit to the electrical contractor, it is more difficult to identify such things as wrappings and crates, and it is impossible to attribute general dirt and rubbish to anyone. But all trash and rubbish should be gathered and removed if the job is to be clean. Cleanup is a vital construction operation. Therefore job superintendents are required to do the following:

1. Insist that each individual trade or subcontractor on the job to periodically clean up their own identifiable rubbish.
2. Insist on the periodic cleanup of general dirt and rubbish. An efficient approach is for the superintendent to do it with his own forces. Prearranged or subcontractual arrangements and agreements for the distribution of costs take care of equalizing cost factors.
3. Insist on the job looking neat and orderly. As well as being essential to safety, a clean job is one of the best advertisements for the job superintendent and his company. He should present a clean and orderly project to the general public. Usually, the impression gained by people when they

see a clean, orderly job is that the contractor and its personnel know what they are doing: because the building in its construction phase looks good, it is assumed to be a quality building. This impression is not erroneous. A clean job shows the professionalism and class of its operators.

4.7 IMPORTANT DUTIES OF THE JOB SUPERINTENDENT

We now examine some specific duties the job superintendent must perform in order to meet his responsibilities. The following is by no means a complete list of his duties but is given as a representative sampling. Most of them are of a continuing nature. The job superintendent is responsible for the following:

1. Determining and maintaining access to the job site. This applies to maintaining the site's physical conditions and storage areas as well.
2. The handling of controversies in labor union areas up to the business agent level.
3. Ordering company materials and equipment used by his own forces on a day-to-day basis. This serves as a central control point for the job foremen to funnel their needs to the yard operation.
4. Ordering testing as required.
5. Handling problems of traffic both on and off the job site.
6. Keeping a job log or record.
7. Dealing with the architect or his representative on job-related matters.
8. Having materials and their deliveries checked for quantity and condition.
9. Coordinating the work constantly with the job expeditor, especially on progress and delivery problems.
10. Suspending (but never stopping) work when necessary.
11. Constantly observing the performance of foremen and spot-checking the performance of the mechanics as well.
12. Acting constantly as the central control distribution point between:
    a. The job and general supervision.
    b. The job and yard.
    c. The job and paymaster.
    d. The job and expediting.
    e. The job and estimating.
    f. The job and labor cost control.

The job superintendent must develop systems for carrying out his duties. The stage of progress a job is in at any time will establish a pattern of repetitive operations. For this pattern the superintendent develops his own systems for carrying out his primary operations of directing and controlling the work.

Sometimes situations arise that do not fit into the pattern of established means and methods of operations. Emergency situations would be one example of this kind of situation; severe or unexpected weather can be another.

4.8 SUBCONTRACTORS AND THE JOB SUPERINTENDENT

The construction operation of subcontracting is of such importance in the total study of construction operations that we devote an entire chapter to it in this book. We also examine subcontracting and subcontractors separately in some detail in the chapters on estimating and expediting operations. Directing and controlling subcontractors is a major responsibility of the job superintendent that needs an expanded explanation in this chapter. Operations connected with subcontracting are so important within the operation of job supervision that it is safe to say that the greatest amount of the superintendent's total time expended on the ordinary job is spent on running subcontractors. In fact, for strict broker-type general contracting all but a very small portion of the job superintendent's time is spent personally directing the subcontractors and their forces in their construction operations; while he has only indirect control over them in their production capabilities and thus indirect control in their profit-making capacity, a job superintendent can, in fact, make the subcontractor's operation a moneymaker or a loser. It should not be surprising then that the subcontractors and their forces regard the individual job superintendent in a special way. This special regard will be discussed in detail in Chapter 8.

Because he has direct control over subcontractors, the superintendent must direct them in a very special and particular manner, because in one aspect they are unique. The job superintendent has the responsibility of directing the subcontractors and their operations on the job, but only in relation to their being independent contractors, and this leads to the special point of this section of the text.

There are some very serious operational drawbacks in the superintendent regarding the subcontractors' forces in the same way he regards his own. Should he regard them the same, even unintentionally, some very serious legal and business consequences to his company could result. Therefore the superintendent should very carefully consider and develop some separate and distinct operational methods by which he handles subcontractors. We repeat here for emphasis that the job superintendent should always have in mind two things when dealing with subcontractors:

1. The subcontractors' employees are not the superintendent's employees.
2. The subcontractors are independent contractors, and because of this, they have their own legal identity and can and will closely guard their prerogatives.

4.9 THE DEVELOPMENT OF A JOB SUPERINTENDENT

How does one become a job superintendent? An obvious but superficial answer is that he probably comes up through the ranks of the construction field operational organization. It is just a matter of time; he learns as he goes along. These answers are partially true, but they do not furnish a complete answer.

The job superintendent must usually direct his learning to mastering the skills required of his position: he observes closely along the way; he listens; he learns how to initiate action and begins to take on the responsibility of handling people; he learns about manual operations, trade operations, and the job operations, all of which form the background of the job supervision operation. As time passes, he learns about these things in more and more detail.

But over and above merely learning the required skills, he must develop "construction thinking" skills that allow him to use in the best way all the things he has learned in order to supervise a job properly. What are these "construction thinking" skills? In part, they consist of the following:

1. *Thoroughness.* He thinks things through; he questions, investigates, and examines things through to logical conclusions.
2. *Adaptability.* He is able to change his operational thinking quickly and easily. He adjusts to conditions on the job and to those over which he has no control. He does not let these conditions govern him.
3. *Firmness.* He analyzes problems, solves them, decides, and orders. Knowing that he is responsible for his decisions, he relies only on good reasoning and judgment to alter or change them once they are made.
4. *Fairness.* He thinks and acts fairly and impartially.
5. *Cooperativeness.* He thinks of construction operations as a part of the total construction process that requires cooperation from all parts of the process in order to build the project.

Some say that the position of the job superintendent in the construction organization is the "cream of the crop." When compared to other organizational positions in operations, it is probably the most independent. It may be the most interesting in that it presents so many varied challenges to the operator, and it is probably the most satisfying in terms of fully appreciating the final product of the construction process—the building. Certainly it is the operational position that most strongly demands good thinking skills that provide initiative and adaptability.

5

Job Engineering

In vain we build the work, unless
The builder also grows.

Edwin Markham

5.1 INTRODUCTION

Many people consider job engineering to be the initial step in training for job supervision. Some construction industry people consider it to be prerequisite training in the construction methods and technology for all the divisions of the operational organization. It is true that the job engineer is in a rather unique position to learn important construction operations, which are requirements for top level operational control and direction. On the one hand, he is exposed to the means and methods of the field building processes and is able to observe them firsthand on the job; on the other hand, his field responsibilities also give him initial training in some of the other processes and procedures involved in office-based divisions of the operational organization. Job engineering furnishes opportunity for training in the broadest range of supplemental construction operations, procedures, and processes. The job engineer learns about them while carrying out his main job responsibilities.

Currently, the construction industry is finding its job engineering raw material among the graduates of engineering, architectural, and technical schools. Whereas in the past the majority of engineers were recruited from the ranks of the trade crafts, the new and complex technologies and techniques in the construction process require the employment of these graduates as a matter of practicality. The construction industry as a whole, however, has not given proper recognition to the position of the job engineer relative to the amount of responsibility it has given him. He has been regarded at times as somewhat of a necessary but burdensome overhead item. But the real value of the job engineer can best be illustrated by asking the operational supervisor and forces on the project the following question: Would you as operators prefer to have a good engineer (regardless of the cost) on the job rather than not have him, in relation to the conduct of your own operations? The answer is usually affirmative.

The most convenient way of defining job engineering for the purposes of this book is to define its responsibilities. Engineering operations on the construction job are carried out by the job engineer. The job engineer may have assistants, depending on the size of the project and the complexity and extent of its engineering work, or for reasons that usually relate to the time necessary for running concurrent construction engineering operations on the project. He may be on the project for its entire duration or he may be moved at any time the general superintendent sees fit.

5.2 THE JOB ENGINEER'S DIRECT RESPONSIBILITIES

The job engineer primarily has direct responsibility for the following:

1. The actual furnishing of lateral and vertical control and measurements for the project both as an initial and continuing operation. He must establish and maintain permanent control and reference points for measurement.
2. The furnishing of usable data based on engineering and mathematical principles where required in all field construction operations at the job site.

He keeps these responsibilities from the time he comes on the job until he leaves it, at whatever stage this may be.

Measurements

Let us examine these responsibilities more carefully. The first responsibility of the job engineer requires the following:

1. To determine the location of the proper furnished data for measurements and to give this data reasonable checks of its accuracy.
2. To determine from furnished information the actual location of the legal property lines.
3. To bring in from the given points both line and grade to the immediate site of the building.
4. To establish easily usable and permanent monuments and targets for building line and grade as permanent points of references at all stages of development in construction operations.
5. To establish building and offset lines where operations require them.
6. To establish existing site topography wherever operations may require it.
7. To establish convenient grade work points for efficient operational use.
8. To develop efficient and quick means and methods of providing layout for the purpose of serving building construction operations at any stage.
9. To establish, by measurement, all information that is important or required for record purposes.
10. To make all the mathematical computations necessary to accomplish the above measurement functions and to record them when necessary.

There must be a development of the thinking process in job engineering as well as in any other division of the operational organization. Nowhere is this more necessary than in the development of efficient and quick means and methods of providing layout for the building construction operations at any stage of their development.

Example. What does the engineer have to consider in providing measurement data for column anchor bolts installed in the top of a pier at 0′ 6″ below finish floor elevation. The pier is on a footing that is at an elevation of 12′ 0″ below finish floor elevation.

First of all, the engineer has to consider the numbers and kinds of layout operations along with every other operation involved in building the anchor bolts into their proper place. So he thinks about the following:

1. The original layout for bulk excavation for the footing.
2. The layout, if required, for close machine-cut or hand excavation.
3. The layout for the formed footing.
4. The layout for resteel dowels on the formed footing.
5. The layout for the pier template.
6. The layout for the anchor bolt template on top of the pier.
7. Lining the piers and bolts.

He must think about both line and grade for each of these operations involved, but, in addition, he must also think about how they will be done at the same time or along with the digging, formwork, pouring, and resteel operations. Whatever method is developed for providing the layout and serving all the construction operations involved, the engineer is primarily responsible for giving reference line and grade so that the construction operators doing the actual building operations can use them with the least degree of trouble and cost. He is responsible for checking the work after installation. What thought and analysis does this require of the engineer? It is clear that he needs to know in what sequence and direction the job is to be dug. He must analyze in which direction disposal of excavation is to be handled if taken off site, and he must realize the possible obstructions created by leaving it on site. He must think about the clear line of sight in both grade and line layout, offsets and their need for frequent checks, and the vulnerability to his set reference in relation to all operations on the site. He must also consider how to furnish references or actual line and grade at a moment's notice. All of this thinking is necessary in order to have the anchor bolts ultimately installed in their proper position.

Engineering

The second main responsibility of the job engineer is the furnishing of data that requires the application of engineering and mathematical principles for conversion into usable information for construction operations at the job site. These job requirements for data can be so numerous and varied that we will give only some specific examples of fairly common requirements.

Example 1. The job engineer is often required to make mathematical calculations from the contract drawings of coordinate reference points for his own layout purposes or to enable the trade foremen to lay out their trade operations. Coordinate points of this type are often required when circular or oval construction is required or when, for example, a spiral curve layout is involved. He must furnish this information to help those who are unable to calculate it themselves.

Example 2. The job engineer is involved at the very least in checking the structural capacity of concrete centering. He can be responsible for its entire design, including framing details. He can also be called on to justify one or another system of forms as to their structural abilities versus their respective costs. Whether or not the trade foreman details the centering work or even if a formwork supplier furnishes the design and details it, the job engineer should at least check the design system and its detailed component parts.

Example 3. The job engineer is often required to systematize measurements of installed work in order to record its status at any stage of production. A common example of this type of recorded measurement would be with camber in horizontal concrete structures before pouring, after pouring, and after stripping. The engineer may or may not be responsible for designing camber into the formwork, but he will probably be responsible for recording its subsequent deflection.

These three examples illustrate the need for applying engineering and mathematical principles and thinking about building technology, systematization, and the means and methods of construction. But some engineering operations require consideration of one other important thing: in Example 2, for instance, the job engineer must be aware of the cost as well as the structural validity of systems and methods. Because he is forced to consider cost here, he learns concrete framing construction processes and procedures more thoroughly. The opportunity to thoroughly learn this and many other operations is excellent training for this field operator.

5.3 THE JOB ENGINEER'S INDIRECT RESPONSIBILITIES

The job engineer works for and answers to the job superintendent. This relationship creates some indirect responsibilities for the job engineer that result from the superintendent's responsibilities. The job engineer is expected to meet his direct responsibilities for measurement and informational data no matter who he is working for or on what job he is operating but his indirect responsibilities are usually delegated to him by the individual job superintendent, as he sees fit. In most cases the job superintendent has the responsibility of performing his duties relating to directing and controlling the job. Sometimes, however, it is necessary, simply because of time factors or because the superintendent might be so inclined, to have the job engineer perform them. Some of the indirect responsibilities given to the job engineer might be as follows:

1. Handling and checking working and shop drawings for job-related purposes.
2. Doing all work-completed quantity surveys.
3. Knowing in detail the terms and conditions of subcontracts or any other similar information that will put him in a more valuable backup position to the job superintendent.
4. Preparing and keeping the superintendent's estimates of work completed for payment purposes.
5. Compiling the job log or making out the superintendent's reports.
6. Making and keeping the "as-built" drawings.

7. Field control of the schedule.
8. Field recording of changes and keeping the actual labor and material field costs.
9. Safety and conducting safety meetings.
10. Recording and keeping the record of job back-charges.
11. Maintaining the status and currency of the general contract documents, drawings, and specifications (particularly bulletins).

It is important for the reader to understand that the job engineer may only potentially be made responsible for these things. The job superintendent, for varied reasons, may want to maintain the responsibility for any or all of the items on the list.

It may be, for example, that the job superintendent wants exclusive and direct control over making the job log. He may want to maintain the status and currency of the drawings himself to keep abreast of any changes; he may feel that a constant review of the drawings makes him more familiar with the job. But in any case, since the job engineer is directly responsible to the job superintendent, he serves the superintendent as requested.

5.4 SPECIFIC DUTIES OF THE JOB ENGINEER

The job engineer has certain specific duties to perform on the construction project that are necessary for him to meet both his direct and delegated responsibilities. These specific duties include the following:

1. Caring for instruments.
2. Supervising job testing.
3. Supplying usable detailed information.
4. Checking.

Instruments

The job engineer has the specific duty to be constantly aware of both the condition and accuracy of his instruments. The tools of his trade are very expensive and relatively delicate, and they must be cared for and properly handled and maintained. The engineer should be keenly aware of the consequences of using inaccurate or improperly adjusted instruments; these consequences can be very serious and costly. The accuracy of much of the work depends on the proper functioning of these instruments. The major point the engineer must realize is the effect that just one use of one bad instrument in one operation can have on subsequent operations and work. An elevation reference, for example, brought in from a given bench with a bad instrument may not have too much effect on relative elevations in an individual building on an open site, but it could have the most serious effect on the relative elevations of an addition to an existing building on a closed site. The inability to close a traverse with a poorly adjusted instrument may have little effect on an open-site job but is of serious consequence in addition work on a closed site. The engineer cannot afford to be anything less than certain of the condition and accuracy of his instruments.

Testing

The engineer must at least exercise supervisory control over testing procedures, samples, and facilities on the job site. If, for example, he does not take the samples of concrete himself, he must see to it personally that the procedures for taking and keeping the samples are correct, that the samples are handled properly in the testing process, and that facilities are available for slump measurements and the curing of the samples. It is best for the engineer to consider himself the operator of the entire testing operations, from directly supervising the sample taking to reporting test results to the people who need them.

Supplying Usable Detailed Information

The job engineer has the duty of converting complex engineering and mathematical concepts so that they may be understood by the trade foremen and mechanics. These complex problems must not only be resolved by the engineer but, in addition, he must be able to make them operationally useful. The mechanic who builds must in some way be made to understand the measurements resulting from the engineer's calculations and solutions. One of the primary duties of the engineer is to furnish information and references quickly and accurately, but he is also required to make it useful and understandable. This duty falls within his main responsibility to furnish information, which we expand on here to emphasize the need for his being able to furnish it in detail.

Checking

One of the job engineer's main duties is to be a checker. He checks many different things, from the measurements of the work installed to conditions for the installation of future work. Accuracy and efficiency of operations calls for him to place substantial emphasis on his duty of checking the work. In construction operations a prior operation almost always affects the succeeding one; therefore the prior work must be in its proper place before succeeding work operations can proceed. Once the building starts, either all mistakes or incorrect work must be replaced with correct work or the subsequent work adjusted to it.

If the checking of installed work is not done, this will result in continuing and more extensive incorrect work. The engineer should develop thinking habits to suit this requirement of checking the work, considering primarily how to correct or coordinate subsequent work to the incorrect work, and develop efficient and quick ways and means with which to do this.

5.5 THE JOB ENGINEER AND CRAFT FOREMEN

It is important for the reader to understand the relative positions of the craft foreman and the job engineer as far as job measurements are concerned. Obviously, the job engineer does not actually perform every measurement on the job. Time would not allow for such a thing. Generally, there is a point at which the trade foremen take responsibility for correct measurements. The question is, "Where is this point?" If it could be stated that the job engineer's responsibility runs only up to the furnishing of reference measurements, this would make the answer simpler. But this is not the case. Usually, an arbitrary point must be established where one operator takes over responsibility from the other. The job engineer, for example, is stationed only on one end of a survey line or grade. The reference point to be established is on the other end of the line or grade or somewhere on it. The job engineer can give a mathematical location of points and man the instrument, but the physical placement of and measurement from these points must be done by someone else. Many times actual physical measurements involve the engineer's supply of reference measurement data, but the actual production of work from this data is handled generally through or by the trade foremen. It is therefore essential that the job engineer establish early in his development personal rules, guidelines, and thinking processes relating his duties to those of the craft foremen.

The job engineer must establish a rule for himself and the individual trade foreman about where his responsibilities end and the foreman's takes over. The specific point where this transformation takes place is not ordinarily formally established either by the job superintendent or some other operational authority. This being the case, the job engineer should make it a rule to establish the point himself. In most cases this has to be done with each craft foreman.

Example. The mason foreman and the carpenter foreman are not likely to be of the same age, disposition, abilities, or personality. The job engineer must furnish reference to both these foremen at different times and phases of the job. The mason foreman may be extremely sure of his abilities and take great pride in doing his own measurements from the most elemental reference points on the project. There are many reasons why he might insist on so doing: it may simply give him satisfaction or he may feel it gives him better control over his mechanics. On the other hand, the carpenter foreman may insist on almost all of his measurements being performed by the job engineer. For him, reference data must be furnished in finer and finer detail as he sees fit. These demands on his part can be based on many practical considerations: the foreman may be unsure of himself, he may not have the required knowledge to do more than the most simple layout, or he may be just plain lazy.

This example forces the job engineer to make respective rules and guidelines when dealing with each foreman. This is particularly necessary when the job superintendent has not set the limits of responsibility.

It is quite likely that the job engineer will be a young person. Trade foremen, if not older, are generally more experienced in their trade. The young engineer must face the fact that he cannot offer experience in specific trade operations to the same degree as they can. He can, however, offer them his engineering skills and clear thinking processes. There can be a reciprocal benefit to each party and to good operations in setting and maintaining proper attitudes about the worth of each operator in the total production operation when the limits of responsibility are clearly established between them.

5.6 THE DEVELOPMENT OF THE JOB ENGINEER

Looking and Listening

We have stated previously that the job engineer has the golden opportunity to learn how a building is put together operationally. He is exposed to difficult primary operational processes more directly than any other member of the construction operational organization. He experiences firsthand how many of the separate construction operations are handled and controlled.

There are several personal qualities and abilities the engineer should develop that will give him competence and self-confidnce not only in engineering operations but in learning the entire construction process as well.

The two primary personal habits he should develop are those of visual observation and listening. Observing means watching the way other people do things. This means watching how they act and react, how their machines work, how sequencing works in construction operations and so forth.

Listening means simply that. It does not necessarily require participation in a discussion or agreeing or disagreeing on what is being discussed; rather, it means being aware of how people think and act through what they say. Let us examine these points.

When a simple construction operation such as driving a nail or shoveling concrete is to be learned, no more than one careful observation is necessary for a person to learn many things about these operations; perhaps he or she can learn all of the necessary things from one observation. These observations are observed, noted, and stored away. The two manual operations mentioned are only a part of a concrete framing and pouring operation. In these operations there is much more to observe, however: both these and similar kinds of simple operations in the resteel, cement finish, and mechanical trades are observed. As more and more of the simple operations are observed within more and more complex operations, it becomes necessary to observe even more. This, in turn, leads to learning even more complex construction operations until the total operation is learned. Noting form design, its cost, and its building, for example, in the total concrete framing and pouring operation leads to the ultimate knowledge in this area of production operation. A single experience is then compared to other previously observed form designs and their building, and this leads to real experience learning.

The same observation, notations, and comparisons are made with the other trade operations. Time and the mechanics of the operations are observed. The tradesmen themselves are observed in their motion and efficiency. The interrelation of the trades is observed. The storage of materials is observed, as well as their movements to and from the location of installation. The sequencing of the trades is carefully noted along with their most effective use. These and all other kinds of things are learned by watching them take place; they are not easily forgotten.

From learning through observation about the production operations required to build the entire building comes knowledge of the effects of each operation on the entire progress of the building. Thus the concepts of scheduling are learned. Completions, quality, and the cost effects of the operations are understood more clearly. Observations are made until the functions of a total specific operation are learned—the simpler operations are remembered and applied almost automatically to new construction situations.

As experiences multiply for the engineer, they are (even though they become more and more complex) noted, analyzed, compared, and filed in his mind for future use.

Because of the great number of things the construction operator can observe, he must be able to store away only the valid experiences for reuse in his operational thinking process. He must do this until he thoroughly learns the proper construction operations and those that have been done better than others. There is a need for caution here, however.

The engineer must never believe that the construction operations he observes and experiences are necessarily the correct or only ways of performing them. He should never consider that the construction operations he observes, once learned, satisfy his knowledge of operations. To do so would be stagnating. As the young, inexperienced operator becomes older and more experienced, those things he has learned through observation only make his operations easier to perform. No amount of experience will ever complete his knowledge of the total construction process.

Developing Two Specific Skills

There are two significant areas of skill development the job engineer should gain from his experience of learning, and he will learn them naturally and easily if he keeps his eyes and ears open:

1. He begins to learn through experience the concept of costs and their importance in the entire construction process.
2. He can begin to learn about the handling and management of men.

Cost Concepts

These two skills listed above are essential to complement the job engineer's thinking processes in construction operations. Take the first—the area of cost. The engineer can learn, for example, the following:

1. All other things being equal, saving time means saving money. Efficiency and quickness in operations contribute to low costs.
2. Costs follow closely the progress of the job. Progress results from both proper operational organization, processes, and procedures. Job progress almost always clearly reflects true job costs.
3. Passing time has cost consequences. Therefore time, in some way, must be planned and controlled.
4. Some operations are simply more time consuming, more complex, more expensive, or more risky than others. Some of these operations must be done within the limits imposed not only by operational planning ability but by

the factors that govern planning within the operation itself. An example of this is seen in concrete operations: no operational planning can set the limits on the concrete's physical characteristics of, say, weight, plasticity, or required curing time. Indeed, it is these factors that set the limits on planning. The engineer learns that costs can be vitally affected by such limitations, particularly if they are not carefully considered by the operator.

5. Cost is dependent on productivity. Productivity depends heavily on planning. Overall productivity of manpower is based on something more than the personal characteristics of the people involved. No mechanic or operator can be expected to work productively in a planned fashion on unplanned operations. Nonproductivity cannot always be ascribed to poor attitudes of the tradesmen when, in fact, the real fault lies in their improper direction and management resulting from poorly planned work.

Learning these and similar facts about costs comes from experience. Admittedly, some of them are so essential to the understanding of the entire concept of costs that it might be said they will occur to the operator automatically if he really thinks about it. This is because even when he does not purposely look for them in an operation, his operational behavior is going to be influenced by their effects. The operator is better off if he is fully aware of the importance of the seemingly obvious factors that affect cost.

This is not to say that the only way to learn about cost concepts is through experience; it is to say, rather, that some essential and useful thinking abilities which can be applied to operating with the cost concepts of the construction process can be developed naturally through experience.

Handling Men

Although knowing how to handle forces is a management function that can be best learned by experience, we believe that it must at least be mentioned in any study of construction operations. It is taken up here because it is convenient. This is not to imply that handling men is the primary or even secondary responsibility of the engineer; this is not so. But job engineering is the entering phase of the construction operational field organization and, being such, it emphasizes the need for all field operators to know, as early in their development as possible, how forces are handled. The following general comments do not attempt to tell the operator the processes and procedures of the construction management of personnel; they merely give the reader some idea of how he must consider handling forces as these affect construction operations.

First of all, the operator must consider people as human. The way in which they perform their particular operations is affected by this simple fact. They can get cold, for example, and this will make them perform physical operations at a different rate than when they are comfortable. Another example of human factors affecting operations is the case where a man's state of mind at a particular time will seriously affect his attention to his work, which in turn might affect his safety consciousness.

Humans must be considered both in groups and as individuals in operational thinking processes. Does, for example, the slowdown of one tradesman affect the production of others? How do group thinking and group action relate to each other? The answers are not easy, but solid operational thinking requires them.

Realistic and earnest thoughts must be given to the human side of construction men. This is a practical thing to do; but there must also be some point at which consideration for these things in purely operational thinking must stop. The construction process, with its objectives, demands that people work and be productive. The operator must determine that at some point consideration of human factors must be balanced against the realities of the process. While no decent person would compromise with job safety, for example, the simple fact is that construction work takes place up high or underground, both of which can create dangerous situations. The dangers and their possible consequences can be minimized, but they cannot be eliminated entirely as a practical matter.

The operator, for example, must learn, as one small aspect in handling people, who is afraid to work up high and who is not. He does this, of course, with careful consideration of safe operation; but, also, he will do it for reasons of productivity and thus economy. The young operator can learn through observation how to determine such human characteristics as fear, reluctance, laziness, carelessness, and so forth. Then he must adjust his operational thinking towards conducting operations successfully, avoiding characteristics that would cause adverse consequences.

The young operator obtains most of his learning about handling people through the careful observation of how others with more experience handle them. He sees how the superintendent, for example, deals individually and collectively with his foremen, subcontractors, suppliers, architects, clients, and others. He observes the broad range of the various attitudes and approaches that can be used with people in order to have them work and be cooperative. He sees that the job superintendent will treat different people in different ways and adjust his personal methods of handling people to suit his disposition or a particular situation at hand. He learns to become an actor, a talent that is sometimes an important tool in handling people, and he learns how to gain and maintain respect. He learns the rudiments of discipline and disciplinary action as well as theories on compensation and reward and observes all kinds of people acting and reacting under stress and pressure. All of these things he picks up and stores in his experience file, and he remembers and uses them as his responsibilities increase within the structure of the operational organization. There is no better position in which to start learning how construction people are handled and managed than in job engineering.

6

Expediting

Time and tide wait for no man.

Geoffrey Chaucer

6.1 DEFINITIONS

To expedite means "to accelerate the process or progress of; to facilitate; hasten; or quicken." In this book we use the word *expedite* to describe the purpose and functions of a major division of the construction operational organization; one who expedites is called the expeditor. We use the word *expedite* for this description rather than such terms as *project management* or *contract administration* because we view expediting as an operational function and not as a management function.

Expediting is the operational division that, through its operator, the expeditor, facilitates the building of the work included in the construction contract. It is a support operation: to some extent this operation offers support to all the other divisions of the construction operational organization, and most paarticularly to job supervision.

The expeditor is the person in the construction operational organization who, immediately after a low successful bid and until final completion of the contract, expedites the office work required by the construction contract in order to facilitate the building of the project. The expeditor's relationship to the job superintendent is one of a subordinate in matters of field processes and procedures; the job superintendent's relationship to the expeditor is one of a subordinate in matters of the office procedures and processes required to implement the performance of the contract. The two most closely related operational divisions in the production and subcontracting subprocesses are job supervision and expediting. In many ways the expeditor and the job superintendent are co-partners in the entire construction of a project. Each has their own distinct responsibilities and duties to perform, but both of their operations are directed specifically toward gaining the objectives of completions, quality, and cost.

The expediting forces in the operational organization are generally regarded by the industry as being office forces, even though many expediting operations may take place at the job site. Expediting is considered an office-based operation because its functions require the expeditor to simultaneously take intermediate positions between the job and the contractor's own office, between the client and his agents, and between the contractor and the subcontractor/supplier forces.

Those who choose expediting as their field of operations will find some particular facets of this operation to be quite unlike those in the other operational divisions. First, of all the divisions in the operational organization, expediting gives the quickest introduction to the operator of the greatest number of major responsibilities. The expeditor, upon starting his job, is immediately faced with decisions and judgments on matters of major importance to the project: he is immediately involved in the subcontracting process, he starts dealing directly with owners, architects, and engineers, and he very quickly starts fully conducting many of the operational processes and procedures described in this book.

Another exceptional facet of this operation is the requirement of dealing with many different people at different levels. Expediting requires dealing as an intermediary, arbitrator, or colleague with a great variety of people, which sometimes calls for some extreme adaptations of personal behavior. It calls for putting on many faces, depending on who or what is being dealt with. To be successful an expeditor must become an excellent judge of the character and personality of the many diverse people involved in the whole construction process. He must learn to appreciate the special interests of the persons and groups he deals with.

The expeditor is the operator closest to the executive and management areas of his company. He does not have the same intermediary as the job superintendent does in having the general superintendent operating between himself and management. This means that from a personal and professional standpoint he must be more careful in performing his operations properly, because he is under the direct observation of the management of the company. On the other hand, the personal benefits of being under this direct observation are obvious in relation to this operator's opportunities to move into top management.

Before we take up the expeditor's responsibilities, we must reemphasize a point discussed several times throughout the text. Responsibilities are assigned to the various divisions of the operational organization based on the perceptions of which particular division can most efficiently and logically handle the assignment.

These perceptions are guided by certain criteria that are based on an analysis of which operator is most likely to be involved with the responsibility over the longest period of time on a project, and which operator is in the best position, operationally, to handle it efficiently from a physical standpoint. We could not object should the reader use some other criteria for making his choice; we can only repeat what we have stated before—that it is essential that he or she realize that each responsibility we have assigned to an operational division in this text has to be carried by someone within any construction contracting organization. To examine the point a bit further let us examine the following.

Take the subcontracting process and its operations. The subcontract can be executed by the estimator, the expeditor, or an entirely separate purchasing division in the construction organization. Each division would have some legitimate claim based on a logical analysis for performing this operation: the estimator could claim it on the basis that he is the closest to the subcontract operations at bid time and immediately thereafter; a purchasing department could claim it on the basis of the efficiency of having central control over all purchasing; but the expeditor can claim it because he deals with and uses it constantly and throughout its life on a project. He takes on the responsibility of subcontractor payments, submittals/approvals, changes, and so forth as part of his common operational duties. Thus our perception is that he should make and administer the subcontract. His continuous use of this document, in our opinion, logically requires him to take the overall responsibility for the entire subcontract process.

We recognize that some contractors in the industry today split the control of subcontracting operations among both the expediting and estimating divisions of their organization. Here we only suggest that the continuity of use and the need for administration calls for the operator who is best able to give more time to perform the operation.

6.2   THE MAIN RESPONSIBILITY

The Objectives

The main responsibility of the expeditor is to direct his operations to the three main objectives of completions, quality, and costs. Of these three main objectives, his operations have the most direct effect on completions. He works toward this objective by expediting the work of the subcontractors and material suppliers. It is his responsibility to see primarily that they are initially prepared to perform their work on the job in their proper and timely sequence. It is the job superintendent's responsibility to see that the site is prepared for their work. This means that the actual areas of work, the planned places for the storage of materials, and the proper sequencing of consecutive and concurrent job operations at the job site are the job superintendent's responsibility. It is the expeditor's responsibility to see that the subcontractors' preparation for performance has been cleared for movement onto the job. It is the expeditor who must initially fit the individual subcontractor or supplier into the schedule of operations from the beginning of the contract. This means that he must take the subcontractor through all of the submittal/approval processes and that he must pay the subcontractor and supplier properly and on time; but most of all, it means that he must simultaneously and continually guide the subcontractors in the performance of their subcontracts in order to insure proper performance of the general contract.

The expeditor must be constantly aware of the interrelation of subcontractor completions and the completions of the work of his own forces. He is the operator who fits all the various construction operations into the whole operation as far as the general contract is concerned.

In expediting completions, the expeditor has a function that is unique to his position: to push the other party to the contract (the owner) and the architect/engineer in their responsibilities designated in the construction contract. This is an extremely important, sensitive, and sometimes difficult part of his job. It demands thinking processes that are somewhat different from those the expeditor uses with his own forces and the subcontracting forces. His contractual position, for example, with an owner is not the same as with the subcontractor, yet he must fit the owner's functions into the building of the project just as completely and efficiently. One of the main skills the expeditor must develop is his ability to adapt and adjust his thinking when dealing with all of the various segments of the total project forces.

The objective of cost is not primarily the expeditor's responsibility, although his expertise and thoroughness in the preparation of the subcontractor's and supplier's forces for their respective operations undoubtedly have a great effect on the project cost. The expeditor makes some significant contributions in meeting this objective, for example, in areas such as changes in the work or in supplemental personal assistance to the job superintendent's task of directing the various subcontractor's operations on the project.

There is another area in which the expeditor has an important effect on the total project cost: "buying out" operations completed during the subcontracting process. It is essential here that the expeditor develop and adapt his thinking to the legitimate "picking up" of money or extra services from the subcontractor or supplier. This generally calls for a very careful analysis of the contract documents and a real knowledge of what the total work on the project is going to require.

As for the third objective, the expeditor has no direct control over quality work; this is the job supervisor's responsibility. In some operations that are the expeditor's responsibility to perform, the expeditor can, however, assist the job superintendent in maintaining quality work and workmanship on the job.

Example 1. The expeditor is responsible for the operation of payments. If the job superintendent is not satisfied with the quality of the work of a subcontractor or supplier, then the expeditor can, in his payment procedure, force the issue with the subcontractor. This means that he takes forceful action in withholding payments (within the proper payment operation procedure) to obtain quality work.

Example 2. The expeditor, in negotiating the subcontracts with a subcontractor who is known to be careless, can emphasize the requirement of quality on the job. This makes it necessary for the expeditor to obtain a commitment from the subcontractor, at the time of issuing the subcontract and as one of its emphasized conditions, that he will do his work with quality.

Example 3. It is necessary for the expeditor to be aware of the consequences at close-out time on the job of poor-quality work. Long or recurring punch lists are an indication of poor-quality work or workmanship. Since the expeditor's direct responsibilities include the operations of close-out, it is his duty to make every effort to minimize these operations. He can do this by constantly stressing to all his subcontractors and suppliers the need for quality construction; if he does not, he will be the operator faced with the most work at the end of the job. This is not only because the poor work must be corrected before final payment, but because the processes and procedures of correcting the work become extremely complex in relation to the actual work to be done within the various contractual relationships in the close-out process.

Example. To correct an incorrectly built wall may involve the work of many subcontractors. There being no privity of contract between the various subcontractors makes it extremely difficult to have the wall corrected in terms of determining the various responsibilities and costs involved.

To sum up this section on the expeditor's responsibility of meeting the objectives of all construction operations, this particular organizational division obviously extends further into the entire construction process than any of the others. Expediting clearly illustrates the diversity with which a division's operations are directly and indirectly applied to gaining the main objectives of completions, quality, and costs.

6.3 OPERATIONAL RESPONSIBILITIES

The Identification and Award of Subcontracts

After the estimator selects the apparent low subbid for inclusion in the bid price, the question often remains as to whether it is the true low or responsible bid. It is at this point that the expediting operator begins a close relationship with the subcontractors and suppliers that will continue until the end of the job.

The operation of identifying the true low subcontractor must be based on the expeditor's knowledge of what the bid documents (and later contract documents) require the subcontractor to do within his trade division of the documents. The expeditor works closely with the estimating forces at the start of the subcontracting operation. He must extend his knowledge beyond the general scope of what is required by the documents into greater refinement of the specific requirements of the technical specifications. Since he will, before the project is done, have to extend his knowledge of all the contract documents, the subcontracting operation is a good place to start using this operational skill.

The reader should understand clearly that the identification process and procedures discussed here are those which are considered legitimate and ethical. The legitimate ways of determining low subcontract bids are many and varied, but the construction operator should be very careful about stepping over the rather fine line between honest and dishonest practices in this area of operation. This point is discussed further in Chapter 8.

The awarding of the subcontract is discussed in Chapter 8. It suffices to say here that this operation is probably the most important in meeting the cost objectives of all the operational practices of the expediter. Why? Because the award of the subcontract in its proper amount is the operation that has long-lasting effects on the project. Aside from its importance in the establishment of the total general contract price, the percentage of subcontracted work on construction projects today forms the major portion of the work. All phases of the subcontracting process, including the determination of low responsible subbidders, requires precise study and perseverance. It is necessary to do the subcontracting in a thorough way and this requires the expeditor to give it his full concentration, using all of his skills and talents as a construction operator. The time required for this operation is that time the operator needs to do it correctly.

Cumulative Tabulation of the Contract Price

Cumulative tabulation of the contract price is an initial operation in the production and contracting subprocess of the construction process. Perhaps it is unfortunate that the term *contract price* has to be used here, because it might confuse the reader with other usages of the same term in construction contracting, but it is used in a specific sense when identified with the subcontracting operation. It is used in the sense that it is the cumulative but changing amount of money owed at any time on both the general contract and subcontracts because of changes or modifications in the work under the general contract.

Contract price in this context means simply the price (with any and all modifications to it) that is actually to be paid by the owner to the general contractor, and thence from the general contractor to each of his subcontractors and suppliers including payment for any and all work not included in the original contract. The term is applicable and used in both general and subcontractual relationships.

Let us suggest a general rule to help the expediting operator expedite this operation in the subcontracting process.

The contract price, from the time that all of the subcontracts and purchase orders are awarded, must be brought to a certain sum of money, and it must be kept current in relation to changes in it at all times during the course of the project.

The expediting operator must be aware at all times of the following:

1. The total price of the general contract with its modifications.
2. The total price of each subcontract and purchase order and the sum total of all of them. When approximations of quantities and their unit prices are necessary (in concrete or resteel quantities, etc.), their individual estimates must be carefully made and recorded.
3. The total cost of all changes to the original subcontract price of each of the subcontractors' and suppliers' contracts.

The expeditor must develop a system to tabulate the contract price that gives accurate and complete information of the contract price at any point in time. With subcontracts a system is generally developed by using a subcontract form as the recorder, to which are added or deducted the changed amounts to the sub-contract price. Thus in every case the last total subcontract amount owed must be reflected in the latest subcontract change order by the staement: *"total" contract price including this change* = ———. By using this system, the expeditor simply glances at the last subcontract change order to get the total contract price owed to that subcontract (see Appendix A for illustration).

The total contract price for the general contract then follows from the sum total of all of the subcontract amounts taken from all of the subcontracts plus all other costs of the job. By using the suggested form of subcontract and adapting it for use as a change document (as described in Chapter 12), the operator is able, at any point in time, to find that part of the total general contract price that con-sists of the subcontract prices.

The finalized general and subcontract prices form the basis for final payment from the respective parties. Unit price subcontracts and purchase orders at close-out time are usually brought to a lump sum amount, because all of the work at that point has been completed. At this point the expediting operator must be able to tell everyone interested in the project the amount of money left to be paid to each and every party on the job, including what is owed to his own company.

Processing Bonds, Permits, and Licenses

The expeditor secures all bonds, permits, and licenses for the project. It is his operational task to provide the owner with the contract bonds and process all ac-tions on the bonds whenever they might require operational input throughout the life of the bond. This is discussed in Chapter 14. Unless higher executive action is required on contract bond matters, the expeditor is the operator who deals with the bonding company and its agents.

The expeditor secures the permits required for carrying out the work and han-dles any requirements in permit administration or related problems. The build-ing permit is generally the most important instrument of this type used on the job. The physical inspections required by the permit are handled by the job super-intendent, violations, payments of permit fees, and so forth are handled by the expeditor. The issuance of a permit generally depends on meeting required criteria in the design of the building and thus often requires the expeditor and architect/ engineer to work closely together in securing it.

Licenses of varying types may be required on the individual construction project. It is a good idea, even when these licenses and their allowances and restrictions are closely field related, for any paper work involved to be processed by the expeditor, for accounting and record purposes.

The new expeditor will have several problems in developing his operations to handle these things. Mainly these problems demand the right amount of time for securing and processing the paper work. The operator will find, for example, that in many cases the securing of a building permit requires time all out of proportion to its importance in the production of the job. He will also find out that building inspectors come in all sorts of different packages: some are slow, some are fast, some are reasonable, and some are not. The expeditor will serve himself by turning over as many of these problems as he can to the job superintendent, while still reserving final operational control for himself.

Procurement of Insurances and Proof of Insurance

Every contract requires insurance of one sort or another and every subcontract should specifically require the subcontractor to furnish proof of insurance coverage. These points are discussed fully in Chapter 14.

Operationally, contract insurance is handled by the executive and expediting forces in a construction firm. The executive selects the insurance company and decides which type of coverage his company wishes to buy. Blanket or combined coverages on the traditionally required construction contract insurance are common.

The expeditor takes on the matter of insurances from that point and performs all the further operational functions that are necessary to meet contract requirements. He must transmit the policies or proofs to the owner and perform the necessary operations to keep his own company's insurance up-to-date. Dealing with the insurance company or its agents when an accident happens is his responsibility. He must develop specific processes for these operations.

Distribution of Information

The responsibility of distributing information is very extensive in nature. It is safe to say that all documentation for a construction contract on a project, without exception, should at least be brought to the attention of the expeditor. In almost all cases the distribution of documents is the responsibility of the expeditor. It is necessary to establish central control for the distribution of all project information. It should be the rule that all documentation on the job pass through the expeditor's hands. In this way, at least one person has knowledge of the complete paper work involved in the job.

Distribution of Contract Documents

It is the responsibility of the expeditor to distribute the contract documents to the subcontractors and suppliers, which must be done with the greatest thoroughness after a close analysis of their respective needs. All necessary, complete, and up-to-date documents actually needed by each subcontractor and supplier must be given to them.

The normal procedure for the distribution of contract documents is as follows:

1. The architect/engineer provides the required sets of documents according to his obligations under the contract. When additional sets are required, the expeditor is responsible for obtaining them from the architect/engineer.
2. The contractor's expeditor receives the documents, checks that they are complete and up-to-date by referring to the signed contract documents and sorts them for distribution.
3. The expeditor transmits (with a transmittal form; see Chapter 11) to each subcontractor or supplier the proper number of sets of documents.
4. The subcontractor receives the documents and repeats operation 3 above with all of the project subcontractors and suppliers.

These steps are followed for all succeeding issues of new or modifying contract documents of any sort.

In determining the needs of each subcontractor or supplier, the expeditor must consider the costs of reproduction and the operational time it takes him to sort and separate the document sets.

Example. A mechanical contractor who has several subcontractors will require not only sets for himself in his own operations but also sets for his subcontractors and suppliers. These people may have their own subcontractors or suppliers as well. The proper number and completeness of the sets of drawings, or parts of sets, actually needed by the mechanical subcontractor are determined by the expeditor and the subcontractor.

On the other hand, it is very unlikely that a door and frame manufacturer, painter, or hardware subcontractor will require any more than those documents that apply specifically to their divisions. These subcontractors require mainly the contract schedule drawings and specification schedule documents plus perhaps a few of the detail sheets.

The performance of subcontracted work depends operationally, as well as contractually, on the subcontractor or suppliers having all of the required contract documents in their possession.

Distribution of Communications

Both the general contract and subcontracts require specific and separate channels of communication. Therefore it is necessary that all communications clear a central operating point in the entire construction process, this point being the expeditor. The expeditor is responsible for distributing all information furnished to all who might need the information. Since it is obvious that each communication cannot or even should not be distributed to everyone, the expeditor must analyze and use discretion in this operation. The expeditor learns quickly where this or that communication should go or not go. Systems must be developed for record-ing and distributing each *type* of communication.

Submittals/Approvals

Shop drawings, brochures, samples, reports, and many other documents must be submitted back and forth between the expeditor and the architect/engineer. The operations required by this obligation of the contract are handled primarily by the expeditor. A full discussion of the submittal/approval process is given in Chapter 11.

Here we must discuss the nature of this operation and its processes and proce-dures in relation to the expeditor and his personal capabilities. The young, inex-perienced expeditor must be given encouragement in this process: after doing this operation once or twice it becomes a tedious, repetitious, and time consum-ing operation. It must be emphasized to the reader that this operation is an abso-lute practical requirement in the process of constructing buildings today; there is no other reasonable way to coordinate the many facets of the physical construc-tion of the typical modern building or structure. It is not only a matter of the ap-proved documents providing proof of compliance with contract, but a necessity for the efficient operational control of the entire construction process, given that the requirements of economy and restrictions in time are mandated by the mod-ern industry. The component parts of the building must be built in their proper scheduled sequence.

The economics of the construction process today do not allow for mistakes or their correction in the physical construction of a building. The construction con-tract recognizes this fact and either directly or by implication requires the sub-mittal/approval process to expedite the work.

Nowhere else in construction operations can an operator, by carefully develop-ing his operational processes and procedures according to systematic thinking practices, make it easier or harder on himself than in this operation. The expeditor must minimize the undesirable features of the submittal/approval process through developing efficient methods and at the same time giving it the attention it de-serves because of its importance in the construction process.

Payments

The importance of the operation of payments is recognized in this text by treating it in a separate chapter. The discussion of this operation is an exception to the exclusion of financial operations from this book.

Payments to and from all of the parties involved in the construction process and its subprocesses are the heart and soul of the construction *business*. The con-duct of the whole construction process is based on the recognized, formal, and traditional ways the product is exchanged for money. A construction firm as a business cannot survive unless its payments due are adequate and on time. The operator who is primarily responsible for the payment process is the expeditor. It is true that he is assisted by the other parts of the operational organization, but he is the initiator and governor of the payment operation throughout the entire build-ing sequence of a project; he formalizes all of its processes and procedures. While the construction contract itself gives the format for the payment procedures, it does not specify the specific operational procedures of payment except in the most general terms and leaves this up to the parties in the contract.

Since this operation is discussed thoroughly in Chapter 13, we wish at this point to state a general rule of payment operations for the expediting operator.

Both final and, in particular, periodic payments must be processed on time and in their entirety; no other operation can preempt this operation when it is time to perform it. The expediting operator must always put all other things aside in his operational responsibilities in order to accomplish this task. Only the most extreme circumstances allow for an exception to this rule.

Representation

The expeditor is the representative of the contractor to all of the various inter-relating parties to the general contract and subcontracts. He is, for example, the prime and direct representative of the firm to the subcontractor in matters involving money, contract, or work in dispute. At job meetings he is the co-representative of the contractor along with the job superintendent.

The expeditor is the contractor's representative who is responsible for all notices, protests, claims, and matters of litigation. The executive will decide initially to act on a claim or start litigation, but from that point on the expediting operator is both the chief processor and negotiator in the claims situation and the prime supplier of information and data for the attorney in litigation.

When representing the firm on some matter, the expeditor must be sure of his authority and position, having clear-cut authority delegated to him by his superiors. It is not always necessary that they give him unlimited authority, but the limits of his authority should at least be defined. This is necessary for the expeditor to operate effectively and efficiently in this area of his responsibility.

Operationally, the expeditor should have the last word on the interpretation of the contract documents as far as subcontractors are concerned; his authority over them should not be hindered by his superiors, because of the confusion that could result. It is also clear that his authority to interpret the contract documents from the contractor's standpoint should extend to maintaining that interpretation with the architect/engineer and the owner.

Keeping Records

The expediting operator is the center point of all communications. In addition, he is also the final depositary and keeper of all records on the job; *all* documents should pass through his hands. The total record of the project depends on his judgment of the importance of all of the correspondence and documentation collected during the course of the work.

The expeditor must develop thinking processes that allow him to make judgments about what to record and what not to. From this development follows the formation of systems, methods, and operational procedures for examining all the various records and incorporating them into a sensible job record.

Even though the actual communications, correspondence, logs, reports, and so forth may be sent or received by some other operator within the operational organization, the expeditor should be the operator who collects and correlates them. A superintendent's report or log should, for example, finally end up in the hands of the expeditor, at least in copy.

The systems and subsystems the operator develops to meet his responsibilities of keeping good records are important to efficient operations: they must satisfy his own need to save time and be convenient to use and should also be adaptable to the use and interpretation of other members of the construction organization. The system of listing of submittals/approvals and their status, for example, must be capable of being used by all, but the list is usually formed by the expeditor to suit himself. His own system of filling these same documents, when finally approved, is an example of a special subsystem developed for the operation of keeping records.

Closing Out

A direct goal of an expeditor on a construction project must be toward closing out the job as fast as possible. This responsibility requires that the operator's thinking be directed toward getting his firm its total money in the shortest time possible. This attitude must always be based on setting minimum (but reasonable) time allowances for the performances of all close-out operations required in the contract, including his own operations.

The contract requires certain performance functions for close out that the expeditor is responsible for. These requirements call for the development of specific operations to perform them, which will be more fully discussed in Chapter 10.

Here our primary concern lies in emphasizing the operator's recognition of the importance of the close-out operation to the profit-making and future business abilities of the firm. The close-out operation is especially difficult to perform for several reasons. By this time the operator has expended much time and effort in all of his other operations to build the building. At this stage the building is physically complete; now comes an operation that is seemingly insignificant relative to the time and effort already spent. But it proves to be otherwise. The expeditor finds that the details involved with this operation can be frustrating and time consuming. Just the fact that he may be tired of the job and ready to start on something new is enough to cause impatience with this vital operation.

But the operator must adjust his thinking and attitudes to realities. As frustrating and time consuming as these operations are, he must still perform them exactly and completely in accordance with the contract, which also states that full payment is not due until these operations are fully performed. The last dollars due on a construction job are generally those that contain the profit and future operating monies of the contractor. The operator must do everything he can to get final payment as fast as possible.

6.4 SPLIT OPERATIONAL RESPONSIBILITIES

There are operational functions in the production and contracting subprocesses of the construction process that must be split between various segments of the construction operational organization. We discuss split responsibilities in operations here because the expeditor plays a significant role in many of them. The following are three major operations requiring input by more than one operational division of construction organization:

1. Scheduling.
2. Certain areas of operations with insurance.
3. Certain areas of changes and modifications to the contract.

To illustrate the extent of the split in operational responsibility possible on a project, we should examine a specific area of the change operation [item (3) above].

Example. A bulletin is received, which is processed into the contractor's office by the expeditor. The expeditor handles the subcontractor's quotations on the bulletin work, and the estimator quantifies and prices the bulletin work to be done by the contractor's own forces. When the proposed work in the bulletin might affect the job itself (effects which may be obvious only to the job forces), then the job superintendent is responsible for making these effects known. When all three of the operators mentioned perform their tasks, all information is gathered by the expeditor and his quotation is based on all the information he receives from the other two operators.

Each of the three operators in this example have separate responsibilities in the operational process of changes in the work. Their efforts must be utilized and directed towards the accomplishment of the whole operation of quoting the bulletin. The expeditor operator must coordinate his work with and into the work of the others in order to gather a complete quotation within a certain time frame. Each of the operators have a semi-independent function: the expeditor does not tell the estimator how, for example, to quantify and price the work; he gathers the information and puts it in a suitable form for quotation.

Scheduling, insurance, and change processes and procedures are dealt with individually in this text. It is necessary, however, to mention scheduling and insurance here just to emphasize their importance.

The expeditor makes the schedule; when this schedule is used, it becomes the responsibility of both the expediting operator and the job supervisor to coordinate it to their respective operational responsibilities on the job. It is particularly important for the expeditor to objectively view *all* the operations necessary to perform and maintain the schedule. This means that the expeditor must fairly judge the relative importance and priorities of his own operations when they are opposed by the job superintendent's operations. Which operations should have priority between, for example, the delivery of brick (the expeditor's operation) or providing space for it on the job (the job superintendent's operation)? Sometimes questions like this pose extremely difficult problems; situations arise where two competent operators, both trying to do their respective jobs correctly in relation to the schedule, come into conflict with each other—one of the two must yield. Generally, it is the expeditor's particular responsibility to resolve this kind of situation. Except under extreme conditions of immediate need for one operation in the schedule to take precedence over another, it is the schedule maker who should decide on the priorities. The example above illustrates how split responsibilities in a single construction operation must be adapted to the total operations—in this case, the scheduling process.

Split responsibilities for operations in an operational process call for the adoption of intercompany rules and procedures to insure the coverage of vitally important operations. For example, in the insurance process both operators (the expeditor and the job superintendent) *must* require the proof of insurance and its maintenance from any and all parties working on the project. Each operator meets this responsibility in a different operational way, but it must be emphasized that both operators are responsible to see that an uninsured party performs no work on the job site.

6.5   EXPEDITING THE SUBCONTRACT

For the expediting operator the expression *expediting the job* means primarily expediting the subcontractors and suppliers. Expediting the subcontractors means pushing them; it means driving them by one means or another to perform their work on time, correctly, and in accordance with the wishes of the job superintendent. The expeditor is in a way both a master and a servant to the subcontractor.

When a job superintendent is unable by any of his means to get the subcontractor forces on the job, he must call on the expeditor to try to use *his* means and methods. For example, when a concrete supplier's dispatcher knowingly or unknowingly shorts a job on trucks during a pour, the expeditor is expected to exert pressures to solve the problem. Or when hard feelings occur between the job forces and the subcontractors and suppliers, the expeditor is expected to exert pressures to mediate and ease the tension—supposedly in an objective manner. The ideal attitude for the expeditor to have is to take such situations in stride as their being a part of his job. He must develop individualized means and methods of meeting these expected occurrences.

In expediting a job, it is not simply a matter of the operator just telling a subcontractor to follow the schedule but, rather, a matter of convincing him of the advisability and profitability of following the schedule and the directions of the job superintendent. The subcontractor must be shown the benefits of cooperation and, sometimes, the logic in coordinating his work with others.

If the reader should question these statements, let him be assured that certain subcontractors either never learn these things or deliberately choose to ignore them. As the circumstances warrant, it becomes the nearly constant task of the expeditor on a construction project to have hard and soft approaches, quiet and loud manners, friendly and unfriendly attitudes, and compromising and uncompromising views when dealing with subcontractors.

The expeditor and the individual subcontractors must get to know each other. This happens through the deliberate affirmative action of the expeditor, not by chance. It is vital for the expeditor to learn and remember certain things about subcontractors in order for him to operate with them most efficiently: it should be taken for granted that the ordinary subcontracting company is basically made up of solid, intelligent, and reasonable people; it is also necessary for the operator to realize that subcontractors have their own interests to foster and protect and that his company is not their only customer. Granting this, there are other significant things the operator must know about each of his subcontractors:

1. The size of the entire subcontractor's organization, including its field forces.
2. The financial capacity of the subcontractor.
3. The subcontractor's history of performance. This includes his quality of performance and timeliness. This knowledge can be gained through prior experience or through inquiry in the trade.
4. The attitude of the subcontractor as to his individual interests when they are opposed by other interests.
5. The real authority in a subcontracting company in matters relating to operations and decisions. In many instances it is the owner of the company.

The expediting operator should govern his own operations by adopting the following personal attitudes toward subcontractors.

1. The expeditor must be persistent with the subcontractors, getting them to the area of the work and, when there, to perform with quality and cooperation. The expeditor's persistence must be based on the attitude that his subcontract is the only contract in the subcontractor's volume of work (even though he knows this is not so). The operator must insist on being served first on the subcontractor service list. A paraphrase on the old adage "the early bird gets the worm" is "the bird who squawks the loudest gets the worm," which can certainly be applied to this construction operation.

2. The expeditor must set and maintain a positive tone for the job. By tone we mean a positive attitude of cooperation between the job superintendent and the individual subcontractors and the subcontractor as a group. The job superintendent is responsible for setting this tone too, but he directs the performance of the individual subcontractors and the group only when they are on the job. The expeditor operates with all of them on and off the job constantly throughout the project, before, during, and after they are on the job.

Sometimes a particular job superintendent, by his behavior and attitudes, can thoroughly disrupt the subcontractor's cooperative attitudes with other subcontractors and the contractor. From the point of view of the expeditor this is unfortunate, because he must deal with both the subcontractors' and the superintendent from the beginning till the end of the project. The expeditor should clearly recognize, however, that the superintendent, even though he caused a particular disruption, may still be doing his job competently and thoroughly. In such a case the expeditor should adapt himself to the handling of the subcontractor, realizing that the job superintendent's actions are planned and purposeful.

The expeditor, as a general rule, must promote cooperative attitudes on the job, even if it means confronting the job superintendent. The expeditor must make it clear to the job superintendent that disruptions in relations with subcontractors are not productive and that any differences over this point must be resolved for the benefit of the job. A clear understanding of the respective roles of both operators is necessary.

3. The expeditor must be patient with the subcontractors. Impatient attitudes,

no doubt, have value at specific times when dealing with subcontractors, but as a general rule the expeditor must work within the actual (not desired) limits of the subcontractor's abilities and capacities.

The various subcontractors are actually preselected for the expeditor. While he may himself award the subcontracts, it is usually the case that only low sub-bid prices govern the selections. Capacities and abilities are rarely considered at bid time. They are given consideration in awarding the subcontracts, but only very broadly when compared to the lowest subbid prices. The expeditor must learn to operate with what he has.

4. The expeditor must be money conscious with the subcontractors; in other words, he must be careful in operating with them in all matters involving money. The expeditor is the intermediary between the subcontractor and the financial department of his own company and it is his duty to see that the subcontractor is paid no more or no less than what he is entitled to. The expeditor, for example, handles all processes of back charges involving subcontractors, as well as all claims for monies between the various subcontractors on the job. The individual money interests of each and every subcontractor and supplier are the expeditor's direct concern and responsibility.

5. The expeditor must be proper with subcontractors. He must be decent and considerate with them, as well as fair and honest in his dealings. He must be reasonable, just and objective and must guard against favoritism.

Obviously, the above criteria are a hard order to fill, but the construction operator will find that the development of his thinking about them relative to the whole expediting process will serve his operational ability. The expediting operator must be persistent, patient, money conscious, and proper with his subcontractors. Efficient performance of the expediting operation depends on refining these attitudes.

6.6 EXPEDITING THE GENERAL CONTRACT

Expediting is an operation that implements the construction contract; the instruments of implementing the contract within the contract itself are the conditions—general, supplementary, and special. The expeditor, when expediting the contract, operates to implement the conditions of the contract within his area of operational responsibilities. Expediting the contract then becomes a somewhat different operation from expediting the job.

The operator must develop operational procedures, means, and methods that serve to obtain the proper and full performance of the contract according to its terms. This means that, in addition to expediting the job, he must not only expedite all of the parties directly involved in the contract, but also the parties that may be only indirectly involved but necessary to its performance.

Parties involved directly in the contract are the owner, architect/engineer, construction manager, subcontractors, suppliers, and sub-subcontractors.

The construction contract creates the necessity for expediting operations that are indirectly related to the production of the job itself. The contract requirements for bonds and insurance, for example, contribute nothing to the production of the building; but the operations necessitated by them call for expediting operations to facilitate their purposes and uses in the contract. There are significant differences in the approach taken to expediting the contract from that used in expediting production of the work.

Example. Expediting production through subcontractors is based on a direct contractual relationship between the subcontractors and the general contractor. Expediting the contractual obligation of the owner to have his architect perform properly in the submittal/approval process is based on an entirely different contractual situation. Because of this difference in contractual relationships, the expeditor is faced with adapting his procedures, means, and methods to each of the two situations. Expediting procedures must be developed to suit the situation.

Other than subcontracting, the most important area of operations to the expeditor is in his dealings with the owner and the architect. The operator is not expediting the subcontract but, rather, the general contract. There is a significant difference between operating with these types of contracts. The main difference is that the expeditor, in the subcontract is in much the same position as the owner is in the general contract. In the general contract he is not the controller; he is the controlled. Besides this, the expeditor is controlled through the contract not only by the owner but also by his agents. The general contractor has no direct contractual relationship with the architect, yet the architect's operations must be expedited in order to fully implement the peformance of the contract.

Obviously, it is more difficult to obtain the performance of the architect than that of the subcontractor from a purely contractual standpoint. Nevertheless, expediting this performance is as important to the job as is expediting the other when the total project picture is viewed.

It then follows that the more remote the party to be expedited is from the contract, the more difficult it is to gain his effective performance. Certainly, where there is no contractual relation at all, it is a difficult task for the expeditor to demand performance. A building inspector, for example, sometimes becomes an important factor in the total job production; his timely performance must be facilitated by the expeditor. Building inspectors can at times be very unpredictable, uncooperative, or obstructive for any number of reasons.

How the expeditor operates with direct and indirect participants in the construction process depends on many different factors. It is necessary for him to find out, for example, how architects operate in general. He needs to know this to determine the acceptable performance of the architect he happens to be working with. Some of the duties and responsibilities imposed on the architect by the contract can have very broad ranges of interpretation as to their adequate performance. The time necessary, for example, for the architect to process the submittals/approvals would be an example of a contract obligation contained in the architect/owner agreement, which extends itself by implication to the general contract. The time it takes an individual architect to do this operation must be balanced against the time that architects in general take to do it.

The range of possible expediting operations flowing from the general contract is very broad. Some operations necessitated by a general contract on one job might possibly never occur again in the expeditor's experience. On the other hand, there is always the possibility that some particular general contract will call for the development and use of a new operational procedure, means, or method that can be used in future contracts. Operations connected with expediting the operations of architects, for example, are commonly required by contract, but operations involving contract claims are liable to occur much less frequently, and operations involving termination are even less common. The reader must realize, however, that submittal/approvals, claims, and terminations are all conditions of the contract that require certain specific construction operations. There are some main theoretical points of the general contract the expeditor must learn in this part of his operations:

1. What a condition of contract is.
2. What performance is.
3. How the contract condition and its performance relate to each other.

Two brief definitions of the terms *condition* and *performance* are given here to assist the reader in his understanding of the interrelationship of these two important facets of the construction contract.

1. The term *condition* means anything called for as a requirement before the performance, completion, or effectiveness of something else.
2. *Performance* is the act of execution or accomplishment.

Generally speaking, the expeditor must always remember that the required performance under the conditions of the general contract is that which is needed to reach the three main objectives of all construction operations, except that his objective of completions requires completion of the contract conditions as well as completion of the work itself.

6.7 PATTERNS

It is improbable that all of the composite desired personal characteristics and qualifications for any operational position will be found in one person. The ideal construction operator probably does not exist. Of all of the operators in the operational organization, however, the expeditor must possess the largest number of common beneficial traits that are necessary to conduct successful construction operations. Let us examine this point by comparing the different operational positions.

The expeditor must be exact in his operational thinking, but not as exact as the estimator. He must be a diplomat, whereas in most of the estimator's operations there is little need for this talent. The job superintendent must be able to run his men, but his approach to this task is quite different from the one the expeditor must take in handling all sorts of people, especially independent subcontractors with their special interests.

One area of special concentration for the job superintendent in his operational approach deals primarily with handling craft tradesmen. These tradesmen, as individuals, somewhat follow behavioral patterns that are common to all building craft tradesmen. The job superintendent must then specifically learn these patterns in order to direct and use them in his operation.

On the other hand, the expeditor must learn the behavioral patterns of the many more varied parties that he deals with in his operations. In many cases these patterns are vague: owners, architects, engineers, and insurance company agents have different patterns of behavior or attitudes about the contract and the construction. There are broad ranges of patterns in the subcontracting forces as well, resulting from the size, makeup, and internal direction of the individual subcontractors. Patterns of behavior can result from the financial capabilities of a subcontractor or the importance of a particular trade division within the subcontracting forces in the construction process. It is important that the expeditor learn these patterns of behavior and attitudes to operate effectively.

Example. The patterns of attitudes and behavior of an automotive manufacturing company as an owner are not going to be the same as those of a municipal corporation owner. The auto manufacturer is going to be quite restrictive about time factors in any construction he buys. On the other hand, this same owner may become quite liberal in the expenditure of money when his own production capabilities are affected by construction. The public owner may have quite the opposite thinking and behavior on these same points.

The patterns of behavior of the various architects that are encountered in building may also be different. Individual architects may practice differently, but they still remain architects. For example, they may disagree upon considerations of quality workmanship in building versus aesthetics: one architect may be more interested than another in the aesthetic qualities of a building rather than in its functional abilities, some may have a more balanced interest, while some may have an extreme interest in function, and their attitudes on the technical aspects of a building may vary accordingly.

In dealing with subcontractors, there is another behavioral pattern for the expeditor to observe carefully. Patterns of attitudes and behavior of the mechanical, electrical, and painting subcontractors, for example, are liable to be extremely diverse. Their different attitudes on the necessary prerequisites for coming on and off the job, the size of the crew required, and so forth all fall into patterns, which are different for the various trades; there are even different patterns of behavior between different subcontractors in the same trade.

The way most expeditors learn patterns is by analyzing and speculating about the interests of each individual or group they deal with. This process becomes a constantly expanding analysis of the interests, co-interests, and counterinterests of all of the parties involved in the entire construction process.

All parties in the construction process have a common interest—the building of the structure for money and its utilization, from which spring all kinds of legitimate (and sometimes illegitimate) self-interests. It is the expeditor's task to analyze, separate, discard, and emphasize all of these special interests toward the goals of completing the building at a profit for the builders, profit and reputation for the architect/engineer, and use of the building by the owner.

Developing and working with the required thinking processes for distinguishing patterns are time consuming but necessary. Actually, the development of these processes is not necessarily done in a deliberate way, and it is greatly aided by experience.

The expeditor cannot ignore these patterns of attitudes and behavior. If he does, he will find that his own planning and practice in operations will become cumbersome and unproductive. When he is faced with refusals or reluctance to cooperate in the construction process that are based on these attitudes or behavioral patterns, he will find them extremely difficult to overcome. The sooner he learns about them, and adjusts his operational thinking to them, the more efficient and productive his operations will become.

6.8   TWO BASIC SKILL REQUIREMENTS

In this chapter the discussion of the expeditor's duties takes a different direction from the discussion of the duties of other operators in the construction organization. The duties required of the expeditor are discussed here in the sense that the he is obligated to improve his skills in some very general areas of operational functions. Specific functions are discussed in several other places.

The expediting operator has a duty to learn the following:

1.   To communicate effectively.
2.   To properly delegate portions of his time to operations in relation to their importance and priority in getting the project done.

The expeditor's duty to gain greater skill in these areas is for his and his company's benefit. The reader must understand that the proper performance of these general duties is dependent upon the operator's own view of their importance in operations.

Communications

A construction operator must be able to communicate with those he deals with; he must be able to communicate effectively, both verbally and in writing. No other operational division requires more writing than expediting. Writing properly then becomes a major learning objective for the expediting operator and the pencil becomes his main operational manual tool.

What is a better indication of a person's abilities to communicate than his writing? These abilities show more about the basic mental discipline and capabilities of a construction operator than any other single construction operation. No single construction operation illustrates more clearly careless thinking processes and hence operational incompetency than does sloppy, incorrect writing; improper writing demonstrates the operator's inability to perform the basic function of communicating with others. In other areas of the construction operational organization it is possible, we suppose, to get by without extensive development of this skill, but even in these areas it is only barely permissible. In any case, this lack of discipline should not be boasted about or viewed with complacency by either the construction operator or his company. In the expediting operation there is no valid excuse for lack of writing skills. A construction company should carefully examine an operator who, through indifference or laziness, does not consider the proper performance of this operation seriously. Indeed, such an indifferent attitude cannot be tolerated by any organization that views itself as being professional. Writing improperly is simply not professional and it reflects poorly on the reputation of the construction firm, as well as its employees.

When developing skills in verbal communication, the operator should develop a wide range of methods of proper verbal expression. There are times, for example, when peculiarities of language might serve to facilitate conversation, but even in verbal communication it is important that the operator give exact and clear impressions. This includes giving a correct impression of himself as a professional to others. While the use of profanity or obscenities, for example, might have their place at particular times and places, it is doubtful that their routine use really serves a valid purpose in conducting operations.

The reader should consider the vast areas for developing skills in verbal communications. It will serve him well to undertake and continue to develop such skills, because their usefulness never ends in the field of construction operations.

Delegation of Time

During the course of a project there are days when there is simply not enough time for the expeditor to do all his work. His time is often dependent on other people's availability—not theirs on his. His time is often governed by their schedule and the expeditor must therefore schedule it to accommodate subcontractors, suppliers, architects/engineers, and others, and at the same time he must perform all of his other necessary operations. When the expeditor is handling multiple jobs, the number of problems with time increases.

It is a major duty of the expeditor to carefully analyze the time he has to spend on each task on each job. He must take care that he allows time for situations and circumstances that can arise unexpectedly. Experience will help solve the expeditor's problems in scheduling and using his own time.

6.9 COMMENTS

Each division of the operational organization is essential to the successful building of work within the construction process, but it is safe to say that the expeditor is the operator who forms the hub of the wheel in construction contracting. Many diverse operations are required from this operator and many of the operational problems of other divisions find their way to his desk. The contract, its performance, and the relationships of the contracting parties within the construction process constitute the focus of a large part of his functions. This operational division combines the greatest number of significant construction operations within the shortest segments of time. Even the inexperienced expeditor is forced to assume a great deal of responsibility for operations that may affect the main objectives of the operational organization.

Certain of the expeditor's operations are given special treatment in this book. In this chapter we have tried to view expediting in a very general way. Those expediting operations treated elsewhere are payments, changes, close outs, subcontracting, scheduling, and claims. These are so extensive and specialized in nature that they cannot be adequately explained within a single chapter.

7

Estimating

*Sagacity and a nameless something more
—let us call it intuition.*

Nathaniel Hawthorne

7.1 DEFINITIONS

Estimating is the initial construction operation in the bidding phase of the contracting subprocess of the construction process. An estimate is an approximate calculation of value; it is an approximate computation of the probable cost of a piece of work, made by a construction contracting firm that proposes to do the work.

The operator who estimates is called an estimator. In the construction operational organization he is the operator in charge of the estimating process of the individual project for bid purposes. His estimate of the cost of the work forms a part of the bid or "offer" in the contracting subprocess. The estimated cost of the work plus the markup comprise the amount of the bid. The bid price is proposed in a proposal (offer) form to the owner. When the proposal is accepted by the owner, the construction contract results.

In many construction firms estimating is viewed by the company as a departmental function: the entire department is considered to be the estimating force on all individual projects, with no one estimator being totally responsible for the entire operation. When figuring all except the smallest jobs, estimating is usually a team effort, because of its very nature. The restrictions on time required to perform the various operations involved, particularly as bid time approaches, rarely allow a single individual to do all of the required operations properly.

The estimator is that person in a construction operational organization who takes off quantities of various items of work from the contract documents, prices each item as an integral part of the total work, and then summarizes his prices in order to arrive at a "cost of the work" price. Markup, which consists of a desired amount to cover general overhead plus desired profit, is added by the executives of the firm to form the total bid price. This bid price is then proposed to the owner, in many cases with a bid security. The bid is a firm offer, usually made for a set period of time, that requires the owner's acceptance to form a contract.

For practical operational purposes one estimator in the estimating operational organization is usually put in overall charge of an individual project to be bid. Other people are added to the operation as it becomes more complex and as bid time approaches. Generally, the estimator will be assisted by another person who takes control over the operations of subcontractor invitations, prebid subcontractor liaison, and the subbid tally on bid day. As bid time approaches more people may be added to the operation, so that on bid day a major part of the contractor's whole office organization becomes involved in the operation of bidding the individual job.

One who estimates must consider himself to be first and foremost in the line function of a construction organization. Other parts of the operational organization might object to this statement, but the simple fact is that the continuing existence of the construction company depends primarily on its estimating ability. Field forces, knowing full well their importance in the production subprocess, will probably disagree with this assessment, but here, again, simple logic will show that while a construction firm might function without the directly employed field forces, a field organization cannot function without successful estimating operators.

7.2 THE PERSONALITY OF THE SUCCESSFUL ESTIMATOR

Required Traits

The nature of the operation of estimating requires those who wish to become successful estimators possess certain personality traits and special thinking abilities. Most of the personality traits must be inherent; only a few of them can be developed.

When faced with the task of describing the required personal attributes of the successful operators of the various divisions of the construction organization, those of the estimator are by far the most difficult. It is even more difficult to describe the relationships of these various traits to each other. Nevertheless, we will attempt this description, because it is necessary and valuable for the reader to understand the personality of the successful estimator. If the reader understands who the estimator is and how he thinks, it becomes easier for him to become acquainted with the operation as a whole, including its special processes and procedures.

No other construction operation relates as closely the personality of its operator to its function, procedures, and processes. This operation also requires the person to fit a specific mold more than any other construction operation.

We must first define some general terms so that when they are used in the following discussion of the required traits and thinking abilities of the successful estimating operator, the reader will understand their meaning in that context.

Optimism is the tendency to have the most hopeful view of matters or to expect the best outcome in any circumstances. *Realism* is a tendency to face facts and be practical rather than to be imaginary or visionary. *Sagacious* means being keenly perceptive and discerning, shrewd, or farsighted in judgment. *Meticulous* means being extremely or excessively careful about details. *Analysis* is a separating or breaking up of a whole into its component parts so as to discover their natures, proportions, functions, or relationships. *Guessing* means to form a judgment or estimate of something without actual knowledge.

These definitions may seem to be a strange way of introducing the reader to the estimating operator; they do not appear to be related, but each word has its meaning as it applies to this operation and to its operator. When one looks for a definition of the operation, it is quite simple to name its component parts and physical functions; however, when one tries to describe the ideal person for this operation, their characteristics are seemingly contradictory in nature. For example, the estimator must be an optimist as well as a realist. Good sense and keen judgment are not exclusive traits of the realist, and the optimist does not have an advantage in his positive views on production, nor is he better at guessing unknowns than the realist. For the estimator to be both optimistic and realistic means that he takes an optimistic view of the unknowns based on a realistic analysis and appraisal of the knowns, thereby being more likely to produce low bids.

Guessing versus being meticulous also seems to present a similar contradiction, but here, again, it is a matter of viewing both thinking processes as essential to the operation when properly applied. The fact that it is desirable for the estimator to be meticulous in the quantity survey and unit pricing, for example, does not mean that he should necessarily be as accurate in pricing the general conditions.

Or, if one holds the view that only meticulous analysis and realism govern good estimating, one should consider the common practice, performed by the executives at markup time, of adding or subtracting substantial dollar amounts to or from the price of the work. It would be foolish for the estimator to believe that his meticulousness in figuring formwork to the exact square foot or in making extensions of unit prices to exact figures is of any value at all when an entire item of work he has estimated can be totally eliminated by a percentage point or two drop in the markup. If, after all, markup determines winning or losing the job, what then is the need for all of the exactness in this operation?

Meticulousness is necessary to minimize the possibility of loss in the determinates on the job in order to allow for more latitude in guessing the indeterminates. Thus it is vitally important that parts of this operation be highly systematized and meticulously quantified. The care and skill required in pricing known quantities exactly are of prime importance in arriving at the estimated cost that is given to the executive for markup. The estimator who believes that markup alone determines the success or failure of the bid is simply wrong; the markup is added to the estimated cost, but its amount may be based on many other factors of business or competition that he is not aware of. The executive must be able to rely on the accurate cost of the work figures. It is unrealistic for the estimator to assume that upward or downward movement in the markup is the sole reason for winning or losing the job: denying the possibility that his estimate could have caused the same result.

An estimator who does not or cannot develop the traits and thinking abilities that lead to becoming a low bidder does not remain in this field for long. Let us now examine how these traits and abilities affect just one important aspect of construction estimating operations—competition. The estimating operator must have a competitive nature, which he should carefully nurture, and he must be influenced by competition, but only to a reasonable extent.

The statement that the estimator with the most optimistic attitudes and views has the best chance of success is likely to be strongly objected to by those who believe just the opposite. Our contention is based on the simple fact that the ordinary contractor, when bidding in competition, must be *low* on his bid when he wants to or has to be. We repeat here that a construction firm must have work to exist. Ideally, it must get the work in a somewhat sequential order and at specific times. Because of the way the bidding competition works in the industry, it is not possible for a contractor to be sure that he will get the job he wants at the desired time.

The surest way for a contractor to get work is to be the low bidder in competition. Estimators should emphasize optimistic attitudes and exercise positive thinking in their approach to the job at hand. Consider the effects of having negative thinking and attitudes when estimating. What good does it do an estimator to be so hesitant about being wrong that it causes him to consistently be second bidder on needed work? Though he may have the satisfaction in thinking he is right and the lower bidder wrong, he does not have the job. He must be able to give the benefit of the doubt to his field operational organization, to expect maximum productivity rather than minimum productivity, or to reasonably anticipate good rather than bad weather; he must equally be able to think that something is possible rather than not, to hope that the temperature will only reach 40°F rather than −10°F, or to estimate that the job can be finished in 10 months instead of 12. When good projections seem possible, it is better for the estimator to keep this possibility in mind rather than turning to negative projections based on pessimistic thinking and attitudes.

7.3 THE THINKING SKILLS OF THE SUCCESSFUL ESTIMATOR

Artist or Scientist?

We have already mentioned various traits that we believe are necessary for the successful estimator to have. Here we look at the development and use of certain thinking skills that are necessary for him to become successful.

Is an estimator an artist or scientist? This question is important because its answer gives clues to the primary thinking skills required in estimating operations. An artist finds certain mental abilities of more value to him than the scientist does when functioning in his profession. Other skills and abilities serve the scientist better. It is not suggested here that being an artist or a scientist eliminates the importance of anyone utilizing the desirable abilities and skills of both professions. It is a matter of the degree to which these special skills are developed that is important.

J. S. Mill wrote, "Art in general consists of truths of science arranged in the most convenient order for practice instead of the order which is most convenient for thought." Art is knowledge made efficient by skills, and science is systematized knowledge based on general truths and principles. William Stanley Jevons has said that science teaches us to know and art teaches us to do. The most relevant point to this discussion is that an artist is one who professes and practices an art in which conception and execution are governed by imagination, visualization, and intuition.

It is important that the estimator consider himself first as an artist, because his profession indeed meets the criteria contained in the definitions of art given above. He must compare and analyze the scientific abilities required in his tasks in relation to those abilities required of the artist. For example, his use and thus his required knowledge of mathematics in his tasks do not extend much beyond those things learned in grade school. On the other hand, his ability to imagine and visualize this or that building operation is absolutely essential in the pricing process of the estimating operation. He must be imaginative, able to mentally visualize processes, and he must use whatever intuitive powers he might have to their fullest in the estimating operation.

Guessing

Guessing is a part of estimating. The estimator must guess at the unknowns. We emphasize here that guessing means forming a judgment or estimating something without actual knowledge.

Most of our readers have heard the term *educated guess*. This term is used at times by construction industry people in relation to the estimating operation. Supposedly it is meant to be the result of some sort of special ability a successful estimator must have. Besides the term being somewhat of a contradiction in itself, it must be understood by the operator that a guess is still a guess no matter what qualifications are given to its definition.

Intuition

Intuition is defined as the immediate knowing and learning of something without the conscious use of reason. In guessing the unknowns in the construction estimating operations, we do not know of a more valuable mental characteristic for an operator to have than intuition. We suggest that the deliberate use and development of the intuitive power of the mind is a vital and necessary part of the estimating thinking processes. Perhaps this will be difficult for pragmatic construction operators to accept. To alleviate their doubts, it can only be suggested here that these operators try to use some other method based on specific knowledge, statistics, or whatever else to help them in the process of guessing; they will find that the application of scientific thinking processes will not be of any practical value, because guessing correctly requires artistic abilities.

Imagination and Visualization

When the imagination is used in estimating, it is used to form a notion of how construction operations of all sorts are to be done when constructing the building. In fact, in the bidding process the building is constructed in the estimator's mind, and the *imagined* cost of this construction forms his estimate. Imagination sets the ideal sequence and completion of the project. Tempered with realism, using the imagination in assessing the determinates and good guessing techniques for the indeterminates is what gives substance to the estimate.

Being able to see in one's mind the different and varied construction operations involved in the work items to be done on the project furnishes, in most cases, the fundamental basis for pricing the items correctly. Visualizing the operation itself and seeing how it is to be done by the components of the construction process are essential in pricing the work. Using the imagination is vital when, for example, historical data cannot be used in pricing, either because it is not trustworthy or not available.

The development of these artistic thinking abilities and skills first of all requires thinking about them carefully. Then practice makes perfect. Experience, based on visual observation, fills any gaps in the ability to imagine and to visualize. Intuition is by far the most difficult essential mental facility to describe, much less to give guidance for its development. Indeed, we must admit the possibility of intuition not being a learned mental ability.

The results of the use of intuition in estimating shows up very clearly when the estimator finally gains self-confidence in making reliable judgments on the knowns, and, most certainly, when guessing about the unknowns.

Pressure

Sometimes it is said that an estimator is born, not made. This is not wholly true, because any interested individual having good thinking habits and common sense coupled with imagination and the ability to act intuitively can become an estimator. But there is one more natural trait in addition to those discussed previously that is necessary for the estimator to have—the ability to withstand pressure. Most knowledgeable construction professionals will agree to this. The ability to withstand pressure required in estimating can rarely be learned by a person; it is of such major importance in estimating that perhaps it is legitimate to say that an estimator must be born with it.

The best way to describe the pressure involved here is to say that it results from the estimator's realization of the necessity of having a low but reasonable price. This pressure can result from the estimator's feeling of responsibility to the many other colleagues who are depending on his skills in this operation, or it may be related to his pride in being a winner and his personal feelings about his worth to the company.

The major cause of this pressure is the relatively short amount of time the estimator has to perform the considerable number of operations required. The bid is like a bet at the race track, but only winners pay off; there is no "place" or "show." The estimator realizes, considers, and sometimes dwells on this, and as he does, the pressure begins to mount and continues to do so until bid time.

It is not at all uncommon for a bid price to amount to tens of millions of dollars. As a rule, 95% of this figure results directly from the estimator's own operations. The estimator cannot help but feel that even if he has done his work conscientiously on the knowns, the unknowns, whose costs are also included in the figure, are merely the results of his judgment and guesses. This by itself would be enough to worry about, but, in addition, he is forced to work within close and ever-constricting time limits. He must be fast, accurate, thorough, and on time; but, above all, he must not be unsure of his work, otherwise he puts the whole company operation in jeopardy. It is not necessarily the case that the company's survival will depend on a single good or bad estimate. Even mistakes in the bid price can be corrected—even after submission (although, with increasing actions being taken on bid security these days, such relief is becoming rare). It is the self-confidence of the estimator that is the most important factor in the time-pressure situation; indeed, his confidence in doing just simple and elemental operations such as adding or subtracting correctly becomes essential. Pressure causes mistakes. Pressure causes loss of confidence if it is not handled properly and affects the performance of the estimator in almost all of his functions.

Thinking clearly and thoroughly about complex and serious matters is difficult in itself and the nature of the construction bidding process makes the additional pressure factor unbearable for some people. For example, otherwise good thinkers can become extremely upset when their concentration is interrupted. On bid day it is common for the estimator to be constantly interrupted in his thoughts on matters of serious concern; this is almost inevitable, even when the company policy for bid day procedures is designed to protect against such interruptions.

Pressure causes many different problems for the estimator, and it is usually the factor that separates those who are able to perform in estimating operations and those who are not. Many otherwise simple and uncomplicated operations can be affected by pressure.

The mathematics involved in an estimate are not by any means complicated. Although the extent of the numbers and different kinds of estimating operations involved is great, were it not for pressure considerations, these operations could be handled by almost any person of ordinary intelligence. Simple calculations or the distribution of subcontractor bids (in themselves simple operations) are often completely mishandled. The amount of confusion and resulting uncertainty that can occur under the pressure of the bid process is usually a problem of major proportions for all estimating personnel. Some can handle it and some cannot.

The problem of gathering all of the components of the estimate together within a specific time limit would seem to be easily solved by developing good systems of estimating operations. In most construction firms good systems are ordinarily developed to minimize the pressure. The problem is that many parts of the estimating operations are not directly controllable at all times by the estimator, no matter what system is used. It may be, for example, that the estimator would like to see the subbids come in with some semblance of order and timeliness so they can be easily plugged into an organized system, but this rarely happens. The estimator must adjust to whatever situation arises, and this puts great pressure on him. What, for example, can cause more consternation and trouble in bid day estimating operations than the submittal of an unsolicited combined subbid that is not broken down? Quick and forceful action must then take place on this kind of subbid. At the same time the estimator is concerned with an unlimited and unpredictable number of other concerns. (The unsolicited combined subbid will be discussed at greater lengths in Chapter 17. We mention it here only as an example of pressure-related situations which must be dealt with that cannot be handled by even the best of organized systems.)

Pressure is always present in the competitive bid process. What can the estimator do to minimize its effects?

Firstly, he must organize estimating and bidding procedures and fit his operations into them. Secondly, he must distribute work loads to people he can rely on, keeping to himself those overseeing operations he feels are most critical to producing the successful estimate. Thirdly, he must see that the work he preempts for himself is finished in as short a time as is practical so that he may be truly the overseer on bid day. For example, can there ever be a valid excuse for extensive amounts of work being done to the estimate of the work of his own forces near bid time? Never. All this work should be finished at least one day before bid day. Even with this precaution, the estimator will be faced with enough adjustments to that part of the estimate on bid day. Fourthly, he must arrange things in their order of importance in making his estimate into a low and responsible bid. Certainly, a $1000 item in an estimate cannot be as important as a $100,000 item. When there is the possibility of only one subbid being available on a major item, it is more important to worry about having more than the one price than it is to concern oneself about small differentials in multiple subbids on an item of lesser impact. These and other methods require the estimator to give priority to the most important things in order to minimize the factor of pressure in the estimating process.

Can the estimator who is simply unable to adjust to pressure be adaptable to estimating questions? It is not very likely. Can experience help this person? Possibly; however, a construction firm can rarely afford the luxury of allowing such an estimator to practice with its present or future success. The reader looking for a preferred place to start in the construction organization must realize the mental strain and nervous tensions involved in estimating and decide whether he is capable of handling them. He may find the answers to his questions by comparing his ability to handle the required mental and emotional adjustments to pressure in estimating to the ability of a medical student to face the sight of blood or pain, or to the lawyer's ability to adequately express himself.

Psychologically, those people who cannot accept the atmosphere of the gambling situation will have a hard time with the pressures involved. Those who cannot stand being wrong, or who dislike even situations where they can be accused of being wrong face difficulties in this operation; those who will not accept help or delegate responsibility to others will not fit either. All of these things, if they are a part of a person's makeup, will just contribute to unneeded additional pressures in this operation.

The Estimator in the Team

So far we have looked at the estimator as an individual and estimating as an operation performed by him individually. Next, we must examine the estimator as a member of the team within the construction operational organization. We have already stated that time considerations, relative to the number of operations to be performed on a project of any real size, automatically call for making estimating a team operation. In most instances, it is impossible for a single person to perform these operations efficiently or even at all.

Different types of team organizations are used in the industry to estimate and bid work and usually they are established to fit the office organizational structure of the firm. These team organizations may change as the organization grows or because the nature, type, or size of the work estimated requires it.

In most cases it is necessary to consolidate and coordinate the operations of estimating the work to be done by the company's own forces and of figuring the cost of the general conditions of the contract of the project into one main division of work. The other main division of the work of the team involves bringing together adequate numbers of responsible and representative low subbids for inclusion into the estimate and the bid. These are the two main divisions of the operation of estimating the cost of the work, to which markup is added in order to arrive at the complete bid price.

These divisions of operations, however, require one person to take central control; this person is called the estimator-in-charge. The estimator-in-charge is responsible for arriving at the final total estimated cost of the work for proper presentation to those who will mark it up. The estimator-in-charge has central control over the total team on bid day, which usually includes a major part, if not all, of the entire organization. Generally, on bid day the estimator-in-charge puts the finishing touches on the estimate of the work of his own forces; in fact, he does so right up until shortly before bid time. The estimator-in-charge generally takes on the work of the first main division of the operation; the other main division is headed by the person who is identified as the subcontractor man or subbid organizer or tallier, who handles the subcontractors from the invitation to bid through the submission of their subbids on bid day until just shortly before bid time. At bid time the entire estimate comes under the control of the estimator-in-charge. The tally and evaluation of low responsible subbids for their inclusion into the estimate write-up of the estimator-in-charge on bid day is the main responsibility of the subbid tallier.

7.4 THE ESTIMATE FORM: ITS PARTS

Each estimate has four major parts:

1.   The general conditions.
2.   The estimate of the work to be done by the contractor's own forces.
3.   The subbid sheets.
4.   The summary sheet.

Most construction firms use estimating set-forms, which vary a great deal in the amount of detail they include. The general conditions estimating form is usually a printed company set-form that lists those conditions most commonly found in any set of bid documents (i.e., AIA 201, National Society of Professional Engineers Consulting Engineers Council, or others); in many instances the items on this set-form are inferred from the requirements of the bid documents. These documents' provisions do not always directly call for a price of the work involved; nevertheless, it is necessary to price work items in order to give a complete cost of the work (see Figures 7, 8, and 9).

The estimate of the price of the work done by the contractor's own forces is written up on a printed set-form (see Figure 10). This form allows for the proper distribution of quantities, and labor and material pricing, with a column for the totals of labor and natural prices.

The separate items of work are handwritten by the estimator on this set-form, as is the list of the subcontract divisions of the bidding documents. For convenience, the same form (Figure 10) may be used for the subbid sheets.

The summary sheet is usually a set-form that brings together all of the totals of the rest of the estimate sheets for markup and the tabulation of the final proposal price (see Figure 11).

The estimator-in-charge usually handles all the estimating of the general conditions and of the work of his own forces. He also personally handles the summary sheet on bid day; it is his task to summarize the total estimate. The summary sheet set-form lists the various items of importance to facilitate the judgment and decisions of the executives responsible for the markup.

ABC AND ASSOCIATES, INC.

GENERAL CONDITIONS	QUANTITY	UNIT	LABOR	UNIT	MAT'L	TOTAL
TITLE				SHEET NO. 1		
OWNER				JOB NO.		
LOCATION				DATE		
ARCHITECT				ESTIMATOR		
PERSONNEL						
Superintendent						
Assistant Superintendents						
Layout Engineers						
Engineers' Helpers						
Timekeepers						
Clerks						
Stenographers						
Tool Room Men						
Waterboys & Supplies						
Traffic Men						
Watchmen						
Fire Patrol						
FIELD EXPENSES						
Temporary Field Offices						
" Labor Shanties						
" Toilets						
Heat & Light Temporary Buildings						
Storage Yard Rental						
Parking						
Office Supplies						
Office Equipment & Furniture						
Temporary Telephones						
Signs						
Progress Photographs						
Additional Plans & Specifications						
Mileage						
Subsistance						
Raise In Wages						
Payroll Insurance						
Pensions						

Figure 7. A typical general conditions set-form write-up.

7.5 THE BASIC OPERATIONS OF THE ESTIMATING PROCESS

The required basic operations in the estimating process are as follows:

1. The entire work is written up in categories of work.
2. The categories listed are itemized.
3. The items are quantified.
4. The quantities of work are priced.
5. Subprices are gathered and recorded in a tally.

6. The total estimate is summarized into a "cost of the work."
7. The total cost of the work is marked up.
8. The price is proposed to the owner.

During the entire course of performing these eight operations the estimator is constantly reviewing and analyzing one major aspect of the operation that affects the total estimating process—the scheduling of the work. It follows from the nature of the estimating process that the aspect of schedule must be constantly reviewed and reanalyzed in the estimator's mind: it is changed, modified, amended, or expanded throughout the performance of all eight of the basic estimating op-

ABC AND ASSOCIATES, INC.

PG.	GENERAL CONDITIONS	QUANTITY	UNIT	LABOR	UNIT	MAT'L	TOTAL	
	FIELD EXPENSES [Cont.]							
	Travel Time for Trades							
	Building Permits & Fees							
	Tests							
	Builders Risk Insurance							
	JOB REQUIREMENTS							
	Temporary Light & Power Hook-up							
	Electric Current							
	Temporary Water Hook-up & Wells							
	Payment for Water							
	Temporary Stairs & Ladders							
	" Runways							
	" Roads							
	Maintain Permanent Roads							
	Cutting and Patching							
	Barricades & Warning Lights							
	Fences							
	Sidewalk Bridges & Fans							
	Temporary Partitions							
	" Closures							
	" Roof Protection							
	Miscellaneous Safety Work							
	Fire Protection							
	Fire Extinguishers							
	Glass Cleaning							
	Glass Breakage							
	Aluminum Cleaning							
	" Protection							

Title _____ Sheet No. 2
Owner _____ Job No.
Location _____ Date
Architect _____ Estimator

Figure 8.   A typical general conditions set-form write-up.

ABC AND ASSOCIATES, INC.

PG.	GENERAL CONDITIONS	QUANTITY	UNIT	LABOR	UNIT	MAT'L	TOTAL	
	TITLE				**SHEET NO. 3**			
	OWNER				**JOB NO.**			
	LOCATION				**DATE**			
	ARCHITECT				**ESTIMATOR**			
	JOB REQUIREMENTS (cont.)							
	Weekly Cleanup							
	Final Cleanup							
	Rubbish Chutes							
	Stakes & Batterboards							
	Pumping after Excavation							
	Trucking from Yard and Freight							
	Jobsite Trucking							
	Punch List Work							
	Spot Overtime							
	EQUIPMENT SET-UP							
	Hoisting Equipment for Concrete							
	" " Masonry							
	Misc. Equipment for Concrete							
	" " Masonry							
	Small Tools & Surveying Instr.							
	Tarpaulins - General Use							
	Elevator							
	Hoist for Subs							
	Air Compressor							
	WINTER REQUIREMENTS							
	Cold Weather Protection-Concrete							
	" " " Masonry							
	Space Heat							
	Snow Removal							
	Frost Removal							
	SPECIAL REQUIREMENTS							

Figure 9. A typical general conditions set-form write-up.

erations. Let us go into the details of this aspect of operations before examining the individual operations by themselves.

Scheduling

The estimating operation requires scheduling. Some estimators put the schedule down on paper and some do not. But every estimator must schedule the work to some extent on every job he figures; he cannot do otherwise. He must have his own view of how and when the components of the proiect are going to go together.

ABC AND ASSOCIATES, INC.

TITLE								SHEET NO.	
OWNER								JOB NO.	
LOCATION								DATE	
ARCHITECT								ESTIMATOR	
		QUANTITY	UNIT	LABOR	UNIT	MAT'L		TOTAL	
Insurance, Pensions, etc.									

Figure 10.   A typical set-form write-up sheet.

Scheduling during estimating is not the same operation as scheduling after the award of a contract. There are two important differences between the ways in which the estimator and the expeditor approach the schedule. First, the estimator knows he will have no future control over the actual performance of the work; he is simply predicting how the job can be done, not how it actually will be done. Secondly, the estimator must view schedule and time with a certain degree of subjectivity, idealism, and optimism, because he knows he must be competitive. His views on schedule must utilize those necessary traits and abilities discussed previously in this chapter; he must look much more carefully at what is possible rather than what is probable. Were he to do otherwise, he would probably not be

ABC AND ASSOCIATES, INC.

TITLE						SHEET NO. SUMMARY			
OWNER						JOB NO.			
LOCATION						DATE			
ARCHITECT						ESTIMATOR			
							LABOR	TOTAL	
SHT.			SIGNIFICANT QUANTITY						
	ABC AND ASSOCIATES WORK								
4	Form Work								
5	Concrete Work								
6	Cement Work								
7	Masonry								
8	Resteel & Metals								
	Specialties								
	Demolition & Connection								
	Carpentry [F-D]								
	ABC AND ASSOCIATES WORK			Sub-Total					
1-3	General Conditions								
				Sub-Total					
	Subcontracts								
	Sales Tax								
	Allowances								
	Bldg. Permit								
	Mech., Elect., Elev.								
	Sub Contracts to be Assumed								
				Sub-Total					
				Subcontractor Bonds					
				Correction					
				Total					
				Correction					
				Total					
				Correction					
	Due Date			Total					
	Due Date			Bond					
	Bid Security - Check			O H & P					
	Bid Security - Bond			Total					
	Liquidated Damages			BID					
	Cube		Cu. Ft.						
	Area		Sq. Ft.						
	Addendums								

Figure 11.   A typical set-form summary write-up sheet.

competitive. This is not to say that his views are not to be tempered with good sense and reasoning, but he must give a realistic appraisal to the factor of time relative to the sequences of work involved in a positive and not a negative way. He must predict the lowest possible price and, in doing so, consider primarily the fact that time costs money.

The Write-Up

The medium through which the estimator-in-charge exercises the necessary central control over the bidding of a particular job is the write-up. The write-up simply means the listing of the itemized categories of work that are to be quantity-surveyed and priced. As mentioned before, set-form write-ups are used by most companies in their estimating operations. The set-form write-up of general conditions, for example, serves to remind the estimator of the most common items of work to be priced as implements of performance (see Figures 7, 8, and 9).

Next, the work of the estimator's own forces is listed. Some contractors use a detailed set-form write-up for these items of work and others use only the most general set-form. The detailed set-form includes the most common types of items found on most construction projects. Sufficient space on whatever form is used must be provided for the listing of uncommon items of work.

Many contractors, however, insist that the estimator make his own individualized write-up of all the items for each job. The reason for this is based on the theory that the estimator will take more care to thoroughly examine the entire contents of the bidding documents if he must list by hand each item for takeoff.

The next step is the listing of the subcontract divisions of the bidding documents, which is done by the estimator-in-charge. It is a good idea for the person (bid tallier) who takes over the handling of the subcontractors to also perform this operation: this duplication furnishes a check of one listing against the other, thus giving greater assurance that no subcontract division or item of work has been missed. Again, set-form write-ups for the most common subcontract items are used by some contractors, but the same theory as mentioned before applies: an individualized write-up allows a more complete understanding and analysis of the bidding documents, and this results in a better estimate. When filled out, these sheets of paper serve as the central written control of the entire estimating process for a particular job. The documents record the price of the work. These sheets of paper, and only these, record the best and lowest responsible cost of the work. The write-up documents, including the summary sheet, make up the written estimate.

The contents of the total write-up must include the *entire* work to be done on the project, which is specified in the bidding documents. A word of caution is needed here: the bidding documents specify more than just the physical construction of the building; they specify the physical building and all of the supplemental work required to accomplish the construction in accordance with the contract. An estimator may find it necessary or convenient, at times, to deliberately leave items of work out of his write-up, but he must always at the very least be aware of and consider the *entire* work; nothing less will suffice for the proper formation of the written estimate.

Items of Work

The estimator makes his write-up by listing separate items of work, which is generally done according to a system or method of distinguishing individual items for the purpose of pricing. Many times the system consists of a set listing based on historical categories of cost experience. Thus, for example, formwork, wall, beam, and slab items of work will all be found in the write-up under the work category that includes reinforced concrete construction. But in most cases there are special items of work, even within these historical categories, that will call for an even more specific itemization. When there are significant variations in a category of work (such as high or circular wall forms), it will be necessary to list these forms as separate items of work, to be priced accordingly.

In masonry work, for example, it is one thing to lay ordinary face brick in common masonry bond into a straight wall without pilasters, but it is another to lay this same brick in some other bond using a mechanical bond system. Then, too, there are differences between the cost of laying face brick in exterior wall construction and that of laying interior face brick.

There are many variables in all the items of the work on a construction project that call for a different analysis of how much they will cost to produce. Most of these variables will have to be written up as separate and distinct items of work; they cannot be priced from historical data. It is the estimator's responsibility to determine what these items are, to list them, and to give them special attention in their pricing.

Distinguishing and listing separate items of work are as necessary in the sub-contracted work portions of the estimate as in the portion concerning work done by the company's own forces. It cannot be taken for granted that the bidding documents will separate the categories of subcontracted work in the proper way. All the sections of the technical specification documents must be read and under-stood by the estimator in relation to the estimate at hand. There are, for exam-ple, many instances when the document divisions include items of work of more than one subcontract specialty; there are also times when an item of work re-quired for the total construction will not be found in the specifications at all. The estimator must determine these things for himself and list all the items necessary for the complete work. Contractually, his company binds itself to furnishing the entire work. The estimator must make sure that the costs of all the contractual obligations have been covered in the "cost of the work" price.

The write-up of all the items of work in the project demands the close attention of the estimator. With experience he will develop the operational skills required of this task, namely, good concentration and the ability to thoroughly and care-fully read the bidding documents to truly understand what they say. This calls for careful organization of the items of work in the estimator's mind, and this organization should be reflected accurately and completely on the estimate sheets. The tasks of writing up and itemizing the work are the starting point of the estimating operation.

Quantities

The estimator must quantify the work to be done by his own forces, and each item of work in the write-up must be quantified from the bid documents. This operation is called the quantity "takeoff." The unit of work (on which the unit price will be based) is usually designated and listed in the write-up for the con-venience and ease in taking its quantity from the bidding documents. Thus the formwork unit can be established as the square foot, the linear foot, or the board foot, whichever designation is the easiest to take off from the drawings and best suits the estimator in his pricing function. Concrete quantities are traditionally unitized into cubic yards, and concrete finishes into square feet, whereas grouting is usually unitized by the piece or possibly by the square foot. Brick masonry units are generally quantified in lots (per hundred or per thousand), whereas larger units can be quantified by the individual piece. Metals can be priced by the unit (per piece) or by lot weight (per ton). Once established by an individual construction firm's estimating forces, these designated units of work are usually maintained over a long period of time for the purpose of maintaining uniformity in cost recording and reporting and for the assignment of historical values to use in pricing these units.

After some practice, the estimator trainee will become familiar with quantity takeoff procedures. Many esitmators find this task to be the most uninteresting of their various operations: it is exacting, and sometimes repetitious work; it demands concentration, and interruptions in concentration are not desirable. Each estimator develops his own individual style of taking off the work; the form and extent of his write-up more or less set the pattern of his takeoff operation. But the methods used for the tabulation and extension of quantities can vary to a great extent from one estimator to another. Methods of tabulation are individualized to whatever extent is necessary to suit the estimator's self-confidence in quantifying the job properly; any method that helps him to perform this operation thoroughly and exactly within a restricted time frame is valid.

The construction operator should watch carefully for such things as forgetting, misplacing, or mistransposing quantities to the estimate write-up from his tabulation sheets and notes. The size of the quantity usually governs the seriousness of such a mistake, but this is not always the case. Many times items of small quantity can be extremely expensive, and misplacing or forgetting one of them can be more serious than a mistake involving a large overall quantity item of less value. This is a matter to watch closely, because when an item is actually produced, there is no room for making up the loss in cost on a small, expensive unit, whereas in a large-quantity unit of work the cost of missed quantities can usually be adjusted to within the production process.

Provided that the write-up has been done properly, quantity takeoff involves little more than the careful dimensioning of the drawings, using simple arithmetic calculations and extensions. Ordinarily, there is no need for the use of more than elemental mathematical calculations. Some real difficulties can arise, however, because of the nature of these calculations and extensions: repetitious operations sometimes cause carelessness and inattention, which results in errors.

The estimator, as he gains experience, will have a tendency to consider the operation of quantity takeoff to be less than challenging. He should remember, however, that this operation is as essential to good estimating as any other more interesting operation used to arrive at the proper bid price.

7.6 PRICING THE WORK

General Comments

Pricing the work means to assign an estimate of value in dollars to the quantified units of work. This is the heart and soul operation of the estimating process; some say it is the most essential operation of the whole construction process. There are different ways to approach the subject of pricing work. Up to the point of assigning dollars to quantified units, there are certain skills the estimator must develop in order to prepare the estimate for pricing. These have been discussed previously in this chapter. The write-up itemization and quantity takeoff operations both require the estimator to organize his thoughts. But in pricing there is one essential additional thinking process that the estimator must learn and develop; pricing requires, for the first time in the estimating process, thinking seriously about competition.

For the first time the element of competition begins to play an important part in the estimator's thinking processes. Every firm's write-up should include all the work required. Theoretically, the quantities of work should come up the same for each estimator in each competing firm on any one particular project. Indeed, there are quantity-survey firms that furnish the same quantities to competing general contractors as a service. In many instances, unit price contracting methods furnish identical quantities and work itemizations, so that the pricing and markup operations become the only differentials in the estimating process. Even the itemization of work units becomes the same.

In competition, the greatest single differential over which the estimator has direct control and which influences arriving at the lowest cost of the work is the assignment of prices to the units of work done by the general contractor's own forces. But this is by no means the only differential that determines the cost of the work. Competitive estimating also requires the proper assignment of the lowest responsible subcontract prices into the estimate. The estimator must realize that in the closely competitive bidding situation there are usually two other significant differentials, over which he does not have direct control, that will affect the competitiveness of the estimate. Aside from mistakes, these differentials are the following:

1. The ability of the contractor to have the lowest subcontract prices available.
2. The markup that is assigned by the contractor to the estimated cost of the work.

An estimator generally does not have as much direct control over the first two of these factors as he does over the third. In the first case he is depending on the pricing abilities of the subcontractors who are quoting their portion of the work. In the second case the markup is more often under the direct control of company executives who are not as fully aware of the details of the job as he is.

Each factor, or a combination of the three, can make a bid high or low. An estimator does himself and his company a disservice when he consistently blames the loss of a job on the improper handling of the two factors that are not under his direct control. It is his direct responsibility to have the correct estimated cost of the work of his own forces and the correct cost of the contract conditions. Obviously, in high-percentage broker work these responsibilities decline as the trade work of his own forces declines. But even in this type of work the conditions of contract still remain a major and competitively significant part of the cost and within the estimator's direct responsibility for correct pricing.

The reader should remember here our previous comments and not overemphasize the competitive features of pricing the work. They are important, and he must be aware of them, but he must keep them in their proper perspective.

Competition

While we have examined the element of competition and how it affects the estimator in pricing units and conditions, the reader is advised that competition should never solely govern the pricing of units in any case. It should only be regarded as a restriction insofar as it influences the estimator in his pricing the units at their lowest reasonable cost. The estimator's judgment of his own company's organizational abilities and capacities is the prime factor that should govern his unit pricing. To allow a competitor to determine in any way the price of a unit is neither sensible nor logical. Even if it were possible to copy the write-up itemization and quantities of a competitor's estimate, it would be foolish for a contractor to use another organization's unit prices. This is so not only because the estimated costs would then be identical, and thus noncompetitive, but because it would assume that the competition had the same ability to produce all the units at the same price. Such an assumption is not logical. Our premise is that in matters of estimating the contract conditions and the work of his own forces, the estimator must realize that he is competing against other competent general contractors. Trying to estimate the probabilities of the competitor's pricing in these same units, however, is not a wise practice. If competition is to be considered at all, then it should be considered in the markup and not in the estimated cost of the work. And should a more extensive cut of the estimated cost be necessary because of a consideration of competition on the bidder's list, the cost should be cut from general conditions before unit costs are touched. Certainly the reader can appreciate the possible confusion and the probability of error that could occur should unit prices be juggled from job to job in order to satisfy some supposed competitive necessity. This will not work. Unit pricing is the trade secret of the construction firm. It should be considered a major determinant of being properly low on a job bid.

There should be serious reservations about the estimator overemphasizing bidding competition in the estimating process, because this, in effect, allows competitors to do his estimating for him. His estimate should consist of the lowest cost of the work involved based on his best rational judgment on the determinates and his best guess on the unknowns. Great surprises will be in store for estimators who base their judgments on the belief that their competitors are desperately in need of work because they have been unsuccessful in bidding their last few jobs or in trouble of one sort or another. The safest estimate is that which is based on the solid objective analysis of the estimator without his reference to such outside factors. It is far better to avoid wasting money and time on a job where questions about the competition are preeminent by avoiding bidding on the job in the first place.

Methods of Pricing Work

How does one price construction work? There can be as many answers to this question as there are construction firms. In addition, there can even be different ways and means used to arrive at prices by different estimating operators in the same firm.

First of all there are two major ways of pricing categories or items of work that are *incorrect:*

1. Pricing work by "square footing."
2. Pricing work through the *exclusvie* use of "catalog" unit prices. By *catalog* we mean the popular and, to some degree, valuable publications that list and price units of construction work.

We prefer the use of the detailed and developed unit price method of pricing work. Let us offer a brief explanation of the two methods listed so that the reader may compare them.

Square Footing

To square foot an estimate means to quantify the total square footage of the building from the bid documents and assign to it a cost per square foot. This results in a unit price; but only this one unit price is used to form to cost of amount. The assignment of a cost per square foot is usually arbitrary and made without adequate analysis (or any analysis at all) of the uniqueness of the individual project. If the building is a church, a cost per square foot for churches as a general category of building is arbitrarily assigned; if it is a factory, cost of the work for a "typical" factory is assigned, and so on.

The most extreme way of square footing is to simply make a single all inclusive unit for a single project based on its type or classification. An entire industrial building project would, for example, have a unit figure of $30 per square foot as its estimated cost of factory construction. Assigning an area square foot figure to the square footage of the project will result in an estimated price. Thus a 10,000 square foot factory building would be priced at $300,000.

A variation of this method of pricing work would be to establish limits of say $30 to $40 per square foot for an industrial building, with the variance in the unit price depending on the type of building it is. Thus a 10,000 square foot building could be priced at from $300,000 to $400,000, depending on how the estimator views the job and its unique features.

To further expand on ways of using this method of pricing, a firm might assign different square foot values to the various parts of the job, depending on the finishes to be included in each area. Thus an industrial building may be priced out for 9,000 square feet at $20 per square foot for unfinished factory space and $60 per square foot for finished office space. This 10,000 square foot factory building would then be priced at $240,000. As the estimate of the building's unit cost via this method becomes more detailed, it becomes closer to the proper estimate of the construction of the project.

The opposite to the square foot method of estimating is the one we recommend to the reader. We claim that the more the components, categories, and items of work of the project are broken down, the greater the chance for pricing its actual cost and thus for furnishing a legitimately low bid. Rejecting the extreme method of pricing and bidding the entire project as just one unit, we suggest breaking up the components of the project into as many separate categories and items of work as estimating time and good reason will permit.

The greatest flaw of the square foot method of pricing is that it does not allow the estimator to give valid consideration to such things as conditions of contract, the potentials of the unknowns and indeterminates, and the many other unique features of an individual job. Even when guessing within limits established for a particular type of construction (industrial, commercial, institutional, etc.), the estimator has much less of a chance of coming up with a proper figure when using this method of pricing the work.

The reader should realize how difficult it is to be in competition with those who use the square foot method: usually, this kind of competition is found to be either extremely low or extremely high in its bids. It is therefore wise for estimators to carefully consider the bidders' list to see who the competitors are and whether any are "square footers" in order to avoid what could turn out to be a waste of time and money in bidding against them. Just one square footer on a list is enough to make bidding difficult for others.

Square footing can be a valuable tool in the total construction process, architects, engineers, and some particular types of speculative contractors using it as a guideline for budget and financial considerations. But the reader should note that the great majority of successful general contractors do not use this type of pricing method.

The Catalog Method

The various catalogs used for pricing present another problem for the estimator, namely, establishing in his own mind the best method to adopt in pricing work. Catalogs are extremely valuable tools for gaining information for some estimating purposes, but it is not good to use their prices when the estimator should price the work himself; even the authors of these books do not recommend that they be used without reservation. These catalogs do, for example, come in very handy on bid day when no subcontract price has been recorded on a particular item. If there is no plastering sub-bid, or possibly just one suspect sub-bid, then the division of work (plastering) can be quantified by the estimator and the pricing information supplied from the catalog, with proper adjustments made for the job at hand. This kind of situation happens quite frequently on bid day, and catalogs then become extremely valuable. But to price entire projects by using these kinds of books is not advisable.

The real danger in using the catalog method of pricing (or even variations of it) is that the estimator may not give the necessary considerations to a more extensive detailing of the components of the estimate in order to obtain a better estimate of cost. It is not good strategy to rely on these catalogs as a matter of practice.

The Recommended Method of Developing Unit Prices

So far we have discussed methods of pricing the work in general; now we approach the proper method of unit pricing, an essential operation in estimating the work of one's own forces.

There are two basic thinking abilities required in learning the proper method of unit pricing:

1. The ability to price a unit as though it had never been figured before.
2. The ability to use proper and correct information based on valid experience in the unit pricing operation.

The estimator must use a combination of these two thinking abilities: one establishes the ability of thinking in a certain way, and the other discriminating standards when he must rely on experiential information in pricing units. These two thinking abilities lead to the development of specific methods of unit pricing.

It is obvious that when an item of work has not been done before within company operations, no experiential information can exist and therefore methods for developing a unit from scratch become necessary. On the other hand, the estimator must realize that operational time considerations will seldom allow for the figuring of all new and individualized unit prices. Thus empirically developed experiential data must be used, and it must be used properly. Using a combination of both thinking processes in developing methods usually produces the best result in unit pricing operations.

The Labor Unit Price

The essential thinking process needed to properly price the nonhistorical labor unit price is quite simple: the estimator visualizes the number of time units necessary to accomplish a quantity of an item of work, and from this visualization is derived information that fits into a simple formula:

$$\text{the unit price of any item of work} = \frac{\text{cost of the number of hours of all estimated labor forces required to do the unit of work}}{\text{unit of work}}$$

Thus the unit price of a square foot of slab form will equal

$$\frac{(4 \text{ carpenters}) + (1 \text{ laborer}) + (\tfrac{1}{4} \text{ operating engineer})}{100 \text{ square feet of slab form}}$$

and given

carpenters' wages @ $10.00 per hour (base or total wages)
laborers' wages @ $6.00 per hour (base or total wages)
operators' wages @ $12.00 per hour (base or total wages)

therefore

$$\frac{\text{one square foot}}{\text{of slab form}} = \frac{40.00 + 6.00 + 3.00}{100} = \frac{49.00}{100} = 49\cent \text{ per square foot for labor}$$

If then the base unit price cost of this unit price is 49¢ per square foot for labor, and there are 100,000 square feet of slab forms quantified on the job, the estimated base labor cost allowed for the slab formwork will be $49,000. Each and every labor unit of work is handled in the same way. Equipment costs or some other costs closely related to labor will be included in the formula used for pricing the labor unit when special circumstances warrant it.

The Material Unit Price

Somewhat different considerations must be given to pricing the material unit. This unit price can be simply the outright cost of a material permanently installed in the work. Thus, if concrete costs $30.00 per cubic yard for five sack concrete, then $30.00 is the material's base unit price. But if, for example, there are costs of 75¢ per cubic yard for additives and the cement differential is $2.00 per sack, then the material's unit price must be $32.75 per cubic yard six-sack concrete.

However, there are construction materials that are reusable in the construction of the work, even though they eventually become expendable. Take form lumber, for example, and more specifically wood deck slab form material. This material is usually purchased in sheets ($4' \times 8' = 32$ square feet). It is used and reused until it is no longer fit for use as a forming material. When it is cut into pieces, it is still reusable, but not to the same extent as it was when it was whole. Its usability and value is eventually expended.

The estimator must estimate the number of reuses possible for this material, because this determines the unit price to be used in the estimate.

Example. If a $4' \times 8'$ sheet of ply form deck costs $9.60 new, then its unit price at the time of purchase is simply

$$\frac{\$9.60}{32} = 30¢ \text{ per square foot}$$

But if this square foot of form material can be used 10 times, then its unit price becomes 3¢ per square foot as it applies to pricing a unit of slab formwork to be done. If it is determined, for some reason or another, that the material can be used only two times before it is expended, then its unit price would become 15¢ per square foot.

If 100,000 square feet of contact surface of slab form are quantified on a job and the estimator figures 10 reuses, then the estimated cost of the material is only $3,000, whereas with only 2 reuses $15,000 must be allowed for material cost.

When the material unit is added to the labor unit price (see previous example) for the same quantity, then the total labor and material unit price is established.

Example. A labor unit of 49¢ per square foot plus a material unit of 3¢ gives a total unit price of 52¢ per square foot, or a price of $52,000 for 100,000 square feet of contact surface.

The same labor unit of 49¢ per square foot but a material unit of 15¢ gives a total unit price of 64¢ per square foot, or a price of $64,000 for 100,000 square feet.

The significant differences between the total unit price and the total price in these examples should show the great importance of the reuse capabilities of materials in competitive material unit pricing.

There are all kinds of variations in the numbers of tradesmen and trades employed in doing a unit of work. The cost of their labor and the material and equipment costs are, when necessary, introduced into the simple formulas for unit pricing shown above.

The Art of Unit Pricing

The real skill in the art of pricing is the ability to imagine how ideally a unit of work can be done, which results from the ability to visualize the mechanics doing the unit of work. The use of these skills constitutes the correct way of pricing construction work.

The reader may comment that visualizations or imaginations result in a price that is just a guess anyway, but he or she should realize that the estimator, by going through these thought processes, is establishing reasonable limits for the guess; he is setting limits on the time needed to perform the work. An estimator with common sense, even if he is short on actual operational experience, can reasonably assess the number of mechanics and the time it should take to do an item of work. Relying on his common sense, the estimator can narrow the limits of his price until he reaches what he feels is an acceptable rate of production. This gives him reasonable numbers to insert into his unit pricing formulas, which are well thought out, making the guess more reliable.

At some point in the unit pricing operation the estimator must consider the indeterminates. At the start of this chapter we emphasized that a successful estimator should maintain an optimistic attitude about the possibilities and dismiss any pessimistic thoughts about what might occur under an adverse situation. In the unit pricing process the estimator must be an optimist. If his judgments of the unit price are based on good sense, then he should not let the possibility of adverse unknowns unduly influence his unit pricing operations. If the estimator does not adopt a positive attitude, his work will not result in competitive unit prices.

We emphasize here that good logical reasoning based on the knowns combined with a positive but realistic approach to the unknowns is more essential to the estimator than operational experience in unit pricing operations. Indeed, we believe that extensive operational experience can sometimes hinder good unit pricing rather than help it. This is because many times prior experience of observing the performance of a unit of work leaves misleading impressions about the possibilities (or impossibilities) of doing the unit of work for other than actual cost experience. The actual experience of such a unit price that is not consistent with the actual cost, however, should not lead to the supposition that the estimate's unit was necessarily bad; it may be that the unit of work was performed badly. Unfortunately, one who has experienced seeing an actual unit of work performed well or poorly may be left with the indelible impression that the actual unit cost must be the correct unit.

Historical or Experiential Data

Using historical or experiential data is essential to the whole estimating process. When pricing a unit, the estimator must consider the use of *available* proper and correct historical information based on valid actual historical experience with the same unit on past jobs. There is no question of the value of using experience gained with actual unit costs in estimating operations, but the construction estimator must be aware of all of the potential problems when he considers using historical data.

Historical data is obtained by recording and reporting the actual time and money spent on a unit of work. For a unit of formwork, for example, the carpenter foreman will report in his time book the number of hours that his own men and any supplemental labor actually spent on the work unit on a particular day. These actual hours are then converted by the timekeeping personnel and the payroll department into money actually spent on the unit of work. This, in turn, is converted into the actual unit price of the work (usually, at the end of each pay period).

Example. The total quantity of wall form (item #5c) on a project is 100,000 square feet. The estimated labor unit price allowed for the work is $0.875 per square foot.

The total quantity of wall form actually built during a specific time period is 1000 square feet. At the end of this time period the money actually spent on the unit (as determined by the hours of labor in the carpenter foreman's time book) for item #5c is $670.00. The actual unit cost is $0.67 per square foot out of an estimated allowance of $875.00.

When the total 100,000 square feet of item #5c are done, a cumulative actual cost will also be recorded for the work. If the actual unit cost remains $0.67, then the total cost will be $67,000. With $87,000 allowed by the estimate, the gain on the actual unit over the estimated unit will be $20,500.

If, when the 100,000 square feet are completed, the total actual cost is $100,000, then an actual unit cost of $1.00 per square foot and a loss on the unit of $12,500 will be recorded.

Whichever actual cost unit results, it forms the unit price for item #5c (wall form) and becomes historical data.

There are two significant factors the estimator must consider before he can use such historical information in future estimates: he must be sure, first of all, that the assignment of time recorded by the foreman was accurate; he must also be confident that the quantities of the work (both cumulative and total) were accurately kept. These two recording operations are never under the estimator's direct control. We stress in other parts of this book the vital importance of reliable and accurate cost reporting and recording; the estimator must rely on their being done correctly because he must integrate the resulting historical data. This calls for an absolutely rigid company policy demanding that the operations of cost reporting and recording be done consistently and correctly by those responsible.

The second significant factor the estimator must consider in using historical data is more complex; it goes to the heart of the estimating process. We have said before that no two construction projects can ever be exactly the same in time or space. If this is true, then neither should the unit prices of their work components be exactly the same. Therefore, when using a historical unit price, the estimator must make adjustments to historical units to suit the work at hand. The previous example of the 67¢ per square foot wall form can truly reflect the cost of that work on a past job, but because of conditions of soil, weather, or water on a new job, it may be totally wrong to use it in estimating the new job. When there are extreme differences in job conditions, quantities, or the capabilities of job personnel, the historical unit may only serve to establish the broadest limits for development of a unit priced specifically for the new work, even though the work may be nearly identical in its physical detail on both jobs.

The estimator must also be very careful about overextending the use of a historical unit price of one type to include a special unit of work of another type. It would be unwise, for example, to think that because 67¢ per square foot is a proven historical price for a straight wall, it would be applicable to pricing a unit of circular wall construction. It should be clear to the reader that such a difference in the shape of the wall completely negates the use of the 67¢ per square foot (straight wall) historical unit. Other differences of this sort may not be as obvious. Nevertheless, all differences in the nature and extent of the new unit of work being priced must be just as critically analyzed in order to estimate a true unit price. The height of a wall, for example, or the number of pilasters required in straight wall construction produces extreme differences in costs, even though both might involve straight wall forms. The estimator must recognize these differences, and if a historical unit has not been developed for such conditions, he must develop a new unit, using the recommended method of pricing units of work.

The above-mentioned example using the wall form unit price extends to all the units of work listed in the write-up of the work of the estimator's own forces. Very few unit prices should be exempt from the estimator's close review and analysis of historical development and validity. To simply fill in the write-up with historical unit prices without giving them individual consideration is not true estimating: it invites mistakes in estimating operations that are extremely hard to rectify. Historical data furnish invaluable information on known actual costs of work, establish excellent limits for guessing, and provide checks for the estimator, but they will not furnish guarantees of actual cost performance on a successive project. If they are used exclusively or to an extended degree in pricing the work of one's own forces, they will not furnish competitive low bids either.

Pricing the Conditions of Contract

The estimator must constantly review the pricing of the conditions of contract from the start of the initial estimating operations until bid time. One must realize that these conditions embody not only all the items of physical work that can be unitized into labor and material categories, but also all items of work needed to implement the performance of the construction contract. These items of work are often merely supplemental to the performance of producing the building.

Many of the conditions of contract are of such a comprehensive nature that their cost must be determined from a point of view that encompasses the building of the overall project; they must be analyzed carefully as to their effects on the time allowed for the production of the project. These conditions require, for example, guidelines for the pricing of the indeterminates of weather, conditions of site, and access, and they encompass the performance of the entire work on the project from start to finish. They are the most difficult things to both quantify and price in the entire scope of estimating operations.

As the estimator progresses through the basic operations of estimating he begins to get a clearer picture of how the job will go together as well as a closer definition of the effects of the general conditions on the work. This refinement of detail enables the estimator to more accurately assess the extent and thus the price of these conditions.

When an estimator becomes proficient in the technique of analyzing and pricing the general conditions, he will have come a long way toward mastering the total operation of estimating, because he has no doubt developed a true sense of what the entire estimating process is all about. He will have learned that the construction process consists of a great deal more than the mere physical construction of a building; he will know the importance of properly pricing these items of the total work, because they constitute such a vital differential in the competitive aspect of the contract construction business. A unit square foot of formwork is the same square foot in quantity for all competitors. Differences in estimating the time of completion, the probability of good or bad weather, the most efficient types of equipment to use, or the significant and special abilities of personnel are just some of the factors that distinguish one competitive bid from another.

All the conditions of the contract require pricing. Nowhere in the estimating process is there a clearer indication that estimating is an art than in the pricing of the general, supplemental, and special conditions of the contract. These conditions call for the pricing of supervision, all supplemental nonproductive labor that cannot be logically included in the various unit prices of trade work, all indirect field expenses, and some of the office overhead expenses occasioned by the requirements of the conditions' provisions; in union areas they include the pricing of the union contract's working conditions rules and sometimes fringe benefits. Conditions of contract also include the pricing of the individual special job requirements, the equipment to be used in common on the project, and the unknowns of weather and conditions of site.

When each type of condition is examined closely, it is clear that the estimator must rely on the basic fundamental concepts of pricing work as if it were new, as described earlier in this chapter. An even greater reliance on the fundamental concepts is required if the conditions are supplemental and specialized. A special condition, for example, is less adaptable to the use of historical data than is a supplemental condition, and a supplemental condition requires greater use of fundamental pricing concepts than does a general condition. All pricing conditions must rely heavily on procedures that approach the project at hand as a unique product, rather than on those that are based on historical data. The conditions of contract truly characterize the estimate of each and every separate construction project.

When pricing conditions, the estimator must give extensive thought to two significant elements inherent in the construction process:

1. The element of time.
2. The element of the unknown (or indeterminates).

The estimator must always first establish his perception of the time needed to do the work of the entire project before assigning time to individual general conditions. This is true regardless of whether or not a time of completion is specified in the bid documents. It would be foolish, for example, for an estimator to regard an unrealistic specified time for completion as being a limit on his pricing of the general conditions. Most construction projects, however, do allow for adequate time, even though in some cases it would seem impossible to price the job competitively and perform it in the time allowed.

In pricing the conditions, the estimator is not only concerned about the total time allowed but also about how the various operations of the job fit into the total time and into significant portions of total time. He cannot price weather, for example, without predicting how the various construction operations he is pricing will fit into segments of time where various extremes of weather are probable. On long-term projects the price of minimizing the effects of weather conditions versus the need for the completion of specific operations scheduled in periods of bad weather becomes extremely critical in producing a competitive bid price.

In addition to balancing the need for computing individual operations during weather cycles within various time segments, the estimator, when pricing conditions, must consider the cost of weather conditions over the total allowed time as well. He must do this in order to correctly price such things as job overhead, idle equipment time, and lost labor time. Also, he must consider operations that are not directly affected by weather, such as job cleanup or punch list work, but which can still be indirectly affected by bad weather encountered early in the job, where the time allowed to do these operations is then restricted. All of these considerations of costs occasioned by the conditions of contract must be made with the goal of making the estimate competitive in time as well as in money.

Some of the conditions are quite simple to price, whereas others are extremely difficult. For example, the price of a superintendent can be found by simply multiplying his wage by the entire estimated time the estimator allows (or is allowed) for the project. The estimator can be fairly certain that his good competition will price the superintendent in the same way.

Now consider the estimator's choice in pricing the lost time caused by weather for other supervisory job personnel or equipment. Here it is not at all certain how the competition will consider weather and consequential lost time in pricing an estimate. Extend this same situation to include the effects on equipment lost time, and further complexities arise in pricing. No matter how complex and difficult the pricing conditions become, a good estimate should cover the entire cost of the project, including all lost labor and equipment time as well as productive time.

The estimator, when pricing the conditions, should make good, reasonable judgments based not on the probabilities of good or bad weather (or any other unknown), over which he has no control, but on the costs that might be occasioned by the severity or the extent of good or bad weather. However, wide latitude is allowed the estimator when considering pricing the effects of the weather on the conditions of the contract. The estimator can decide, for example, to either enclose the structure and heat for cold weather (an additional cost) or prepare for a job slowing down (an additional but different cost); in each case he must judge the relative costs. Figuring these relative costs can be an extremely difficult task in the competitive setting of bidding.

When analyzing and pricing the unknowns, the estimator must recognize one simple fact: they are a potential threat to every construction job. Whether the unknowns are the conditions of the site, the availability of labor, or any number of other indeterminates, it is impossible during the course of a construction job not to encounter some unknown or indeterminate condition. Some of them can be expected, and some will be unexpected. The task of the estimator is to realistically examine their effects and the extent of the effects if and when they are encountered. As mentioned before, it is not realistic for an estimator to overemphasize the effects of the unknowns to the extent that they become prime determinants of the cost of the work; this is not practical. But it is his task to recognize their potential and their possible effects so that he may constantly keep them in mind during the course of pricing the work. He must learn to study the effects of unknowns thoroughly and realistically.

Example.   The estimator assigns a certain time period within which to perform a certain operation that could be affected by rainfall. He then has a choice to make when he prices the work in the operation: he may consider that during the entire time span it will rain every day, or he may go to the opposite extreme and consider that there will be no rain at all. Using his best judgment on the possibilities, the good estimator must allow for some rain during the course of the operation. But his responsibility does not end with this simple choice. Beyond this, he must consider not only how rain will affect the one operation, but how it will affect all concurrent and particularly subsequent operations relative to the completions on the project. This study of effects is the most imporatnt consideration in the pricing of unknowns.

Estimating and the effects of indeterminates on time and costs is the most crucial part of estimating the conditions of the contract. While the conditions of contract usually constitute only a relatively small percentage of the overall cost of a construction job, some professionals consider their costs to be the single most important element of risk in pricing in the entire estimating process. One can spend an entire career in estimating operations and still feel uncertain and lacking in one's ability to price conditions properly.

7.7   THE ESTIMATOR AND THE BIDDING DOCUMENTS

The estimator has certain important tasks to perform on each individual project he is assigned. Each project demands that he work through its specific bidding documents; it will not do for him to view the bid documents for one job as being the same as those of another. The survey of the estimator's tasks listed below should furnish the reader with a guide to his approach to estimating procedures on any project he may consider:

   1.   The estimator must gather all the contract bidding documents that are necessary to price the work in its entirety. Emphasis should be given to obtaining the *complete* set of bidding documents, which must be the most current, and they must be kept current during the bidding period.
   2.   The estimator must study the Instructions to Bidders. The title of this document reflects exactly what it is, and it is an essential guide to the bidding process. While there are many important parts to this document, there are two of major significance. Firstly, the estimator must check the time given for the submittal of

the bid specified in the Instructions to Bidders and gauge the time and labor necessary for estimating operations within this time period. He must organize his time accordingly and make certain that the entire company organization is able to accomplish the estimating work involved within the allowed time. Secondly, the Instructions to Bidders contains the bonding requirements for the project. Since it takes time to secure and execute bonds, it is essential that the estimator make those responsible for securing the bonds in the organization aware of the exact requirements as stated in the Instructions to Bidders.

3. The estimator must thoroughly examine all of the conditions of contract. This may seem to be an unnecessary and repetitious task for the estimator, especially when he has become familiar with the set-form specifications usually contained in the ordinary construction project's bid documents. Even those owners who develop their own sets of conditions usually specify provisions that are very similar to those in the set-form conditions. It is essential, however, that the estimator read and analyze all of the conditions carefully and in their entirety; he must always look for unusual or specific conditions on each individual project he estimates.

4. The estimator must examine the proposal form, note it, and set it aside for reference and completion. This is the document that formally expresses the contractor's offer to the owner. It is usually furnished by the owner in the form he chooses, and he solely determines its contents. The general rule in construction practice is that the contractor must respond to exactly what the proposal form specifies, allowing for no deviations in furnishing the full information it requires. The proposal form should not be qualified, modified, or extended upon in any way in the competitive bid situation. It contains the words of offer, the time allowed for acceptance, the proposed price, and many other terms of importance, and is usually submitted to the owner in duplicate or triplicate. It also contains places for signatures and sometimes allows for attestations; it clearly identifies the offerer and the offeree and the date of the offer. The proposal form, when filled in on bid day, constitutes the offer.

The proposal form can often be a very complex document to execute, particularly when viewed in the light of the pressures and excitement inherent to the bid day processes. This document usually requires, in addition to the base bid price, such information as the following:

1. The unit prices for extra or additional work.
2. The percentages of markup for extra or additional work for both the contractor and his subcontractors.
3. The fees for the assumption or assignment of other separately bid contracts.
4. The time for completion stated either as specified by the owner or as offered by the contractor.
5. The listed alternate prices in any form or to any extent desired by the owner.

These are some examples of information the owner can ask for in a proposal form. The fact is that the owner may list on the form whatever he wants the contractors to make an offer on. The estimator, since he is the operator who finally determines all the required prices, percentages, or completion times to be offered, must know exactly what the proposal requires, and he must be aware of each of the requirements from the beginning of the estimating process to its end on bid day.

5. The estimator must carefully examine the technical specifications. Here we provide a word of caution to the reader: an estimator who has read many technical specifications in the past tends to regard them as sufficiently similar in content that he might begin to lessen his emphasis on his examination of the documents. Here again, carelessness or presumption can have extremely expensive consequences.

The reader should take note here of how carelessness or presumption with any of the bid documents can result in bad estimates. There is as much danger of adding more work than is called for in the documents to an estimate as there is in missing segments of the work. While the consequences of overestimating may not be as serious as underestimating, the result is simply a bad estimate and a waste of time and effort. Attitudes that allow carelessness in the examination of bid documents cannot be tolerated in a successful estimating operation.

6. The estimator must examine and review the drawings. This task is more or less accomplished during the quantity takeoff operation, whose ordinarily repetitive nature allows the estimator to gain a substantial knowledge of the drawings in the relatively short time during which he uses them in this operation. However, when a split responsibility occurs within the takeoff operation or in pricing the various parts of the estimate, each operator involved should be familiar with at least the drawing documents involved in his part of the takeoff. Preferably all the estimators involved with pricing should know all of the drawing documents concerning the work being priced.

7. The estimator is responsible for including in the bid the cost of all the addenda to the bidding documents. Addenda create a peculiarly difficult situation at times, either because of the traditional system used to handle them or because of their late arrival into the bidding situation. The addenda system is similar in function to the bulletin system (see Chapter 12), with one important exception: it often does not allow adequate time for the affected bidders to make proper investigations of the work included in them. The investigation and pricing of the addenda must be done within the allowed bid time. The estimator must always make sure that he has included all work in the addenda, including subcontractors' work, in the bid price. He must, at the very least, make his subcontractors aware of the issue of new addenda as bid time approaches; when there are late addenda, this sometimes becomes an extremely difficult, if not impossible, task.

Nevertheless, when the proposal form requires, the price of the addenda must be included. The problems of late addenda are discussed in more detail in Chapter 17.

8. The alternate prices to the base bid (when they are requested by the owner) can be an extremely difficult feature of the bid documents to handle. As the number of alternates involved in a proposal increase, so does the complexity and difficulty of the whole estimating process for the project. The peculiar operational difficulties encountered with this portion of the proposal form are discussed at greater lengths in Chapter 17. It suffices to say here that the inclusion of alternate prices requires the estimator to alter his procedures from the initial write-up and takeoff processes to the final fill-in of the proposal form. All alternate prices requested must be provided in the proposal form in order to make a truly unqualified responsive bid.

9. The estimator must give unit prices to the extent and in the form required in the proposal form. There are some special precautions the estimator must take when providing requested unit prices for adding or deleting work. The most important thing is to not forget the costs of overhead and profit that must be included in the unit prices, particularly when the unit prices are stated as being additive or deductive.

10. The estimator is not usually responsible for obtaining the bid security from the surety company, but the bid bond and other forms of security are bid documents. Since he must present the bid security with the proposal form on bid day, he must concern himself with this facet of the bid process.

The estimator is also frequently called upon to make early and fairly exact projections of the cost of the work in order for the firm to obtain a bid bond. These early projections are a necessity when the bid security is a certified check.

7.8 PRODUCING A PROPER PROPOSAL

The estimator in charge must perform the following operations in order to produce a proper proposal:

1. Completely and thoroughly examine the bidding documents.
2. Write up all of the items of work.
3. Take off the quantities of work of his own forces.
4. Price the quantified trade work of his own forces.
5. Constantly analyze and reevaluate the extent and pricing of the conditions of the contract.
6. Initiate, correlate, tabulate, analyze, and include the low responsible subcontract bids into the write-up.
7. Include fringe and insurance costs ("burdens") on the priced work.
8. Include applicable taxes on the taxable items.
9. Tabulate, compute, extend, and check all the items in the write-up sheets thoroughly and in good time.
10. Develop the unit prices required in the proposal form.
11. Transfer all extended and checked figures in the write-up sheets to the summary sheet of the estimate in good time.
12. Check and compare figures against solicited or nonsolicited subcontract prices on work that matches the work of his own forces estimated by himself.
13. Tabulate the summary sheet and have the total cost of the work price ready for markup in a reasonable time before the bid is due. (The executives of the firm usually perform the markup operation.)
14. Add bond premiums and permit fees.
15. Complete and determine the final base bid figure from the summary sheet.
16. Compile the final alternate prices and determine the completion time if these things are required by the proposal form.

7.9   THE ESTIMATOR AND THE SUBCONTRACTOR

General contracting operators tend to rate subcontractors according to their specific relationship to them in their respective operations, thus the field and expediting personnel usually evaluate subcontractors on the project based on their prior
experience with each of them. Estimators must regard subcontractors as being of
prime importance to successful estimating operations; from their point of view
the subcontractors must produce the lowest reasonable subbids possible and, just
as importantly, furnish these low subbids to the estimator's firm. It bears repeti-
tion to say that in the modern construction process the percentage of subcontracted work on any job is so substantial that the combined prices of the work
subcontracted often determine the success of a general contractor's estimating
operation. While each part of the estimate must still be properly figured, it cannot be denied that the greatest percentage of every estimate is made up of
the subcontractors' prices. The estimating department of any contracting firm
must then see to the development, encouragement, and fair and equitable treatment of all the responsible subcontracting segments of the industry.

It is an accepted "fact of construction life" that no estimating force, and thus
no general contracting firm, can be successful unless it has the confidence and
support of the subcontractors it consistently deals with. Confidence and support
does not mean showing favoritism or going beyond the limits of sound businesslike practice, but it should be obvious to the reader that callous, unprincipled,
dishonest, or unfair dealings with subcontractors will gain neither their support
nor lower their subbid prices. This is not to say that there can be no hard bargaining or plain talk in subcontractor relations, but respect must be shown for
the fact that subcontractors are specialists in their field who presumably know
how to price and do their work. They must be shown that their abilities and interests are regarded seriously, with confidence and trust, by the general contractor's operators.

Sometimes there are problems associated with general contractor–subcontractor relationships within the bidding and estimating processes: some are occasioned
by innocent actions either on the part of the subcontractor or the general contractor; some are caused by deliberate attempts by one of the parties to misuse
the accepted system of the construction bidding process to their own advantage.
The two most common incidents of the latter type occur when "bid peddling" and
"bid shopping" practices are used by the general contractor's operators. We do
not propose to discuss these practices at length: to do full justice to their causes
and adverse effects would take too much space. While these practices cannot be
justified, they are certainly prevalent in the present-day construction industry. It
is common for the various segments of the industry to deny that their personnel
indulge in these practices; everybody seems to be innocent. To engage in an argument about the existence and common use of bid peddling and bid shopping
would be a waste of space. It suffices to say that, human nature being what it is,
trying to do anything effective to eliminate these practices is practically useless.
The reader should carefully note, however, that those construction firms that commonly indulge in bid peddling and bid shopping are, for the most part, short-lived
in the industry.

Subcontractors should be invited to offer their prices to the general contractor. They should be given as much information by the estimating department as is practical, and, if possible, provided space in the general contractor's office for the use and examination of the bid documents and offered time to ask for and review answers to their questions. They should be treated with patience and an understanding of their own particular problems in their specialties and in the bid process. Solicitation of the subcontract bid with reasonable consultation or interpretation of the bid documents between the general contractor and its subcontractors is probably the surest way of obtaining a low reasonable subbid price on bid day. Solicitation, however, does not mean that only solicited subbids are received on bid day; the unsolicited low responsible subbid is just as welcome.

The unsolicited subcontract bid, however, can cause serious problems for the estimating forces, particularly if it is received very late in the allowed bid time. These bids, while they are not unwelcome, cause difficulties when they involve inclusions and exclusions to the divisions of the trade specifications that the subcontractor has not previously discussed with the general contractor. A subcontractor can also cause serious problems when he combines work of trade divisions, particularly when the subbid is unexpected and not broken down into comparable portions of the work. This problem will be discussed further in Chapter 17.

It is wise for the general contractor's estimating department to actively solicit bids from subcontractors well before bid day. The invitation can simply be sent out on postcards specifying the project, the trade division solicited, and the date desired for the bid to be submitted in the general contractor's office, with other informational data as required. In most cases the subcontractors are aware of the project through their reading of the published trade reports, which give the information of interest to the subcontractors, including the list of general contractors bidding. An invitation gives them special notice and emphasizes the interest of the general contractor in the project. Formal acknowledgment of the receipt of an invitation is rare, but the general contractor, if he adheres to this practice, can have some degree of assurance that he will receive an adequate number of subcontractor bids on bid day. Invitations should always be sent when highly specialized subcontractor work is involved in the job or when only a few suppliers handle a particular product or specialty.

There are several other advantages to working closely with subcontractors during the bidding time. It is a very rare case when an estimator can pick up all of the ambiguities and contradictions contained in the normal bidding documents, particularly in the trade divisions of the specifications. The individual trade subcontractor, however, is usually watching for these things when estimating his work; when he discovers them, it is likely that he will give this information to the entire list of general bidders on the project. This serves to even the competitive advantages (or disadvantages) among the general contractor bidders.

7.10 DEVELOPMENT OF THE ESTIMATOR

Does practice make perfect in estimating operations? In the operational function of estimating the construction operator must realize that only a minimum amount of practice is necessary to perfect most of the more menial performance functions.

As stated before, there is very little need for the use of higher mathematics in estimating operations. Quantifying the work can be learned quickly by an intelligent person. Even the processes of studying and learning the bidding documents and the write-up become extremely familiar to the estimator after a little practice. Except for the never-ending need for practice in pricing operations, the operation of estimating does not require extensive practice. It is of no value to either the individual or the organization to estimate work merely to furnish a bid proposal; being second bidder does not count for much in the construction process as a general rule. It is a waste of time to estimate a project for practice.

The most valuable practice the estimator can use is in placing and keeping himself within a competitive frame of mind. His entire function is to acquire new work for the company, and he must always be aware of this. He must win when he has to win.

Too often too much emphasis is placed on the need for exactness in estimating and sometimes there are rather foolish expectations of perfection in the bidding process. The estimate resulting in a "perfect" bid proposal price happens by coincidence, not by design. The purpose of the whole estimating operation is for the company to be awarded the job. Over a period of time the quality of the products of the good estimator are proven by the company's success; some of his estimates are going to be better than others. An estimator may have figured a low bid correctly or he may have figured it incorrectly; this is one of the known risk factors in the construction business, and if the estimator believes he has done his best, he should leave it at that. However, the estimator should never figure a job indifferently; this may become a very bad habit. He must always figure competitively.

Many times a person entering the industry, when selecting his preferred operational function, tends to reject taking on the operation of estimating. The prevalent opinion is that its tasks are boring or uninteresting, and therefore this rejection is somewhat understandable. But the person should balance this point of view against the fact that bid day is a day of unmatched excitement, pressure, tenseness, relief, and, hopefully, gratification in this business. The operation of estimating totally illustrates the gamble and risk inherent in the construction business.

Part III

Common Operational
Processes and Procedures

8

Subcontracting

"It ain't the individual nor the army as a whole
But the everlasting teamwork of every bloomin' soul."

J. Mason Knox

8.1 GENERAL COMMENTS

This chapter deals with the subcontracting process that takes place between the general contractor and the subcontractor in the production and contracting subprocesses of the construction process. It is an operational process that consists of two basic subprocesses:

1. Negotiating the subcontract.
2. Forming and executing the subcontract itself.

For the construction operator the subcontracting process must be carefully learned, because the subcontract is a fundamental and essential device used extensively in the construction process. The construction industry requires that subcontracting be the major part of the contracting subprocess. Subcontracted work makes up some portion of almost every construction project built today. It should be clearly understood by the reader that the major portion of all contract construction work performed in the world today is built under the direct control of subcontractors; they perform such a significant portion of the volume of construction work that it is reasonable to state that subcontracting is by itself the single most significant operation in the production subprocess of the construction process. Thus how the subcontract works and, more importantly, how subcontracting is done must be understood by every operator involved in the construction process.

When the general contractor's operators consider the amount of time they will spend in the subcontracting process, they will realize how important it is to learn about it thoroughly. If, as is common today, a construction project is 80% subcontracted, operators may draw their own conclusions as to how much of their daily time will be spent on subcontracting operations. Because of the extensive use of the subcontract, construction operators must develop thinking processes that result in formulating useful and efficient operational processes and procedures if for no other reason than that it will save them time for their other operations.

Operational subcontracting processes and procedures can very easily be formalized. It is the nature of subcontracting and its extensive use by industry that both demands and permits formalization; the industry has adopted traditionally accepted formalized practices and procedures for subcontracting. That this process can be formalized to some extent makes it quite different from other construction operational processes, which can only be informally handled.

Subcontracting consists of one party (the general contractor) selecting another party (the subcontractor) to perform and be fully responsible for a specific portion of the work required by the general contract. Subcontracting is used by the general contracting construction industry as a deliberate device for contractually transferring legal, financial, and operational responsibilities from the general contractor to the subcontractor. This device has allowed specialization in the majority of the physical building processes. Its use has resulted in the concentration of skilled mechanics by the various segments of the subcontracting forces in the industry and has allowed for the distribution of the financial burdens of the individual construction project.

Subcontracting makes the subcontractors the primary employers and controllers of the men and materials used on the ordinary construction job. The general contractor is a step removed from the direct control over both the men used and the materials installed on the job. It then follows that the general contractor should be more interested in the processes of subcontracting rather than in the control of labor, materials, and equipment and their quality or quantity on the job; he must be continually and deeply interested in the operations of subcontracting itself.

Construction operators, therefore, must concentrate their efforts on learning the formation, implementation, and administration of the subcontract. They must develop operational procedures and processes whereby the subcontracts and subcontractors can be easily and efficiently expedited and handled. If, for example, a project has some 60 subcontracts involved in its construction work, then 60 valid, complete, and binding subcontracts or purchase orders must be made and used on that project.

For the reader's information and convenience it should be noted that the words *subcontract* and *purchase order* are synonymous for the purposes of this book and they are used interchangeably.

8.2 THE SUBCONTRACTING PROCESS

Negotiating

The subcontracting process begins after a general contractor finds he has offered the low bid to the owner. The procedure is the same with each division of the work that is to be subcontracted and it starts in the estimating phase of the bidding subprocess. Usually subcontracting proceeds as follows:

1. The subbids are confirmed.
2. The subbids are examined, compared, and correlated.
3. The low subbid is determined in relation to its
   a. Responsiveness.
   b. Completeness.
   c. Price.
   d. Conditions.
4. If there is genuine and honest confusion in the contractor's mind about anything in item (3), then those subcontractors creating the confusion are called in to clarify their bids.
5. Negotiations with the determined low bidding subcontractors are undertaken.
6. The subcontract is formed and executed.

Examination and Comparison

As the various subbids are received in the general contractor's office on bid opening day, they are assembled in some order to be used in the bidding operations. Usually the subbids are segregated by trade division and tabulated in the order they are received. The subbids generally include the scope of the work involved, a reference to the specifications by the trade division, and the price of the work.

Example.   An acoustical ceiling contractor will call the general contractor and say, "This is Acme Acoustic Company, Charley Smith; we are bidding acoustics treatments division 8D, plans and specs, tax included, for $80,000. Will confirm."

This statement takes little time to communicate but contains a great deal of bidding information. The bid tabulator might then record in his tally sheets this and similar bids on acoustic ceilings as follows:

Acme, p&s, t. i.	$80,000
Gold Star, p&s, no tax	$78,000
C&G, p&s, t. i., special ceilings excluded	$57,000
J. C. Smith, p&s, no tax	$73,000

We will examine the procedures for bidding and bid opening day more thoroughly elsewhere in this book, but the bid tabulation operation is where the subcontracting process begins. On bid day the estimator will select the apparent low responsible subbidder, whose figure will be used in the contractor's bid price. If the estimator determines, for example, that J. C. Smith is low, then the price of $73,000 will be used in completing the general contractor's bid. Should the estimator determine with good reasoning that C&G is low, then $57,000 will be used in combination with some other subbid covering the excluded special features. There can be many combinations of subbid figures that will amount to the lowest price and include all of the work.

Negotiating Discrepancies in Interpretations of the Bidding Documents

In many cases it is possible to make a clear-cut determination of the low subbid from the bids received on bid day, which makes the operator's task easier in the subcontracting process. But even when the low bid is clear-cut, it is still necessary to sit down with the apparent low bidding subcontractor and fully and exactly determine the mutual obligations to which each of the parties intends to commit itself in the subcontract. It is the general contractor's operator's task to make sure that the subcontractor fully understands the work which the general contractor expects him to perform. While both the subcontractor and the general contractor have read and interpreted the bid documents, they may have reached different conclusions on, for instance, the extent of the work included. Any discrepancies between the general contractor's and subcontractor's impressions must be clarified before contract. It may be that there is a sincere and legitimate yet different interpretation of the plans and specifications by one or both of the two parties. These differences may result from just a word, a phrase, or an entire section of the bidding documents. A full and complete discussion of the respective parties' impressions of their responsibilities is a valuable and practical operation in the subcontracting process. Such a discussion should result in a better working relationship between the parties at subcontract time and throughout the life of the subcontract.

Even when the low subcontractor bid can be absolutely determined from the bid itself, there may still be items to discuss and negotiate between the subcontractor and contractor. It is true that the bidding subcontractor usually bids "plans and specs"; however, this in no way takes care of certain requirements that must be included within the subcontract and which are not covered at all in the bidding "plans and specs" or contract documents. The following are a few examples of subcontractual requirements of a general nature that are not ordinarily specifically detailed in the bid documents:

1. Payments and payment procedures.
2. Shop drawings and sample submittal procedures.
3. Bulletin and change order procedures.
4. Cleanup, back charge, and damage repair procedures.

Negotiating the Complete and Responsive Bid to Subcontract

Not all of the obligations running in the general contract between the owner and general contractor are necessarily applicable or useful in a subcontract. If the parties assume that the general contract somehow binds the subcontractor to the general contractor in exactly the same way or to the same extent that it binds the general contractor to the owner, they are mistaken.

Negotiations to subcontract constitute a construction operation within the subcontracting process. When the subbids are not complete or wholly responsive, the negotiating operation is absolutely essential. It may be that negotiations are the *only* way to establish the actual low subbidder in some category of work. When a group of subcontractor bids in the same trade division are not in conformance with each other, some extremely complex problems arise for the construction operator in the negotiations process. The general contractor has already made his offer to the owner, the major portion of which is composed of subbid prices. Time and pressure considerations do not allow for the complete analysis of all of the subbids on bid day at bid time. When the low subbids cannot be ascertained on bid day, they must be determined before the award of each individual subcontract.

We are not suggesting here (in order for the general contractor to stay within his bid price to the owner) that illegitimate or unethical means and methods be used to accomplish this. Neither is it suggested that the contractor's own mistakes in the bidding process be passed on to his subcontractors. What we do say here is simply that within the nature of the bidding process itself there arise legitimate and understandable confusions and misimpressions that can and should be corrected in the negotiating operation in the subcontracting process. If negotiations are handled properly, many of these differences can be straightened out to the satisfaction of both parties.

Negotiating Out Ambiguities, Contradictions, and Duplications

The bidding documents are often ambiguous, contradictory, or contain duplications. When the various subcontractors bid on them, they usually bid on the work of their own trade without reference to the entire bidding documents. Work that is duplicated within the bidding documents will often result in different subcontractors in different trades including the same work in their respective bids. Such a duplication can be legitimately and justly adjusted in one or another subcontract in the negotiation to subcontract process. Duplications in the bid documents are an example of an area where negotiations before subcontract benefit the general contractor without doing harm to the subcontractor.

These negotiations result in the "pickup" of money from the process of subbids. A pickup made on a sound, reasonable, and just basis can be a very profitable construction operation. The construction operator should develop special skills and methods of performing the pickup operation. These skills are primarily mastered through the careful reading, study, and thorough understanding of the bidding documents during bid time and after the award of the contract but before the award of the subcontracts.

Contradictions and ambiguities in the bidding or contract documents, however, raise much more complex issues for negotiations. In some cases the operator must be careful here as far as ethical considerations and good business practices are concerned. First of all, certain types of discrepancies are generally very hard to find in the bidding documents during the bidding process. Secondly, when they are found, the question always remains as to how deliberately, or how honestly, they were handled by both the contractor and the subcontractor on bid day. There can be distinct competitive advantages to both the subcontractor and the contractor in either taking a particular action on or ignoring these discrepancies on bid day. Thirdly, there remains the difficulty of coming to a just determination on these discrepancies even when they occur by accident. The following situation very often occurs when contradictions and ambiguities show up in the bidding documents during the bid process.

Example.   A subcontractor's price has not made provisions for discrepancies. The subcontractor's price is a part of the general contractor's price to the owner. If the subcontractor has not provided for a discrepancy, it still remains an item of cost. But who bears the cost? The contractor or the subcontractor? From a practical operational standpoint there is much to be said for putting the full burden on one or the other of the parties. The fact is, however, that if there was no deliberate intent to deceive on the part of either party, then the only solution is for them to negotiate on the discrepancies to their mutual satisfaction.

Usually, it is impossible to fully satisfy both parties in such a situation. One or the other must usually bear an unequal portion of the resulting cost and expense. It is valuable for the general contractor's operator to look for trade-offs—to the possibilities of minimizing the subcontractor's costs by giving him relief in some other part of his subcontractual obligations. For example, in exchange the subcontractor can agree to furnish within his bid price additional but not necessarily more costly services free of charge to the general contractor. An excavation contractor, could agree to maintain temporary roads with equipment he has on the job site as a trade-off for considerations given to him on discrepancies during the subcontractor negotiations. On the other hand, the general contractor could agree to furnish temporary services at or below their cost to the subcontractor. Many times such arrangements made in the course of negotiations with subcontractors are extremely beneficial to both parties without either having to assume a significant cost burden.

Ethical Considerations

When either the contractor or the subcontractor has deliberately mishandled a discrepancy in the bidding documents, different problems arise. Should a subcontractor, for example, who has deliberately mishandled a discrepancy be dismissed from consideration for the subcontract? The answer would be an unqualified yes if there were not more serious matters for the general contractor to consider. The main consideration is that the general contractor has used this subcontractor's quoted price in his own bid price. Should there be a great discrepancy between the erring subcontractor's price and another quoted price, this becomes a serious problem for the general contractor. Under such a circumstance it is not practical to dismiss the subcontractor. Ordinarily, the subcontractor can be awarded the subcontract for the work, but not before negotiations take place based on the idea of having him bear a major part of the cost of his incorrect behavior. Negotiating on problems like this is a perfectly legitimate practice if deliberate intent on the part of the erring subcontractor can be established.

Where the situation is reversed and the general contractor has either mishandled or missed a discrepancy in the bidding documents, the subcontractor should be entitled to a negotiation process as well. Here the general contractor's operator must exercise discretion, fairness, and good judgment in solving the problems of cost involved. To act unfairly and unethically in this situation is an easy trap for the general contractor's operator to fall into. He has the advantage because he decides to whom the work will be subcontracted and the subcontractor knows this. The operator's tendency to use this position to his advantage should always be tempered with good sense. He should be influenced by such things as future business considerations and the fact that if he takes advantage of the situation he is likely to have a very dissatisfied and distrustful subcontractor on his hands for a good length of time. A reasonable rule to follow in such cases is that while it is not necessary for either the contractor or the subcontractor to take the whole burden of the cost of the discrepancy, certainly the major part of the burden should be borne by the party in error.

Methods

We emphasize here that negotiating with subcontractors should be considered an extremely important construction operation that should be done in an orderly and thoroughly developed manner. The operator should think and act professionally; he should engender and maintain goodwill and good feelings throughout the process and make the negotiations convenient and pleasant. He should also plan a schedule for negotiating the various subcontracts. The contractor is the initiating party and therefore responsible for directing the negotiations and making them productive. He should hold them on time and in a proper place. The atmosphere and comfort of the setting for negotiations become important factors to consider. If a social rather than a businesslike atmosphere is more conducive to productive negotiations with an individual subcontractor, then such an atmosphere should be provided.

The construction operator should approach negotiations to subcontract in a fair and equitable frame of mind and he should be aware of the long-term consequences of thinking otherwise. It is not productive to have unhappy, dissatisfied subcontractors on a project. Even if the operator feels or even knows he can handle them himself, it is doubtful that he can be as certain that other operators involved in the project can do so as well. The construction operator who learns to properly negotiate with subcontractors gains a good reputation for himself and for the company he represents; on the other hand, the operator who does not conduct negotiations properly gains a bad reputation and brings on very serious consequences for the profitable conduct of his business.

8.3   THE SUBCONTRACT FORMATION SUBPROCESS

The Basis for Form

It is likely that the construction operator will use some sort of set-form subcontract in practice. This subcontract form will have been established by the firm as the form that all of its subcontracts must take. Since it is impossible to show the reader all of the various subcontract set-forms used in the industry, we will show him one that should suit the needs of any subcontracting situation in both its legal and operational aspects. We will illustrate what is required in *any* subcontract. The reader will then more fully appreciate how whatever form he uses works. When he understands what the form requires and why it is formed in a certain pattern, he will better understand how to use it.

First of all, one must establish the intended purpose of the subcontract form, The subcontract, in its form, must satisfy the normal requirements of establishing a basic contractual relationship between the parties, with their mutual and respective obligations clearly set forth; it must function as a simple contract. Secondly, there is something unique about the construction subcontract that must be examined carefully—the general conditions, which are of special concern to the constructor.

The Conditions

The subcontract must, in its form and content, bind the subcontractor to the general contractor to certain obligations that exist within the scope of the general contractor–owner agreement. A fundamental concept in subcontracting that the construction operator should always bear in mind when forming the subcontract is stated as follows:

> The construction operator in the subcontracting process must bind the subcontractor to what the general contractor is bound to in his contract with the owner *and then some.*

from the set-form itself. Spaces for allowing the proper and complete execution of the contract should appear on its face.

2. The form must contain the general conditions of the subcontract on its back face. The general conditions in form must be applicable to *every* kind of subcontract and material purchase order that will come out of the general contractor's office. Therefore much discretion and care must be exercised by the construction firm when developing the set-form of the general conditions of the subcontract.

3. Space must be provided for all supplemental and special conditions. Supplemental and special conditions are those conditions that apply to a particular subcontractor or supplier under a specific general contract. They are special to the project at hand and vary in number and content.

4. Space must be provided for all inclusions and exclusions to the subcontractor's work. The inclusions/exclusions are specific to the particular subcontractor or supplier and to the project at hand. They also vary in number and content, depending on the requirements of the project at hand.

5. Space must be provided for the subcontract price, as well as for the delivery terms and time obligations for the subcontractor's performance. There should be a space on the face of the set form for an easily identifiable intracompany number for the subcontract.

Some of these details can be incorporated in their entirety into a carefully designed subcontract set-form; the others must have space allocated for their easy inclusion into the set-form. Figure 13 shows a blank set-form. Appendix B gives an example of a typical filled in set-form. Figure 12 shows a set-form list of commonly used general conditions, which are usually written in fine print on the back face of the page shown in Figure 14.

Once the set-form is satisfactorily designed, it is printed in bulk for company use for all of its subcontracts. When the set-form is completed and fill-ins occupy the entire space allowed for them on one set-form, additional fill-ins are simply written on another set-form (Figure 13) and consecutively numbered. Subcontract set-forms permit easy and efficient production, and all members of the construction operational organization, including the clerical personnel, can use them. The clerical personnel can quickly learn how to use the set-form, fill it in, make reference to it, and use it in the subcontract's administration.

In the set-form subcontract the following things can be in printed form:

1. The document's identification.
2. The contractor's identification.
3. The general conditions (in fine print on the back of the page shown in Figure 14).

Set spaces can be provided for the following:

1. Subcontractor identification (with necessary supplemental information).
2. Dates.
3. Time requirements for subcontract performance.
4. Methods of delivery.
5. Intracompany contract numbers.

6. The signatures of the parties, identifying both the contractor and the sub-contractor as offerer and acceptor, respectively.

If the general conditions are not in fine print on the back face of the subcontract (Figure 14), then it must be shown clearly that they are an element of the sub-contract and not separate and supplemental to it.

To:		Date	No.
	ABC Inc. General Contractors 100 Main St. Columbus, Ohio	Delivery	
Attn:		As per	
Phone:	A Subcontract		
		Page_____ of_____	

Accepted by:

_____   _____   _____   _____
Contractor   Date   Date

Figure 13.   A typical subcontract set-form (front and back faces).

CONDITIONS

Figure 14. A typical subcontract set-form (front and back faces).

8.4 MAKING THE SUBCONTRACT

Uses and Implementation of the Form

How is the subcontract made using the set-form? The initial purpose of the set-form subcontract is to produce a document that is easy and efficient to use; it is designed to be an understandable document, thus making it efficient. Any participant in the entire construction operational organization must be able to adapt it to their particular use. The office clerical staff must be able to read it and pick out its pertinent points of information accurately and easily. The set-form sub-contract must be a fully operational document, because of its extensive and continued use throughout the subcontracting subprocess.

When utilizing the printed set-form, the operator fills it in to make it a valid, binding subcontract for a particular project. To do this the operator makes written keys, which are individually designed and written for each individual job.

The Keys

1. The first written key is the complete and entire identification of the contract documents. For every individual project this only needs to be done once. The key allows the operator to take the proper special care in describing exactly the entire general contract document within the subcontract. A general rule for the construction operation of writing this key is as follows:

> The full, complete, and entire listing of the general contract documents (the contract documents binding the owner and general contractor) must be stated in the subcontract. The listing must be an exact description of the general contract documents as they are listed in the general contract. It must identify the several documents by name and date and include every single one of them.

This key is called a contract document identification statement and it might read as follows for a fictitious project:

> All in accordance with the entire contract documents for the Lakeview High School, Los Angeles, California, Architect, Smith and Smith, "address," *"their"* job number _____," and, in particular, the Advertisement for Bids dated _____, the Invitation to Bidders dated _____, the Proposal Form dated _____, the Agreement dated _____, the General Conditions dated _____, the Supplemental Conditions dated _____, the Special Conditions dated _____, the Technical Specifications dated _____, and Drawings A_1 through A_{25} dated _____, (A_{24} dated _____), S_1 through S_{12} dated _____, M_1 through M_{16} dated _____, Addendum 1 dated _____, Addendum 2 dated _____, Addendum 3 dated _____, and including Accepted Alternates 1, 3, and 5, each respectively dated _____, _____, _____, and especially as called for in Section _____ of the Specifications (if concrete is Section or Division 3 of the Specifications enter 3, if masonry is 4, enter 4, etc.).

The operator should make no compromises with the extent and accuracy of the identification of the general contract documents to which he means to bind the subcontractor. If, for example, within architectural drawings A_1 through A_{25} there are one or more drawings dated differently from the others, these drawings and their dates must be clearly identified.

2. The second written key is the authority key. It is written once with care and might be as follows.

For material and labor subcontractors or suppliers:

a. "This is your authority to furnish all labor, material, equipment, taxes, and insurance necessary to do . . . ."

For material subcontractors or suppliers:

b. "This is your authority to furnish all materials, equipment, taxes, and insurance necessary to deliver . . . ."

For labor subcontractors or suppliers:

c. "This is your authority to furnish all labor, specific equipment agreed upon herein, taxes, and insurance necessary to do . . . ."

The proper key is then simply picked out from the key to fit the particular subcontract being written: for example,

Concrete material	(b)
Acoustical work labor and material	(a)
Excavation work labor and material	(a)
Resteel erection work labor	(c)

3. The third written key would be the supplemental and special conditions. These conditions are those required by the particular subcontract on the particular project at hand. A key is made of all the conditions that are considered necessary from a careful reading and analysis of the general contract documents. An example of such a key would be as follows:

a. State sales and use taxes are included.
b. Shop drawings shall be submitted promptly in one sepia with one print.
c. Samples shall be submitted promptly in triplicate, properly tagged and identified.
d. Brochures will be submitted promptly and in 12 copies.
e. Subcontractor shall furnish proof of insurance before commencing any job operations.
f. Subcontractor invoices shall be submitted no later than the twentieth of the month.
g. Subcontractor is an Equal Opportunity Employer and agrees to remain so throughout this project, and so forth.

These conditions are written only once. Some will apply to all subcontractors and suppliers on the job and some will not. Thus the excavating subcontractor might have the following listing for inclusion into its subcontract:

(a), (b), (e), (f), (g), and so forth

The mechanical subcontractor might have

(a), (b), (c), (d), (e), (f), (g), and so forth

4. The last written key is the contract price key. This might be written as follows:

a. For the lump sum price of _____.
b. For the unit price of _____.
      Approximate quantity required _____.

Thus the price for acoustical work could be

a. _____

$10,000

The resteel labor price would be

b. _____

$180 per ton.
_____
1000 tons (approximately).

The concrete supply price would be

b. _____

$38 per cubic yard.
_____
10,000 cubic yards.
_____
Five-sack (approximately).
_____
$3.00 differential—add 75¢.

These are usually the four keys that the construction operator must write for every individual project. Each project requires writing different keys: for example, the contract document identification key is never the same for two different jobs; on the other hand, the contract price key will vary little from job to job. Each key must be written with extreme care.

Once these keys are written, the construction operator simply goes through the entire list of subcontractors and materialmen and supplies the typist with the following information:

1. The contract document identification statement key for each subcontract.

2. The project general contract number (7904 representing the fourth contract awarded to the general contractor in the year 1979). The secretary assigns the subcontract number (7904-1, 7904-2, 7904-3, etc.) for each individual subcontract.

3. The name of the selected subcontract or (AB Ceilings, Inc., JCB Excavation Company, etc.) for each individual subcontract. The secretary fills in the location, address, and telephone number of the subcontractor. It is also advisable to identify the person handling the subcontract in the subcontractor office.

4. The authority key for each individual subcontract.

5. The section or division number of the specification (structural steel, 5a; resteel, 3; painting, 9; etc.) for each individual subcontractor or order. This specifically directs one's attention to the work involved in this contract.

6. The work ("the structural steel work," "the excavation work," "the mechanical work," etc.) for each individual contract.

7. The contract price key and number [(a) $500,000, (b) $180 per ton; $47.00 for five sacks per cubic yard, etc.] for each individual subcontract.

8. All of the condition keys [(a), (b), (c), (d), etc.] required for each individual subcontract.
9. Specifically written inclusions and exclusions to each subcontract.
10. Information on the delivery type (FOB job site, COD, etc.) for each individual subcontract.
11. Time requirements for the delivery of performance (as per job superintendent, as per job progress schedule, etc.) for each individual subcontract.

Thus the secretary might see a list as follows:

Excavation 7905-2, (a), Jim Smith Co., Sec. 2, The Excavation Work, (a), (c), (h), (j), Inclusions, (a), $500,000, FOB, as per job schedule.

Mechanical 7905-10, (a), Adams Plumbing and Heating Co., Sec. 15, The Mechanical Work, (a), (b), (c), (d), (e), (f), (g), (h), Exclusions, (a), $2,500,000 FOB, as per job schedule.

Concrete 7905-1, (b), Acme Concrete Supply, Sec. 3, The Concrete Materials, (a), (c), (d), (f), (b), five sacks $47.00 per yard, $3.00 diff—additive 75¢, 10,000 cy. prox., COD, as per job schedule.

The construction operator using this or a similar system can make numerous subcontracts in a very short time. The inexperienced operator will soon realize that when there are many subcontractors with whom to negotiate and come to contract, this kind of system is very useful and efficient.

Inclusions and Exclusions

There is one part of the subcontract that is not adaptable to keying—inclusions/exclusions. Ordinarily inclusions and exclusions result from the parties negotiating something extra in or out of the subcontract.

Example. The cement finish contractor might accept the obligations of the general contract documents as part of his subcontract. Within the technical specifications of the contract documents, however, it says that stripped concrete surfaces shall be of a good finished appearance. (This finishing can usually be accomplished by rubbing and bagging.) It then follows that the cement finish contractor will be held responsible for the appearance of the stripped concrete. But since he neither forms, pours, nor strips the concrete, he must hesitate before committing himself to finishing items over whose building he had no control; neither can he estimate the extent of rubbing necessary during the bid process. Usually he will want this item excluded from his responsibilities as specified under the general contract documents. The general contractor and subcontractor negotiate this item out of the subcontract by simply stating therein that it is specifically excluded.

There are many other similar instances that reasonably call for specific inclusions as well as exclusions to and from the subcontract. A major cause for inclusion/exclusion provisions in the subcontract is when the same work is specified in two or more parts of the specification, or when the specification has obviously not divided the work according to the subcontracting specialty. An example of the latter would be when the general contract technical specifications create difficulties in the assignment of proper trade jurisdictions to subcontractors. Unfortunately, this happens often in the contract documents. Duplications or the improper assignment of work in the documents call for constant vigilance and remedial action on the part of both the general and subcontracting operators when subcontracting. The remedial action is the inclusion/exclusion provision in the subcontract.

8.5   USING THE SUBCONTRACT

Execution and Distribution

After negotiating with the subcontractors and developing the contents of the subcontract and its keys, the construction operator prepares the subcontract for use. Generally, it is prepared in four copies: two of them are signed and dated by the contractor and sent to the subcontractor, who returns one copy, signed and dated by himself. This then constitutes the executed construction subcontract.

These documents are valuable and should be handled as such. They should be kept in a special "contract file," which should be fireproof and theftproof. One should consider the chaos that would result if the contents of this contract file were destroyed or stolen. The contract file should contain the original copies of the general contract, the bonds, the signed change orders, and the signed subcontracts. Usually, these important documents of all the individual projects are kept in the contract file in a set filing system.

As for the other two copies of the subcontract, one is kept by the construction operator who made it and the other sent to the job superintendent. Each of these operators then prepares a subcontract book for reference purposes. An efficient way of doing this is to place the various subcontracts into an easily used loose-leaf volume. The subcontracts are filed by numerical order and indexed for quick reference. Usually, if the expeditor is the construction operator who makes the subcontract document, he will keep and maintain the subcontract book for his particular job. It is a good idea for the job superintendent to keep his book in the same form as does the expeditor. This allows for these two closely integrated construction operators to reference and cross-reference more efficintly, particularly over the telephone.

The subcontract book should be available to all parties in the operational organization who have use for it. The job superintendent, however, should refrain from letting his book be totally available at the job site; whether or not the information contained in it should be available to other job-related personnel and subcontractors is a matter of good judgment or maybe of company policy.

Changes to the Subcontract

Changes to work authorized by the general contract must be transmitted to the subcontractors if their work has been involved in the change. Again, we can only suggest a method that is functional and efficient for handling this process. Another very simple set-form document system can be developed for universal company use in this process. The set-form in Figure 13 is used again and filled in (see Appendix B for an example of a typical subcontractor change order).

The primary purposes of the subcontract change form are to identify and record the following:

1. The reason for the change.
2. The authorization for the change.
3. The total amount of the (individual) change in money.
4. The total amount of change to the original subcontract price.

Example. Bulletin #16 dated August 16, 1980, is issued by the owner and quoted by the general contractor. The work of 12 subcontractors is involved either in deletions or additions to the work and the general contractor's quote will include their prices for the work in the bulletin. When the price of the bulletin is accepted a change order from the owner to the general contractor results.

The general contractor's construction operator determines from the quotations the amount of money to be added or subtracted to or from each of the 12 subcontractors' original price. Thus, for example, the electrical subcontractor under subcontract 7904-34 is entitled to $10,000, which is the sum total of his quotations for all of the items in Bulletin #16. The general contractor, after the issuance of the owner's change order, owes the electrical subcontractor this amount of money when the work included in the change is performed. The original electrical subcontract price was $100,000. The subcontract price after the change is $110,000.

Using the set-form (Figure 13), the operator would then fill it in (see sample in Appendix B). Basically, the subcontract change can be issued as an order to do the work, as is done in the original subcontract, or it can be issued for work already accomplished by the subcontractor before the owner's change order was issued to the general contractor. The sample in Appendix B illustrates this type of fill-in for the subcontract change 7904-34-1, where the 1 at the end of this number means it is the first subcontract change issued to the electrical contractor on project *7904*.

It is obvious that the contents of the subcontract change cannot be keyed using this system. Each change order must be written separately, but this type of system accomplishes the purposes of the subcontract change process in an easily understandable and efficient way. Its main advantage is that if properly handled and kept current, the total contract price owed to the subcontractor is readily available for reference.

The subcontract changes are kept in numerical order in the subcontract books immediately following the original subcontract of each individual subcontractor. The process of distribution is the same for the change as it is for the original document.

The Work Order

When changed work must proceed before authority from the owner can be granted formally by change order, there is usually a process allowed that accomplishes this on a safe contractual basis. This process is called the work order process: the general contractor transmits his request, in set-form or otherwise, that the affected subcontractors perform proposed changes. There are wide variations in the industry as to how this work order is processed. Basically, the general contractor must transmit to the subcontractor the fact that operations must proceed, that they are based on a proposed change, and that the authority for proceeding is based on these facts. Contractually, it is the general contractor, not the owner, who is ordering the work at this point. There are special implications, however, of implied contract in this process. The work order reinforces these implications and is therefore a valuable operational process to learn and systematize.

8.6   THE SUBCONTRACTOR AND THE SUBCONTRACTING PROCESS

Relationships

In this chapter it is necessary to examine the relationship of the subcontractor to the subcontract in more than just contractual form. This means examining the practical application of contract principles to the realities of the construction process.

It is of little value for the construction operator to develop sound, fair, and complete subcontract documents if the subcontractors cannot or will not operate under them in a practical way. It is natural that the construction operator, whether he be a general contractor or a subcontractor, will try to protect his own interests at all times. Subcontracting is initiated by the general contractor, who ordinarily negotiates from a position of strength. His subcontract forms are the ones used in the process, and the subcontractor must usually submit to their contents. The general contractor not only controls the formation of the subcontract but also its implementation and, most importantly, the subcontracting monies throughout the project. But, in spite of all these things, the general contractor's operators must get the subcontractor to work efficiently and cooperatively. Nothing can be worse on a project than even one subcontractor (particularly a major subcontractor) who is dissatisfied or disgruntled. One way of putting a subcontractor in such a state of mind is to mishandle one or all of the stages in the subcontracting process. Each of the stages of negotiation, execution, and using the subcontractor must be handled properly by the construction operator.

The subcontracting process should not be made into a game of trying to outguess or outwit the subcontractor; it should be done fairly and conscientiously by the construction operator. By saying this, we are not trying to establish a behavioral code, but we are attempting to convince the reader that from a practical and business standpoint, correct and proper approaches and attitudes toward subcontracting procedures are necessary.

At the first, we must examine what a subcontractor is. The subcontractor is a construction organization that performs specialty work or supplies specialty items. It is the subcontractor that is the specialist. The general contractor is the generalist. This must be kept in mind at all times by the construction operator when operating within the subcontracting process; in addition, he must not forget that the subcontractor is an *independent contractor* relative to both construction practice and the law.

The Specialist

Because the subcontractor is a specialist, there is always a question as to what the relative positions of each party should be concerning the performance of his special work. The subcontractor's performance of his work depends on and is guided by the general contractor. At times this creates operational difficulties: on the one hand, the subcontractor is the specialist, but, on the other hand, his work falls within the total work, which, of course, is the general contractor's responsibility. How then should the construction operator allow for both positions to function at maximum efficiency? He does this by carefully considering and analyzing the entire subcontracting process.

It would be absurd, for example, for a general contractor's operator to tell an electrical subcontractor how to install a building's wiring system. Not only would this be absurd, but counterproductive and impractical. But the general contractor will most certainly have to tell the subcontractor when the system must be started and completed, and he must do so continuously throughout the project. It would be simple for us to then say that ordinarily the general contractor should tell a subcontractor when to work but not how to do the work. But even this would not be the rule to follow in all cases. It might be much better, for example, for the general contractor to simply tell the electrical subcontractor that he is bound to the progress schedule. Constant direction by the general contractor as to when a subcontractor must perform often leads to compromising positions for the general contractor's operator. It is good practice, therefore, for the operator to bind the subcontractor to only general schedule and time considerations, with formal subcontract provisions. For example, the statement "as per job schedule" or "as per job superintendent" are much better terminology to use in the subcontract than the statement "to start by October 19, 1979, and finish by December 12, 1979." By using the former type of statement, the construction operator can still push the performance of his contract without compromising himself and getting overinvolved in the special problems that the various specialties might have.

It is certain that the subcontractor's operator, being specialists, will know more about the technicalities of their work than the general contractor's operators. This knowledge of the technology of materials and installation is accompanied by experience and skill in the proper utilization of manpower. Here again the general contractor's operator should rely and depend on this experience and skill, with minimal interference. It is rare when the job superintendent really knows the ins and outs of utilizing specialty tradesmen on even his own job, much less within the subcontractor's total operations. While he may feel that he has the right to govern the use of the tradesmen on his job, he has no such right relative to the subcontractor's total operations. The good job superintendent and expeditor should encourage and, at times, even force the subcontractor's management of his own forces. It is not necessary or practical for them to take on this burden themselves.

The Entity

At times the subcontractor must be viewed from a strictly impersonal standpoint. At these times personalities cannot be assigned to the subcontracting entity. We admit to a degree to the unreality of this situation, because obviously subcontracting entities are made up of people. But we do so in points of both business and legal fact, where the reader must recognize that the construction operator has to approach some aspects of the subcontracting process with some degree of impersonality.

The fact has been stated several times in this book that the subcontractor is an independent contractor. There are other features about subcontractors being entities that construction operators must approach on the same impersonal basis as the foregoing central legal point. Let us present just two examples.

Example 1. The size, both operationally and financially, of the subcontractor must be viewed impersonally by the operator because that is a reality over which he has no control. It is of little practical good for the construction operator to "wish" that a subcontractor had abilities to perform and handle its work financially beyond its actual capacity. Judgments made by the operator engaged in the subcontracting process must always be based on facts, not on wishful thinking.

Example 2. The construction operator must establish certain criteria for impersonal approaches to dealing with subcontractors that will apply in general to all of them. The operator must consider the following:

1. The subcontractor knows better than the operator the true capability of its organization and personnel.

2. The subcontractor is an "adult" entity, and the operator is not to presume that special guidance is necessary to protect it from itself. The subcontractor represents itself as a subcontractor, and with this representation goes the assumption that it can take care of itself.

3. The subcontractor is made up of different people of differing philosophies, behaviors, characteristics, attitudes, and so forth. But it is the entity that the operator is dealing with in the end, not the individuals.

4. Operators must at all times consider the common goal of both parties— profit. Keeping this goal in mind allows the operator to reach impersonal decisions, and it also serves to justify improper personal actions that are directed toward furthering his own interests. If the end result of the job is profit for all parties, any bad feelings caused by the construction operator's attitudes will be quickly forgotten. Perhaps profit will not always satisfy individuals in the subcontracting entity in a personal way, but it should, at least, soothe the pain caused by maintaining impersonal relationships when they are necessary to the process.

Personalities

In spite of the fact that the subcontractor must be regarded as an entity, the relationships of personalities form the basis of the subcontracting process. This is true in all of its phases, from negotiations through the completed subcontracts and project.

Personal relationships often furnish the key to good or bad performances in the subcontracting process; this cannot be avoided, nor should it be. This is an aspect that makes the construction process such a vital and interesting profession.

In any local area both subcontractors and their construction operators are usually closely identified with actual people. Often one person truly becomes the subcontractor entity to those with whom he deals. Many times this identification establishes the reputation and quality of the entity. This is particularly the case with small- and medium-sized subcontractors. As the size of the subcontractor organization increases, dominant personalities generally (though not always) play less of an important role in this phenomenon; individual personalities, as they affect the subcontracting process, become more obscure in larger companies.

In any case, the construction operator must realize that the interrelationship of personalities can provide the basis for successful subcontracting if properly handled. For example, it is the personality of the subcontracting negotiators that gains or loses in the negotiation process. Other evidence of this phenomenon is the fact that the general contractor's job superintendent's behavior with the subcontractors and others often directly affects the success of the subcontractors on a project.

The construction operator should always strive for the best possible personal relationship with his subcontractors. Whether the operator is employed by the general contractor or the subcontractor, it is necessary for him to realize the practicality of doing business the easy way.

In personal relationships in the construction process, the following rules should serve as a guide for the operators' proper behavior when dealing with each other (these apply to both general and subcontractor personnel):

1. One should never do things behind the back or over the head of the contact or person immediately responsible for the operation at hand. This means that each party should deal with suppliers, architects, other subcontractors, and most certainly clients and owners in their proper place.

2. One should never try to second-guess the instructions of the authority responsible for the operation at hand. If the job superintendent, for example, directs work to be done in a specific area, that is the area to be worked, and no other.

3. One should never deliberately stop operations on a project without proper notice to all the responsible and involved parties. Even if the stoppage of operations is unavoidable and irreversible, just a notice of intent to do so is of practical benefit to the project and of importance to the people involved in it. Common courtesy demands this.

4. One should always regard other operators as responsible and capable professionals until there is solid proof of the contrary. Mutual respect between the general and subcontracting operators is an absolutely essential ingredient to successful construction operations.

9

The Scheduling Process

Use thee well the moment; what the hour
brings for thy use is in thy power.

Johann Wolfgang von Goethe

9.1 GENERAL COMMENTS

This chapter is admittedly written with a certain degree of subjectivity—certainly more than in other parts of the book. Still it attempts to be a truthful statement. It is written with a sense of humor, in the knowledge that some of its statements may be violently contradicted by some operators in the construction industry. These statements, however, are made most deliberately, and we hope that the reader will realize this and not believe that they are based on an ignorance of the subject.

The variations and sophistication of the types and methods of scheduling are merely mentioned in this book. It is not our intention to discuss their relative merits, either singly or in comparison to each other. Recent developments in scheduling procedures are relevant and valuable to the construction process, but a full discussion of them would be far beyond the intended scope of this book. This is so not merely because of scope or space limitations, but because of the small value an incomplete discussion would have for the reader.

Here we are concerned with the basics that develop thinking processes in the construction operation of scheduling. In this approach the type and method of scheduling are not important. The basic thinking processes, once learned, can apply to any or all types and methods of scheduling.

A construction schedule is a timed plan for a construction project. It is a schedule of operations. Consider the following statements carefully:

1. A schedule is an operational document.
2. Making a schedule is a construction operation in itself.
3. The schedule guides and governs many other various construction operations; it interrelates operations.
4. One cannot make a valid schedule until one understands the operations it concerns.
5. No construction project of any significance produced within the construction process today can be built economically unless it is scheduled.
6. Every part of the construction operational organization uses the schedule to one degree or another.
7. For a construction schedule to be a valuable tool in construction operations it must be usable and used for its proper purpose.

We summarize these statements with the following:

The construction schedule is intended to be a usable operational document, composed by a construction operator familiar with the nature of construction operations, that becomes the most essential tool in the whole production subprocess if properly used.

Underlying our comments in this chapter are two main ideas about the operation of construction scheduling:

1. Scheduling, as an operation, should be based on a basic knowledge of most, if not all, of the operations involved in the entire construction process. This is simply a matter of putting the horse before the cart: one cannot schedule if one does not know what is to be scheduled. To attempt to schedule without this basic knowledge is an exercise in futility.
2. The schedule must be used. It then follows that the schedule must be practical to use, and accepted as such by those who use it.

What method or type of schedule is to be made and how it is to be processed or used are questions that we leave to others more qualified to answer. For purposes of illustration and reference we will mention the "bar chart" or "critical path method" (CPM) methods of scheduling. Had there been any simpler models available to illustrate the concept of schedule, we would have used them.

9.2   THE BASIC CONSTRUCTION SCHEDULE

Schedule is the time fixed for doing something. To schedule means to arrange something for a definite future date. The single most important factor to consider in scheduling is time.

Within the production subprocess of the construction process there is an operation known as construction scheduling. This is a process whereby the construction operator makes a schematic diagram of the necessary operations based on the time allowed in the contract to perform the construction of the building. The complete construction scheduling process involves both making the schedule and using it throughout the life of the project so that the project can be completed on or before the contract time designated.

There is a basic form to any schedule. One of its characteristics is that it is a time chart. The scheduler, when considering time, looks at different time frames. In most cases the total contract time sets the frame for the completion of the total project. Time is usually of the essence of a construction contract and the scheduler is the operator who, after the award of the contract, considers the time of the project in relation to operations. Actually, estimators always schedule as they estimate, but not necessarily in a formal way. The scheduling operator is the first person in the organization who formalizes, with a written document, a plan for completing the contract.

The scheduler breaks the total time allowed down into increments. These increments are most often made to coincide with the payments required by contract, for the convenience of handling these related operations. Thus if the contract specifies monthly payments on the job, the scheduler will break a one-year total completion time down into 12 equal segments; if he chooses to be more detailed, he may break it down further into weeks or even days. The schedule

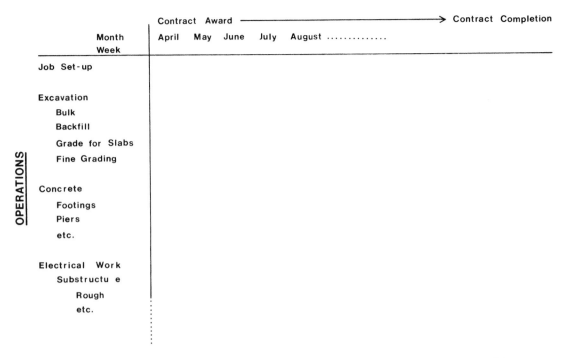

Figure 15. A basic diagram showing time operation function for a construction schedule.

then becomes a time chart proportioned for the convenience of the scheduler into detailed segments of time, which is a simple operation (see Figure 15).

The schedule is also an operations list as well as a time list. When considering operations, the scheduler first looks at the construction of the completed building as one operation. He then segments all the different physical building operations that are necessary to construct the building within the total time frame. He generally does this in a logical sequence of operations, which may be developed individually through his personal experience in physical building operations or simply follow a traditionally recognized listing derived from the trade divisions in the technical specifications.

Thus the listing of operations can be done according to successive general categories of operations:

> Excavation
> Concrete foundations
> Structural steel
> Masonry, etc.

This list follows a logical sequence of physical building operations.

The operations list, however, will usually be more detailed. The amount of detail required in scheduling operations usually depends on the specific demands of the architect/engineer or the owner or it may simply be a matter of convenience for the scheduler and those who use the schedule.

Thus a more detailed listing of categories of operations can look like the following:

Excavation
Bulk
Footings
Backfill
Grading
Fine grading
Concrete
Footings
Piers
Walls
Ground floor
First-floor frame

This more detailed list also follows a logical sequence of physical building operations, but it does not necessarily follow a time sequence; it just lists the kinds of operations involved. For example, fine grading in excavation could not possibly come before footings in concrete.

Depending on the need or requirement for even more detail, the listing can be broken down even further. Thus a detailed listing of the concrete framing of the first-floor operation might be the following:

Concrete, first floor
Layout
Form
Resteel
Pour
Strip

All these lists follow a logical development of physical building operations. Once the operator understands how a list of operations is formed logically, it becomes a simple matter for him to make any listing he desires on any project.

9.3 REQUIRED KNOWLEDGE OF CONSTRUCTION OPERATIONS

We have shown how a basic form is established for a construction schedule. Time is represented in the graphic form and so are operations. Using this form, the operator proceeds to schedule the work, assigning each operation a segment of time within the total time frame. This assignment, of course, constitutes the real task of the schedule-making operation. It is not a simple task; even in the smallest work this assignment can be very complex. In large work it always involves complexities if the schedule is done properly.

Each operation must be assigned a time. Here, as in a few other areas of the construction process, the operator who schedules must know as much as possible about the elements of each and every construction operation that he schedules in the construction process. This does not mean that he must just know about the actual physical building operations involved; he must know about *all* supplementary operations that are necessary to accomplish the work as well.

Let us examine some examples of why the operator must know all about other operations in relation to just the operation he is considering. This will help illustrate the analysis and thinking processes he must develop for the scheduling operational process. Take the concrete framing operation: in order to complete a detailed concrete operations list the scheduler must know how concrete framing construction is done. What, for example, are some of the things he must know about the concrete framing operations? He must know the following:

1. Concrete framing is formed, poured, and stripped.
2. Within these operations are the supplemental operations of resteel, finish, and all electrical and mechanical installations, which must take place in a sequence.
3. The properties of all the materials used for or installed into the work must be taken into consideration in scheduling (concrete is heavy, liquid, and it hardens; resteel is long and heavy, etc.).

The scheduler must also know about the shop drawing submittal/approval process and its operations. He must understand how these processes work, because the shop drawing is the initial step in furnishing the necessary materials for the concrete framing operation. He must know the following:

1. Shop drawings have to be made and approved.
2. Within these operations the shop drawings must pass through several different steps, over which the scheduler has only varying degrees of control.
3. After the shop drawings are made, transmitted, and approved, the products of some of them take longer to make than others.

The scheduler must know about the periodic payment process and its operations for the following reasons.

1. Any portion of the concrete frame he is scheduling is always critical to the completion of the whole frame, and the completion of the whole frame will affect the payments for future sequential work in addition to just the concrete work at hand. The schedule of payments, in fact, is often based directly on the construction schedule.

2. When he knows how the payment process works and how important payments are in the construction process, the scheduler can, if he chooses, deliberately schedule any part of the framing operation (or any other operation) out of sequence if payment for that work is more important at a particular time than doing other work in the sequence of construction.

These examples are just a few basic things the scheduler must know and consider in his operation. They relate to just the production subprocess of the construction process. The reader should realize the great complexities that can arise in co-relating this one operation within an entire schedule where all the other operations must be analyzed in the same way.

9.4 THE BASIC SCHEMATIC

Some General Concepts of Time and Operations

Figure 15 shows the basic form of a bar chart. This is the easiest schematic method of scheduling and it is quite simple in form. The bar for each listed operation must be properly drawn in to make the schedule function as an operational tool. The scheduler must learn some general concepts to guide him in his scheduling.

1. He must first learn the ordinary sequence of physical building operations. This is accomplished by listing these operations in a logical sequence, but the operator must go beyond this listing to see which operations should actually come first. The optimal sequences are not always evident from the listing of the ordinary sequence. The scheduler must remember what the primary purpose of the schedule is and therefore see if it is possible to list certain items totally out of ordinary sequence so that the fastest total completion will result.

Example. Roofing is usually listed well toward the end of the ordinary sequence of an operations list. However, the actual roof completion on the project may be needed very early in the sequence of operations. The scheduler may therefore decide to roof immediately after the frame is erected, and the roofing item would then have to move upward in the schedule listing. In fact, based on the importance of completing the roof installation to the final completion of the project, the scheduler may even rearrange operations that were previously listed in ordinary sequence for the sole purpose of getting the roof done. This same reasoning can be applied by the scheduler to the total enclosure of the building.

2. The scheduler must also learn the complexities of the individual physical construction he is scheduling, bearing in mind all the supplemental operations involved.

Example. The specific operations involved in building a concrete frame are more complicated than those involved, for example, in building footings. Finish work is generally more intricate than rough work. Electrical work may be more complicated than mechanical work. Complications or intricacies in physical operations usually cause time complications and the scheduler must take these factors into consideration.

3. Weather, disruptions in labor, availability of labor for the job, shortages of materials, and water and site conditions are all examples of "unknowns." Un-

knowns cannot ordinarily be scheduled with any degree of accuracy; they can only be scheduled by guess. They are always factors to be considered by operators scheduling work. The scheduler must always take a realistic approach to both their potential and consequences.

A construction operator must know something about physical construction operations before he can properly schedule them; he must also be able to analyze the sequences of operations. He must know the degree of complexity of all the different physical operations and always give close attention to how unknowns can affect the schedule.

9.5 THE ORIGINATOR OF THE SCHEDULE

Who in the construction operational organization is to be the originator of the schedule? This generally depends on the operational division to which the company assigns this task. Since all of the various operational parts of the organization will use the schedule, all should know how it is made and how it functions. Let us examine each organizational division to see which one is the most logical candidate for making the schedule.

The estimator schedules the job when he estimates it. He cannot help but do this; he must, at the very least, price the work according to some schedule based on the time–operation formula. This means that the scheduler, after award of the contract, must take this into consideration when scheduling the work. The estimator would then seem to be capable of performing the scheduling of the job. Assigning this operation to the estimator would at least avoid a duplication of effort. The main reason why the estimator should not handle scheduling is because he usually cannot afford to give the time required to the continuous operations of scheduling and its processes and still adequately perform his main function of estimating new work.

The general and job superintendents' primary roles in the scheduling process are in using and maintaining the schedule, and not in making it. This is not to say that they should have no say in the schedule making, particularly where their skill and experience in physical operations are concerned.

Utilizing these operators' experience is essential when either new construction methods or processes are to be used in the construction or when the schedule depends on a good knowledge of advanced technological or manual processes, techniques, or operations. There are other specific areas of scheduling where the assistance of the general and job superintendents is particularly important. For example, when demolition work is to be scheduled, actual field experience is very important in furnishing information to the scheduler on the duration and hazards of these operations. The thinking processes required in scheduling demolition work are different from those required for building new work. The thinking abilities needed to analyze time, sequence demolition work, and quantify the effects of the unknowns are more likely to be found in the field operator than in the office operator.

Since we have eliminated from consideration the estimator and the field supervision, the expediting operator appears to be the prime candidate for handling

2. Immediately after the contract award submit the schedule for information or approval to the proper parties and press for any responses required and review them. Revise the schedule as necessary.
3. After the submittal/approval process is complete publish the schedule for distribution.
4. Distribute the schedule to all parties involved in the project, making sure none are missed.
5. Make the use and maintenance of the schedule a contractual obligation in every subcontract and purchase order.

The schedule should be sent to subcontractors along with the subcontract when transmitted for the subcontractor's or supplier's execution. It is wise to send it under a cover letter, which should include the following:

1. State that it is the identified project's schedule.
2. Request the subcontractor's or supplier's approval of its portion of work designated in the schedule.
3. Request the subcontractor to examine all operations covered in the schedule, particularly those that are concurrent or consecutive operations.
4. Ask for the subcontractor's or supplier's reasonably immediate approval of its assignments in the schedule.
5. Notify the subcontractor that if he does not reply, his silence will be considered as an approval, and the properly executed returned subcontract or purchase order will evidence this fact.

Of course, reasonable objections to the schedule will necessitate its revision and adjustment until it is acceptable to all the parties. This reevaluation process is not as complicated as it may seem. Even if no objections are raised in response to the letter, it is necessary that the scheduler and his subcontractors and suppliers, the architect and the owner realize that an attempt was made at the beginning of the project to have the schedule coordinated, correlated, and accepted by all. This considerably lessens the number of operational problems involved with scheduling that can occur during the course of a project. This process insures to the greatest extent possible a full and complete understanding of the mutual obligations of all the parties using the schedule.

9.11 DEVELOPING METHODS OF THINKING

In the Formation of the Schedule

There are several similarities in the thought processes required in scheduling operations and in estimating them: both must tend to be optimistic; both must make use of guessing; both involve thinking about time; both require exactness in their product, and thus thoroughness and realism in their thought processes; and both predict and forecast operations that are dependent on analytical and intuitive skills.

This does not mean that there are no striking differences between the two processes. For example, the thinking skills that are necessary under pressure differ greatly from those required in nonpressure situations. Estimating is done under pressure, and scheduling is not. In addition to pressure considerations, there is no real need in scheduling operations for either the same kind or amount of competitive thinking as is required in estimating. This is a striking difference in the two approaches to scheduling.

Whenever the ability to predict is essential to an operation, it is necessary for the operator to at least be aware of the need to develop his intuitive thought processes, should he not already have done so. There is very little doubt that intuition plays an important role in the scheduling process. Knowing upon first glance that something is right or wrong, efficient or inefficient, possible or impossible, and so forth is not only an aid to solving operational problems but also a great timesaver, especially with solving problems that tend to hamper scheduling. When results can only be tentatively predicted, as in the forecasting of operations within segments of time, the use of intuition can be of great service to the construction operator.

The construction process is notable for the amount of guessing required in many of its subprocesses and phases. Scheduling is a prime example of such a subprocess. Scheduling could be viewed as a guess in its entirety if for no other reason than the great numbers of unknowns the scheduler is required to somehow fit into a time frame.

But the art of scheduling most certainly involves making "educated" guesses. To say that the construction process requires guessing in no way means that it is a haphazard process or that it is based on haphazard thinking processes.

The scheduling process demands that consideration be given to many varied knowns and unknowns. The use of statistical and historical data has a place in this process. Giving proper priority to the many different influences and factors and analyzing their relative importance must also be a vital part of the scheduler's thinking. Appreciating the limits and capacities of personnel and organization is another essential. After all the potential negative factors have been examined and reviewed, determining the positive, realistic, and logical sequence of a construction job is the ultimate goal.

In Using and Maintaining the Schedule

When using the schedule, all operators must exercise great determination in implementing and maintaining the schedule. Once the schedule is formed and approved, no small reason should cause it to change; a schedule that is once compromised becomes a less acceptable and applicable document. It is true that the schedule must often be modified or changed to meet the circumstances created by unknown conditions. This is a fact of the construction scheduler's life. Those who use it, however, are usually better off maintaining the schedule until it is obviously unworkable rather than forcing its change because of unexpected difficulties in its use or maintenance or for convenience.

Construction projects must have a schedule, which should be a formed document based on good, sound construction operational thinking, and it must be used and maintained with determination.

10

The Close-Out Process

*All's well that ends well: still
the fine's the crown: What'er the
course, the end is the renown.*

William Shakespeare

10.1 CLOSING OUT AND FINAL PAYMENT

The operation of closing out is the performance of all of the specific things the construction contract requires for the total completion of the work in order for the contractor to obtain the full and final payment of the contract price. In this chapter we will examine the following operations:

1. The punch list and its processes and procedures.
2. Furnishing "as built" drawings and operating instructions.
3. Furnishing proofs of prior payments.
4. Furnishing the various warranties required within the contract itself.

The construction operator has a major goal in mind at the close-out phase of the construction process—final payment. The final payment of the contract price (adjusted for contract modifications) represents the operational goal of all of the divisions of the operational organization involved in this process. It is the last payment made on the project. The operator can assume that the overhead and profit monies for a particular job are contained in this payment. This is also the payment that includes the retention monies withheld from the periodic payments throughout the course of the job, which usually amounts to 10% of the contract price and in most cases contains the amount of overhead and profit figure put on the job in the contractor's proposal.

Needless to say, the contractor is vitally interested in getting the final payment, as are all of his subcontractors and suppliers. This payment often not only furnishes future business potentials and capacities but is also required for conducting the contractor's business in a continuous, orderly, and profitable manner. It must not be delayed for any reason, and most certainly not because the contractor's construction operators failed to expeditiously and properly carry out the closing out operations.

Each division, at some point during the course of the project, will have contributed its efforts towards reaching this goal. At close-out time the expediting and job supervisory divisions are those that contribute a major portion of their time and effort in the closing out operations. Each of these two divisions depends heavily on the other.

Example. "Punch listing" is an operational process within the close-out process. The punch list process is directed initially by the expediting operator, but it is also absolutely necessary that the job superintendent devote a major part of his time to seeing that the work items on the list are completed or corrected. In order to perform his part of this operation, the job superintendent must concentrate heavily on the coordination and scheduling of subcontractors. When corrective work is involved, this is often much more difficult and certainly more frustrating than checking to see that the work items were done correctly in the first place. All the operators must be in close and constant touch with each other in this operation so that the corrective work goes along as smoothly and with as few problems as possible.

10.2 THE PUNCH LIST AND ITS PROCESS

Judgments

The punch list on a construction project is a list of items of work that still need to be done in order to finalize payment and/or that were not acceptably built in the opinion of the owner or the architect/engineer. The purpose of the punch list is to gather these work items into a compact, itemized list of unfinished or incorrect work so that the work may be finished and corrected on time.

Unfortunately, the contract does not define either the punch list or an objective standard of the completeness, correctness, or quality of the work. This lack of direction results in great problems in punch listing, not only for the contractor but for the owner and the architect/engineer. These problems are created by the subjective character of the judgments that must be made on the completeness, correctness, and quality of the work. The degree of subjective judgment required is sometimes very extreme, and many times these judgments arise from the nature of the work itself.

Example. Even though paint shades and tones can be predetermined through samples (using the submittal/approval process), they are still subject to a wide range of interpretations in judging whether or not their appearance on the actual work installed represents what the designer intended. It would seem that approval of the samples should alleviate such a problem; but the problem in punch listing arises because of the individual punch lister's subjective interpretation as to whether the paint shades and tones have matched the samples. If they are not suitable to him, it may not be because the approved selection of the color was not applied but, rather, because the paint was misapplied or the surfaces on which it was applied were not correctly installed. The approved color, when in place, may not look good to the punch lister, and therefore it becomes at least a debatable item on the punch list.

Because an item is on the punch list does not mean that it should be included. The point, debatable or not, is that it is there and must be dealt with as such by the operator, and this takes time and effort. Sometimes items are on the punch list because of its maker's individual conception of acceptable work.

Example. By its nature, the concrete frame of a building cannot reasonably be built to the same measurement tolerances as can finish carpentry or stainless steel items on a building. But what tolerances are acceptable in the concrete frame?

The contractor's idea of allowable tolerances are those to which the building is actually constructed; but the architect may be viewing them in relation to the structural ability or aesthetic results of the construction. This raises the question of who should prevail in a contest when such a debatable item is punch listed. Even if such items are negotiated out of the punch list and do not have to be touched, they are still on the list and must therefore be processed.

From these last two examples it should be obvious that the construction operator must try to keep the number of possible subjective judgments on punch list items to a minimum. He does this by realizing potential problems as early during the progress of the job as possible. He then makes arrangements to settle questions as they arise. By doing this, he establishes the acceptability of the work early in its building.

Standards of acceptable work must be clearly established as soon as work is in a state that will reasonably allow for its being judged acceptable or unacceptable. If unacceptable, corrections should be made at that time that are mutually acceptable as a standard of completeness, correctness, and quality, which then becomes the norm for successive construction. This not only prevents a great number of the same items appearing on the punch list but also furnishes better operational control over the building and the components of all segments of the structure. A subsidiary benefit of establishing standards of acceptability early in the construction process is that they can be applied in the various consecutive and concurrent sequential operations involved in constructing a building.

Separate Contracts

The punch list process is generally much more complicated under the separate contract system than under the single contract system. Operationally it is more difficult for both the owner and his architect/engineer and for the separate contractors' operators. Under this system it is generally the architect/engineer who not only makes the punch list but also coordinates its execution. Not being a building contractor, he is faced with coordinating complex contracting operations. This has serious consequences in punch listing, because the architect/engineer is not generally familiar with contracting or financial problems such as, for example, the subcontractor's size and capabilities or the cost and direction of labor forces. The person responsible for punch list operations on a project should be familiar with the many construction operations that must be performed, and he should be primarily familiar with them from the standpoint of the contracting segments of the construction process. It would be unrealistic for him to assume, for instance, that contractors are immediately available to perform on the items at his call. In addition to this are the problems caused by the complexities of coordinating the work, required at specific stages on many different items of work in a punch list, of multiple separate contractors and their subcontractors. Another major problem is caused by none of the various separate contractors and subcontractors having contractual responsibilities to each other. These problems and complexities show that completing this phase of the close-out operation is clearly a major task. Under the separate contract system it is an area where the system's many advantages must be balanced against these disadvantages.

The only efficient way in which this operation can be accomplished with separate contractors is to clearly define and organize the punch list process as early in the job as possible; this means establishing some type of operational control between the various separate contractors. Operational control is best established by having the contractors agree to the assignments of responsibility and authority among themselves.

Once a method of control is established, the operators in the individual separate contractor organizations best serve themselves and their companies by strictly adhering to it. They must remember that their own individual final payments and financial interests depend on the performance of the close-out operation.

In the single contract system there is little doubt as to who actually controls punch list operations. The general contractor's expeditor bears the most responsibility and his task is a major one. The job superintendent is also a main initiator and pusher in the punch list process: he is closest to the work to be done and responsible for seeing that the punch list work is done. This calls for a close coordination of efforts from both the field and office forces.

The Punch List Procedure

Sometimes the architect/engineer will insist on the contractor making the first punch list. This saves the architect/engineer the task of processing a punch list on many items of work whose responsibility is clearly the contractor's. This demand is reasonable in most cases, because there should be no need for the architect/engineer to waste time listing items of work that are obviously incorrect or incomplete. The first listing is sometimes called the preliminary punch list.

The final punch list is made by the architect/engineer, and usually in his form. Its extent and contents vary with the individual architect/engineer's office's peculiarities.

Example. Some offices will detail each single item of work as a separate item on the list:

Item 78. Paint bottom of door, 173C.

Some architect/engineer offices will generalize as follows:

Item #10—General item of work—Paint all door bottoms not already painted.

Each office has its own way of forming and processing the punch list and each architect/engineer has a different way of compiling it. Some will walk the job alone and list, and others will insist on the contractor's accompaniment. The latter method probably saves time but is not always feasible, owing to the availability of the contractor's field operators. The job superintendent must use careful judgment when making his own or his subordinates' time available for this procedure, because of the obvious advantage of immediately knowing the architect/engineer's opinion of an item of punch list work.

The punch list may be made by the architect/engineer in a continuous operation or it may be a drawn out affair taking several weeks. When he has the list completely or even partially made, he will have it typed at his office (transcribed from his notations) and published. He generally sends one or more copies to the contractor.

The contractor reviews the list, carefully analyzing its contents to determine to whom he must transmit it. Some operators feel that it is necessary to send it to *all* the subcontractors and suppliers on a blanket basis so as not to miss anyone involved. Some operators feel that it is the subcontractor or supplier's responsibility to pick out the items of work that they must do; other operators feel that in order to save time it is best to note the items of work that should be brought to the attention of each subcontractor or supplier. This saves precious time and some future operational scheduling problems, but it takes quite a bit of time to separate and coordinate the responsibilities for work. The contractor's operator publishes sufficient copies of the punch list and distributes them to his field forces and the subcontractors' forces.

The work on the punch list items then begins. This work calls for the close coordination and control of all the forces involved in the punch list work. In many ways it calls for a minischeduling of items that require multiple subcontractor participation.

At the punch list stage of the project regularly scheduled job meetings can serve as a means of expediting the accomplishment of work on the list. At times special job meetings must be held whose purpose is to expedite the punch list work. Those areas of the list that are consistently causing problems are handled in these meetings.

As the items of work are completed or corrected, they are crossed off the list. Items of work are checked for acceptability by the architect/engineer, either by himself or with the contractor's job superintendent. Sometimes the lists are recompiled as they are reduced on a regular basis, and published and redistributed. The punch list process is a narrowing-down process that continues until all the items on the list are satisfactorily completed.

Operational Problems

There are two main operational problems that are commonly encountered when processing the punch list. First of all, the progress of the work passes through various stages to the point where the building is almost a finished product. This has two significant results on the punch list process: first, because the building is being completed, all parties require much higher standards of care and cleanliness in operations than during the beginning stages of construction. The conditions under which work is to be completed or corrected at the beginning of the construction process are simply not the same as those toward the end. One can repaint a wall, for example, early in the building process without too much regard for protecting surrounding surfaces; but at the end of the process the wall's repainting must always be done over and around installed work.

Secondly, as the building nears completion, the owner usually occupies or starts to occupy the building. Many times he starts his own operations in the building, which affects the punch list process because it restricts the availability of areas of work. Because of this restriction, the problem of scheduling and doing the punch list work becomes more complex. A women's restroom/toilet area, for example, is not always immediately available for work after the owner occupies the building. This problem calls for a very close scheduling of the uses of the various parts of the building that both the owner's and contractor's forces must have available for their respective operations.

To avoid or minimize these problems it is wise for the construction operators to meet, early in the process, with the owner to discuss his uses of particular portions of the building, with the idea of specifically pressing for completion of the most critical punch list items.

The second main operational problem of the punch list process has to do with the repeated use of the process by the owner and the architect/engineer to maintain an advantage over the contractor. When a punch list item has been completed and formally accepted, fewer difficulties are involved than when it is completed (but not yet accepted) before the architect/engineer's next review of the building. An item of work may have been done, for example, in an occupied area, but, because of its use by the owner's forces, it has been damaged again. Although good common sense and honest approaches to such situations solve many of the problems, there are often sincere differences of opinion as to who caused the damage or whether the list item had been corrected at all. This calls for establishing definite criteria for the occupancy of the punched out portions of the building.

The real problem that arises from the repeated use of the punch list, however, is when it is used as a tool to delay final payment. When the owner and his architect/engineer use the punch list without cause to deliberately withhold payment, there is very little the contractor can do short of engaging in a dispute with them. There are times when the owner is sincerely dissatisfied with the quality of a portion of the work. It is natural for him to want what he paid for. The problem in this kind of situation involves the reasonableness of the owner's expectations. The construction operator must enlist the aid of the architect/engineer to determine a specified standard of acceptability for the work as early in the process as possible. Most professionals know the acceptable standards of quality; it is necessary for the operator to convince them so that they may, in turn, advise and convince the owner of what is and what is not acceptable construction.

The Nature of Punch List Work

We must closely examine the nature of punch list work; understanding what this work actually consists of will make it easier to accept its troublesome and frustrating features.

Corrective punch list work is repeated work; it is work that has been done once and that must be done over. There is a strong natural reluctance on the part of construction people to do work over, even in spite of the fact that the pay is the same for redoing work as it is for doing it right in the first place. A contractor, of course, never likes correcting work that is installed incorrectly. But it is even more difficult to accept redoing work that has been done correctly. Because of the nature of the punch list process, doing correct work over is a more frequent occurrence than it may appear at first glance. For example, a wall structure can be built correctly in all its segments but one, and if the incorrect feature happens to be the layout, then the wall is still built incorrectly. What are the attitudes, then, of those who have done their work correctly when some feature of the wall appears on the punch list as an item to be corrected? The reader can guess the feelings of the painting contractor, the electrical/mechanical contractors, or others who have done their work correctly and are faced with doing the same work again.

The reader should understand that the construction contract takes very poor account of the legal consequences that can arise out of this situation. The best way for the operator to solve these and similar problems is to depend on his skill and ingenuity in getting the various subcontractors to perform the work and not rely on legal solutions. It is not wise for the operator to rely too heavily on the contract during punch list time, because it does not address these problems from a practical operational standpoint.

It is the general contractor operator's responsibility to educate participating subcontractors on certain aspects of the punch list process. In most cases a reminder or an explanation of the mutual benefits that are derived from completing a project for final payment is sufficient, but sometimes other approaches must be taken.

The operator must emphasize to the subcontractors that the punch list process is simply "a fact of life" in the production subprocess of construction and its performance is a normal obligation of their subcontracts. The major problems encountered in these subcontractual relations involve areas where multiple parties work on composite work items, and these are difficult problems because of the lack of a privity relationship between the various subcontractors.

The Back Charge

A commonly used tool in the punch list process is the back charge. The back charge (or the threat of back charge) is used when a subcontractor has caused an item of composite work to be on the punch list and refuses to pay for its correction. Back-charging consists of billing the subcontractor for the corrective work done by others and then withholding his payments if he does not pay the costs resulting from his faulty work. Usually only the threat of using this procedure convinces the subcontractor to at least be cooperative in the process. The back charge is operationally cumbersome and under ordinary circumstances stands on a very poor contractual legal foundation.

The back charge is therefore really not the most effective tool to use in the punch list process, but sometimes it is the only tool available to the operator. If those being back-charged have even a little knowledge of the law of contracts, they will usually realize the shaky foundation of the back charge. It is rare when a subcontractor will voluntarily agree to a back charge; in most cases he will deny that a basis for it exists. When the back charge must be used, the operator should realize the increase in paper and accounting work involved. In the punch list process it is better to use the back charge only as a last resort. The construction operator who does the punch list should rely instead on his ability to convince all parties in the process to cooperate and do their portion of the list, relying on the theory that completion of the punch list gains the objective of final payment for all of them.

10.3 FURNISHING "AS BUILTS"

The "as built" documents are a designated set of contract drawings that are "marked up" with all of the changes of work which deviated from the dimensions in the actual placement shown on the contract drawings. It also includes the "marking up" on the designated set; the actual placement of work developed from any nondimensioned schematic such as the mechanical and electrical contract drawings.

Example. If a footing is dimensioned in elevation on the contract drawings but must be lowered for soil bearing capacity, then its actual location must be noted on the "as builts." If under-floor drain lines are schematically designated on the mechanical drawings but need to be located elsewhere because of interference with the dimensioned structural foundations shown on the structural drawings, then the actual installed locations of the lines must be noted on the "as builts."

We include the subject of "as builts" in this chapter because we wish to emphasize the need for the construction operator to develop set procedures and designate someone to be responsible for making, keeping, and recording the as built documents in a current condition.

This operation is specified in the construction contract. It is unfortunate that construction operators, as a rule, refuse to take this essential requirement for final payment seriously. Usually this construction operation does not attract the attention of the operator until the owner or his representative makes furnishing "as built" drawings an absolute condition of payment. The operator should be aware that the owner stands on firm contractual grounds when he insists on their receipt before payment.

The furnishing of the "as builts" is merely a transmittal of certain documents from the contractor to the owner through the architect/engineer. It is the actual operations of making and keeping the "as builts" that the operator must carefully consider. It is one task to keep the "as builts" up-to-date with the stages of construction, but it is quite another task to present at close-out time a set of drawings that can only be marked up and compiled from memory.

10.4 FURNISHING OPERATING INSTRUCTIONS

The contract also requires furnishing operating instructions as specified in the documents. These operating instructions can vary from furnishing information on the care and maintenance of the resilient floor material installed on the project to information instructions on complex electronic systems that have been specified and installed. In most cases the owner is fairly adamant about having these instructions as a condition of final payment, which is a reasonable expectation.

The construction operator must make clear to all parties (and particularly to subcontractors) that they must furnish these documents at close-out time as a condition for their final payments. This obligation should be clearly stated in the subcontract. In most cases it involves merely the transmittal of published materials (brochures, etc.); it only takes the operator a minimum amount of effort to secure them and thus furnishes no excuse for failure to obtain final payment on time. The operator should have all of these documents in hand well before close-out time.

10.5   FURNISHING PROOFS OF PAYMENTS

Very few expediting operators will be enthusiastic about the operations required to furnish final waivers of lien and other proofs of payment. This operation is at best tiresome and at worst frustrating, because it causes some very difficult and unpleasant problems for the operator.

The construction contract makes it mandatory, as a condition prior to final payment, for the contractor to furnish proof that he has paid those who have worked on or supplied the project, meaning primarily his subcontractors and materialmen. The final waiver of lien has been established as the formalized process that gives evidence that these people have been fully and satisfactorily paid. When they sign and return the final waivers of lien to the contractor, the owner can be fairly certain that he will not be held legally responsible for any liability to any of these people. At least one copy of the final waivers for each and every material supplier and subcontractor usually must accompany the general contractor's request for final payment from the owner; as well as the contractor's sworn statement that he has paid his debts in full. This furnishes the owner with additional protection.

The construction operator should realize that the general contractor will not (except under very unusual circumstances) receive final payment until he has furnished these documents to the owner; it would be foolish to believe otherwise. Therefore, since the task of furnishing these documents is absolutely required on every construction project, the operator must simply put his shoulder to the wheel and do it.

It would seem that there is not too much to do in this close-out operation other than to start it and keep it going so that all waivers are processed by final payment time. However, on every job there is the possibility that one or more subcontractors or suppliers will refuse to waive their lien rights. Just one refusal is usually enough to thoroughly complicate the process. The owner expects proofs of payment from the general contractor for all work done and material furnished on his project.

What does the construction operator do when a subcontractor refuses to furnish his waiver of lien? The operator must first determine why the subcontractor has refused to do so.

Example.   Let us say, on the one hand, that the general contractor has withheld payment to two different subcontractors during the course of the job. He has withheld payment from one of them because of the subcontractor's refusal to accept and pay legitimate back charges for cleanup. Payment has been withheld from the other subcontractor because the contractor sincerely believes that there are serious deficiencies in the quality of his work. At close-out time this subcontractor refuses to acknowledge items on the punch list but has made his final billing to the general contractor for his full subcontract price.

The general contractor has requested final waivers, and both subcontractors have refused to furnish them with their final billings. The waivers of all other subcontractors and suppliers are in the general contractor's possession and only these two outstanding waivers prevent completing this close-out operation.

The construction operator must view these two subcontractors in different ways. With the first subcontractor he must weigh the value of receiving final payment as quickly as possible. Perhaps he will call the subcontractor into his office and try to compromise on the back charge or drop it altogether. With the second subcontractor the operator has a much more difficult problem and greater risks are involved.

If the general contractor pays the subcontractor his final billings without deducting the costs of the corrections in return for the subcontractor's waiver, then the costs of the corrections will most likely be left to the general contractor because the subcontractor cannot be forced (at least by any money considerations) to pay for them. If there is multiple trade work involved in the corrective work, this can be a troublesome and expensive situation for the general contractor.

Probably the best course for the operator to follow in this case is to first have the work corrected. He must give proper notice to the subcontractor that he is proceeding with the correction of its work and must keep accurate and complete records of his costs. He must also be aware that this course of action is not without its dangers, particularly in the areas of warranties and guarantees, but at this point he really has no other choice.

The operator's next step would be to sit down with the owner and explain the situation, trying to convince the owner to allow an exception in the strict waiver requirements and to rely instead on the general contractor's sworn statement (and labor and material bond should there be one) for his protection.

If the owner refuses the request for some reason, then the operator should ask him to withhold from the final payment some reasonable amount to protect his interests. If this request is refused, the operator can suggest that the general contractor furnish him a specific bond for the full amount of the subcontractor's contract price. If the owner does not accept even this suggestion, then the best the operator can do is try to convince the owner to pay as much of the final payment as possible.

One can see that this kind of problem causes great hardship not only to the general contractor but also to all the parties awaiting final payment on the project. The problem, while being extremely difficult and troublesome to the operator, must be solved as quickly as possible, but at the same time very cautiously.

Operationally, furnishing proofs of payment can be a lengthy process, but the construction operator can minimize the time it takes. If he sets up the operation for the project properly by making initial contact with the subcontractors and suppliers and having someone (an office person) make periodic checks with them, then it should be a fairly painless process. The operator must remember that furnishing proofs of payment depends to a great extent on the clerical operational capacities of his subcontractors and suppliers as well as his own, and he must make allowances for this fact.

10.6 FURNISHING WARRANTIES

The construction contract requires that the entire work be warranted. To warrant construction work means to guarantee to the owner the quality, quantity, and condition of the building or structure. The contractor guarantees the owner that the building has been built in accordance with the contract and that the integrity of the entire construction and all of its parts will be guaranteed for a specified period of time. After the expiration of this period the contractor is no longer held responsible for the condition or quality of the building or its parts, unless the building was improperly built in the first place. Expiration of the warranty period does not release the contractor from the obligations of contract if he never carried them out.

Warranty is given for the entire building or structure, as well as on specific items of work contained in various parts of the technical specifications. Warranty periods vary, depending on what particular component they are given for. Usual time periods are one year on the building or two years on a component such as a built-up roofing installation. Each warranty period is designated in the contract specifications.

There are two reasons why construction operators should pay careful attention to expediting the warranty process: (1) because the furnishing of warranties is a clearly stated requirement for final payment in the close-out process and (2) because the warranty period begins to run from the time the owner is in possession of the warranty. Since the contractual and operational responsibilities hold until the warranty period expires, it is good practice to furnish warranties as soon as possible. Expediting the warranty process decreases the total time for which a contractor is obligated under a particular construction contract. This has significant bearing on the contractor's future business potential, particularly from the viewpoint of the construction firm's bonding company.

The warranty process is operationally quite simple. The contractor, subcontractors, and suppliers are usually aware of their responsibilities in this process. However, as is the case in other formal requirements of the close-out process, it is a task that is at times not taken seriously, and this lack of concern leads to procrastination and carelessness. The warranty process generally calls for the collection and transmittal of the formal documents and normally requires only simple clerical organization in order for it to be done correctly and on time. There is little excuse for it ever being the cause of delaying final payment.

10.7 FURNISHING THE CERTIFICATE OF OCCUPANCY

The certificate of occupancy is a formal acceptance of the building by the public authority responsible for the public's protection when they use the structure. This authority and responsibility is usually authorized by enabling statutes that include provisions for compliance with building and fire codes. The issue of a certificate of occupancy is almost without exception based on the building's compliance with the building and fire code provisions.

Generally the owner or his architect/engineer is responsible for obtaining the certificate of occupancy. We discuss it here because procurement of the certificate sometimes affects the close-out operations of the contract. At times the completion of the work is affected or delayed because of the requirements of satisfying a code in order for the authorized public officials to certify occupancy.

The construction operator should be particularly aware of the role of the fire marshal in this process. The owner, architect/engineer, contractor, and everyone else must satisfy the citations of a fire marshal; there are very few exceptions to this rule. The fire marshal is usually a trained and competent person who knows what he is doing. Therefore, when a fire marshal says to do something to comply with the code, it is wise for all parties to obey. The difficulties of fighting a citation at or near close-out time do not make it worthwhile to refuse to remedy any small problems. Any major changes that are required can, of course, seriously affect the completion of the building and must be dealt with as expeditiously as possible.

While it is usually the architect/engineer's responsibility to see that building's design satisfies code requirements, the construction operator must do everything reasonable to see that the code, permit, and public inspection processes and operations are scheduled in their proper time. It is only practical for the operator to give these inspection operations his close attention during the course of constructing the building.

Expediting and facilitating inspection operations and being helpful with inspectors during the course of the job makes their effects on the close-out process easier to deal with and faster to handle. Surely, the operator prefers to have building inspectors and the fire marshal on his side at close-out time rather than to have them behave in an obstructive or foot-dragging manner. The operator must remember that these people have their jobs to perform. Good inspectors do their jobs thoroughly and conscientiously. In conducting his operations, the operator might as well make an effort to keep them happy and satisfied.

10.8 DEVELOPING METHODS

When involved in the close-out process, the construction operator should think primarily of the importance of his operations to the business potential of his company. Closing out a job quickly and successfully allows the company to take on new business, with some confidence that past business is finished with and paid for. It would be hard for the individual operator to find a more demoralizing factor in the construction process than a series of long, delayed close-outs. As new work is assigned to the operator, his functions are affected by his having to start the new work while still extensively involved in old work. This not only adds to his work load but also affects his mental outlook. When the operator has been involved for a length of time on a project and he begins to feel that it is complete, but it has not yet been fully paid for, the operations that need to be done to get full payment sometimes become tiresome and frustrating. This last major step in the construction process is a long one; nevertheless it is an operation that must be done and the operator will save himself trouble and frustration if he develops methods for speeding it up.

The operator must develop his operational methods according to the realities of the task. Most of the process involves simple operations of collecting material and documenting its transmittal. The operator conducting these operations must be determined to see these operations through. The methods he uses must be fairly detailed and they must be closely followed; it is a matter of developing a method of checking on their status and persevering until the tasks are completed. Determination and perseverance are the main attitudes that the operator must cultivate in order to successfully carry out the close-out operations, and they are the only attitudes of any value when processing the punch list.

The process of closing out is anything but formal in nature, yet it demands the application of the strictest mental attitudes and behavior. The operator must perform his functions and also get all the other parties involved to perform them.

11

The Submittal/Approval
Process

When duty comes a'knocking at your gate
Welcome him in; for if you bid him wait,
He will depart only to come once more
And bring seven other duties to your door.

Edwin Markham

11.1   GENERAL COMMENTS

The submittal/approval process is a creature of the construction contract, born out of the contract performance requirements. The construction operator must view this process as being a vehicle through which the production of the building is made easier and more efficient while satisfying the formal requirements of the construction contract. The submittal/approval process requires the development of special construction operations to serve its purpose.

Basically, the submittal/approval process evolves from the contract performance phase of the contract subprocess; it is also related to the production subprocess. The purpose of submitting data for its approval is to expedite contract performance, which, in turn, expedites production. Data is submitted in order to gather, distribute, and finalize information about a particular portion of the work for the purpose of supplying this information to all of the parties involved in the production of the total building.

Those who receive and use this information are the architect/engineer and the general contractor and all of his subcontractors and suppliers. When combined, this information forms a detailed guide as to how the building is to be constructed and it furnishes a check on the builder's intent to conform to the contract documents; it becomes a record and a reference as to how the building was actually constructed, by whom, and with what materials. This information becomes a commitment, on the part of both the submitting and approving parties, to doing the details of construction in accordance with the data presented in the approved submittals. Other types of submittals include those that are made to show the proposed detailed means and methods by which the contract is to be performed and those that are simply for specific record purposes.

There is always a submitting and an approving party involved in the process and each of their responsibilities are established specifically by or inferred from the conditions of the contract.

The nature of the operations involved in this process are time consuming, and sometimes their boring and repetitive nature make construction operators consider the whole process a waste of time, which is not, however, the case. Practical experience will show the operator that the use of the submittal/approval process is needed to run any job of major significance. It is necessary because it saves time, lessens the possibility of mistakes in construction, finalizes construction details, and thus facilitates adequate planning and scheduling. It is essential for sequencing specific trade operations and allows for the application of before-the-fact sequencing of building operations within physical areas and spaces; it sets the form and specific ways of conducting specific operations and establishes records. It is a very valuable tool to use in the construction process, particularly in the production subprocess.

A certain amount of legal thinking is required when operating with submittals and approvals. The law is not specific in its application to this process, even down to the application of basic concepts of contract law. What, for example, is the legal effect of a shop drawing's submission and approval and the subsequent execution of construction operations in accordance with its data? How does it relate to the whole contract? What rules govern its precedence over the working drawings or other contract documents? Is it a contract document and, if it is not, when does it become one? These are only a few of the complicated and unsettled legal questions that arise from the submittal/approval process. As we said before, the legal considerations of this and other construction operational processes cannot be fully explained in this text because of their complexity and extent. Their legal implications, however, must be addressed by the construction operator and we advise the reader to seek sources on this subject and learn it well.

11.2 FLOW OF EXCHANGE

Generally, the submitted data passes through many hands. The submittal, in its approval stages, passes back through the same hands, which requires the development of specific operational processes and procedures for the specific type of data or item submitted. These operational processes are concerned primarily with the flow and exchange of information. Processes developed for submittals and approvals must result in the efficient and timely transfer of information. Thus in the general contract method of contracting a typical system of flow and exchange would be similar to that shown in Figure 16.

The diagram in Figure 16 represents a set-form of flow and exchange that, once established as a procedure, should not vary on an individual project. Obtaining approval of the submission is the prime objective of this operational system. The subcontractor submits and the general contractor checks and corrects its contents and then submits it to the architect/engineer for his approval. The architect/engineer then checks the submittal, approves or disapproves, and sends it back in the reverse order of the flow and exchange system. If the architect/engineer approves the first submittal, it is sent back in reverse order through the system just once, completing the process; if he disapproves of the submittal, its operations are repeated, going through the necessary resubmittals until approved. Everyone involved in this exchange must at all times be aware of the "status" of each submittal required on the project.

11.3 TYPES OF SUBMITTALS

There are three major types of submittal/approval requirements demanded by the contract documents that will be discussed in this chapter:

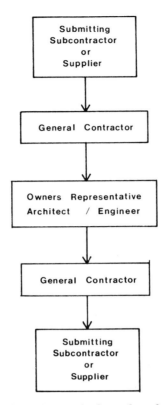

Figure 16.   A flow chart for the exchange of submittals in the submittal/approval process.

1. The shop drawing.
2. The technical brochure.
3. The sample.

The construction operator should first realize that these three requirements are documental in nature; even the sample, which is an actual representation of a specified material, becomes a document.

All of these documents record something and all are used as references and cross-references. They should be accurate and adequate descriptions of the details, procedures, means, and methods of work to be performed or of the item to be produced or installed. The submitting party and all the other parties on the job should be able to rely on this information.

The three submittal/approval requirements take a form, which has been generally established by tradition, as have the accepted general operational processes and procedures for making and using them. In this process, unlike most of the other construction processes, the operator must realize that his individual system for handling submittals and approvals must be developed from the operational procedures established by tradition. For example, the general procedures for processing the shop drawings have been used for years in the industry. The operator need only learn these traditional processes and then develop expediting systems that suit him.

11.4 THE VALUE

As a general rule, proceeding with any construction that uses the submitted materials, methods, means, or other information requires the approval of them by the architect or engineer. This approval allows for the information contained in the submittal to be reviewed and accepted by the design professions, both individually and as part of the total submissions on the project. As the number of submittals increases, so does the need for their integration into a larger and more detailed plan for building the project.

Example. A ceiling is shown on the contract drawings and all the operations involved in building it are directed to making it conform to these drawings. Some of the trades involved in this construction are the mechanical trades (working on the plumbing, heating, ventilating, sprinklers, etc.), the electrical trades (working on the light, power, and sound), and those trades that build and finish the ceiling structure itself (lathers, plasterers, painters, etc.). Generally, a ceiling structure, of whatever type it may be, is built in a production sequence of consecutive and concurrent operations. One trade builds onto or around another trade's work. Sometimes this sequence requires considerable doubling back on and overlapping of operations of the various trades as the item of work progresses.

From the moment it is noted that the ceiling structure is a composite construction, a plan must be made and coordinated to build it in proper sequence at its estimated cost. The first thing that must be determined is whether the ceiling can be built as shown. All the various work procedures of the various trades must fit together. There must, for example, actually be enough space for all the work to fit into the total ceiling, regardless of what is shown on the contract drawings. In addition the work must be accessible, and there must be sufficient space for the trades to work.

The contract drawings do not guarantee that each part of work in the ceiling will fit with the other parts. It is a rarity where there has been complete coordination of all of the items of work in the individual drawings with others in the contract set of drawings. In fact, most times the ceiling structure is only seen as a single line appearing somewhere on an elevation or a section detail of the architectural drawings. Usually the structure of the ceiling must be determined from the information about elevation, materials, and type of construction given in the contract drawing's schedule of room finishes. The mechanical and electrical contract drawings are generally only schematic in form and are therefore of no help at all in dimensioning the work. The designer who makes the contract drawings usually gives little thought as to how the work is to be built in proper sequence into the ceiling structure, and even less thought is given to matters of physical accessibility or cost.

The shop drawing is only one example of the tools required in the submittal/approval process to accomplish both a specific and a general purpose; it helps to satisfy the contract obligations of full and proper performance and allows for the most expeditious and economical way of building the various items of work into the total construction.

The best way of illustrating the value of the submittal/approval process to the construction operator is to ask him to consider building a structure or any part of one without it. He will rapidly come to the conclusion that this leads to chaos, disruptions, delays, and unacceptable cost consequences. He will also see that there are no viable options or alternatives to this process in today's construction industry. The modern, sophisticated, and complicated building cannot, from a practical standpoint, be constructed without a detailed and continuous coordination of all the parts of its construction. In addition to this fact, cost considerations require that the coordination of construction work take place before actually building the structure, and this coordination must continue throughout the entire project.

11.5 THE CONTRACT AND TRADITIONAL PRACTICE

The architect/engineer establishes by contract condition the three common types of submittal requirements and, in a very general way, the contractual procedures for submission and approval. These do not directly specify exact operational procedures; the contract simply says that these submittals shall be furnished through some formal exchange process for approval purposes, and it is assumed that an experienced contractor will know the steps required in the submittal/approval process. Traditionally accepted procedures for each type of submittal have resulted from this lack of direction in construction contracts. We emphasize here that neither the construction operator nor the architect/engineer solely determines the specific operational procedures involved in the submittal/approval process. The following operational procedures and practices used in the process today have been found to be workable and efficient.

1. A certain number of copies of the submittals in a certain form are presented for approval.

Example 1. Shop drawings are submitted in one sepia and one print or in a number of prints.

Example 2. Brochures are submitted in a certain number of copies.

Example 3. Samples are submitted in triplicate, with proper identification directly attached.

2. Submittals can frequently bypass steps in the formal flow of exchange patterns (Figure 16) when it is found to be more efficient.

Example. After its first submittal has gone through the formal procedures of exchange and been disapproved, on a resubmission the structural steel subcontractor will frequently be allowed to deal directly with the structural engineer, bypassing the general contractor and the architect, even though the general contract and the subcontract specifically forbid this practice. This deviation from the formal pattern is allowed for purposes of expediency. This exceptional but traditionally accepted practice generally requires only giving notice of transmittal to the various parties for record purposes. However, such notice must always be followed by the necessary formal requirements of the submittal/approval process for shop drawings.

3. The periods allowed for processing the submittals/approvals are rarely specified in the contract documents except in very general terms. Again, tradition has set time periods for handling various kinds of submittals.

Example 1. Resteel shop drawings are handled within a time frame which is different from that of other submittals required for work occurring later on the job. Strict time considerations set the limits on how long the data can be in the hands of the various parties in the formal pattern without causing delays in the job. Special time limits have been established for processing the resteel submittal, because of the fairly immediate need for this material on the job, and the times for making, checking, transmitting, and approving the submittal have been correspondingly adjusted. Resteel shop drawings can be submitted even before the award of the general contract and the award of a purchase order to a supplying subcontractor in order to save time. The resteel submittal process is a good example of adapting formal procedures to the demands of time. Contractually speaking, this way of operating is absurd, but there is no doubt that it works and has become a commonly accepted practice in the industry. The operator must simply adapt his operations to these practices when the need is evident.

Example 2. A specified concrete design usually has to go through the submittal/approval process. For example, a contract demands that a 28-day strength test for the concrete to be used on the job be made on the test cylinders made from the design. Theoretically, only after the cylinders are tested and found to reach the specified design strengths and the architect/engineer approves the results of the tests can the concrete be poured. But it is obvious that there are few projects that can afford to wait so long for this contractually required design approval. In fact, it is common to pour concrete long before the 28 days are up on construction projects where this requirement is in their contracts. How is this possible? There are accepted procedures of furnishing the test results from similarly designed concrete used in the past that allow the architect/engineer to deviate from the formal procedures of the submittal/approval process. Such a deviation is perfectly acceptable in this process, even though it does not conform strictly to the contract.

4. The approval process is handled both operationally and contractually by the architect/engineer's stamp of approval, which is referred to in the condition of the contract. The stamp generally contains the following notations on its face:

1. Approved.
2. Approved as noted.
3. Not approved.
4. Date.

The contract makes no specific requirements for the form or contents of the stamp notations. The use of the stamp of approval is based on traditionally acceptable procedures for the approval portion of the process.

"Approved as noted" is the stamp that causes the most trouble and confusion when trying to make that status conform with the requirements of the contract. However, this notation has a purpose: it allows the initial steps in producing the work to begin before the formal final approval is obtained. It is a designation that is clearly not in accordance with the contract, but the practice has become acceptable in the industry as a means of saving time while still adhering to the general purpose and intent of the strict contract requirements for approvals.

Example.   Resteel shop drawings are transmitted back through the general contractor to the supplier. Those marked "disapproved" cannot be acted upon in any manner and those marked "approved" can be used for manufacturing and fabrication purposes. However, some or all of the drawings may be marked "approved as noted." These drawings do not, for several reasons, conform strictly to the requirements of the contract documents, but the differences are such that they can be noted by the architect/engineer on the submittals, with the understanding that the submitter will correct them before use. "Approved as noted" means that fabrication may be started on the resteel materials shown on the drawings, and the submitter therefore does not have to go through the formal process of exchange all over again. It is understood that the production and fabrication of the resteel must be done according to the architect/engineer's notations and corrections. It has become accepted practice in the construction industry for the supplier to regard the "approved as noted" status as a permission to fabricate. The operator must, however, still go through the complete operational procedures of the formal submittal/approval process for the purposes of record and reference, but much time is saved by allowing this deviation from strict contract compliance.

Perhaps it seems to the reader that accepted traditional operational practices make the submittal/approval process one of the least restrictive of all the construction operational processes. This is true to a great extent, but the reader should remember that these operational practices are now so solidly based in the industry that they have developed some very strong implied legal contractual complications of their own. In conclusion, it should be apparent that the submittal/approval process, while very broadly governed by the construction contract, is the process that is the most adaptable to practical deviations from strict contract conformance.

11.6   THE SHOP DRAWING

The shop drawing is the visual representation of the details of specific work required by the contract drawings. It details and notes the following:

1.   A component of the work.
2.   The materials used in the work.
3.   The placement of the specific work within components of other work and within the total work required by the contract documents.
4.   The dimensioning of specific work within the scope of the contract draw-

ings in relation to the work of the submitting party and that of others involved in building the components of the construction.

5. The dimensions required to fabricate the individual material items.

6. The sequences of erection and installation of its subject matter, either through direct representation or by implication.

This submittal/approval document is generally a drawing, sketch, chart, or schedule. It can vary in size and form, but tradition has generally established the acceptable forms for each of the various trade divisions of work.

Thus structural steel shop drawings are generally erection drawings, combined erection–layout drawings, details, and schedules of fabrication; resteel shop drawings consist of placement drawings and fabrication schedules; sheet metal drawings are dimensional drawings developed from the contract schematics; and hollow metal shop drawings consist of schedules of dimensions with specific jamb, sill, and head details. Each trade division has its own traditionally accepted forms of shop drawing submittals that facilitates their own operations and serve the overall purposes and objectives of these contractually required documents.

The Formal Shop Drawing Process

The shop drawing process is as follows:

1. Shop drawings are made by the subcontractor or supplier working from information contained in the contract documents. They are normally made on a reproducible transparency (sepia) and reproduced in one or a small number of copies for submittal purposes. The subcontractor sends one sepia and one print of each drawing with a transmittal form to the general contractor.

2. When the general contractor receives the shop drawings, he acknowledges and notes the date of receipt. He then checks their contents both in relation to the entire contract documents and against whatever shop drawings he has that relate to the item of work. After checking and noting his examination of the content of the documents on the document itself, he makes a copy that he keeps and sends the sepia and at least one print (with his corrections on it) to the architect/engineer, under the cover of the company's set-form transmittal. With this print the architect/engineer can make corrections to the drawing without changing the sepia, thus allowing him to keep a record of his corrections.

3. The architect/engineer receives the drawings and notes the dates of receipt and transmittal that are pertinent to his operations. He then checks the document and stamps it with a status of approval. He does this with each individual document and returns them under cover of his transmittal form to the general contractor along with the original sepia and a print marked with his corrections.

4. The general contractor receives the documents and notes their receipt date and approval status. The general contractor's operator must note in his own records not only the approval status of each document but also the architect/engineer's corrections in order to coordinate the work. The general contractor then returns both the sepia and the corrected print to the subcontractor or supplier. The documents for the subcontractor will be accompanied by a transmittal form noting its status at that point in time.

5. If the documents are marked "approved," the process is finished, except for the distribution of the documents to the parties who need them. If the document is stamped "not approved" or "approved as noted," then the entire submittal/approval process must take place again: the sepia is corrected and submitted again through the general contractor for approval. (See the exception to this process for the "approved as noted" stamp discussed previously; it follows the same procedure as is used in the original submission.)

6. When all the transparencies submitted are finally marked "approved," the sepia can be printed by one of the parties in its approved form. The transparencies are kept by the submitting subcontractor or supplier, and the required number of prints bearing the "approved" status stamp are printed for distribution.

7. The prints made from the transparency with the "approved" stamp on them are sent to the general contractor, and he distributes them with a transmittal form to all the parties who need them. These copies are then retained and filed for reference.

The Transmittal

A set-form transmittal that is convenient and easy to use is usually developed by the individual construction firm. The ideal transmittal form records a great deal of information on its face. The form must have adequate spaces for filling in items of a special nature and should be adaptable to any project a construction company might undertake. The properly used transmittal form explains and records the flow of exchange and the status of each submittal.

Architects/engineers and contractors need data that are unique to their individual responsibilities and duties and they must make sure that their transmittal form records this information.

The construction operator should realize that the transmittals will form a time record of the status of the various documents at any particular point in the process, as well as a final record of approval status. The operator must realize that, aside from facilitating the submittal/approval process, the form is also useful in any situation that involves delays, extension of time, and particularly adverse situations where time is a critical issue.

A form that might be used by a general contractor is shown in Figure 17. The subcontractor's and architect/engineer's form can be somewhat different, because they are used in only one direction in the flow of information exchange in the submittal/approval process. The general contractor must develop a form with which he can transmit documents either to the subcontractor/supplier (submitter) or the architect/engineer (approver), whatever the direction of the flow.

The construction operator uses these forms in the way that is most convenient for him. He may, for example, develop an individual filing system that establishes the sequential record of time or the status of submittals and approvals, or he may prefer to transpose time and status to another form designed specifically for that purpose and put the actual transmittal forms into his general correspondence or job files, under whatever system he uses. The operator should be able to tell at a glance the approved status of a document at any time without wasting effort in finding it (see Figure 18).

TRANSMITTAL

```
┌─────────────────────────┐
│        ABC  INC.        │
│   General Contractors   │
│       108 Main St.      │
│  Columbus, Ohio 54112   │
└─────────────────────────┘
```

To: Original Submittal

No.	Submittal	Date	Request	Status this Trans.

Remarks:

By: _____

Figure 17. A sample of a set-form transmittal.

The transmittal form should be adaptable to use with shop drawings, brochures, and samples alike.

Status

Once the document enters the submittal/approval process, it has a status. The status of a document or sample is important and sometimes critical to the conduct and progress of the construction project. As stated before, the "approved" status of the shop drawing is the goal of the whole submittal/approval process; only at

Submittal Name	Date to G.C.	Date from G.C. to Arch.	Date Arch. to G.C.	Date G.C. to Subcontractor/ Supplier	Date Resubmittal	Approval Status	Final Approval

Figure 18.   An approval/submittal record form.

this point can the document become complete, reliable, and usable to everyone on the job. Not all submittals are approved upon first submission.

It is the "not approved" status of a document that requires the time and work of the operator in the submittal/approval process. When the submittal has the "not approved" status, all efforts must be made to get it approved as soon as possible. For example, the operator must see that a disapproved document which is part of a set is pushed through the process in order to make the whole set usable. Sometimes specific documents should be pushed through the process at the expense of momentarily sacrificing the processing of other less important documents for the same project. Such a step calls for analysis and discretion and good judgment on the part of the operator. Just explaining to the architect/engineer the importance of substituting an "approved as noted" designation for a "not approved" one on a submission might allow an entire fabrication operation to proceed without delay. Manipulations of this sort take extra time and effort because they demand a more thorough study of the relative importance of the respective documents to the production subprocess. Nevertheless, the demands of the project may force the operator to take such steps to obtain the necessary information.

To allow the project to progress through its proper sequence, the construction operator must not only be concerned with a specific document's approval status but he must also judge the effects of its status on the entire construction of the project. Therefore he must always bear in mind all the potential operational problems over which he has no control. For example, he must understand that the architect/engineer assigns checking and approval tasks, within his respective office, to people whose speed and thoroughness vary greatly. The operator must be aware of all the production and scheduling processes with which the subcontractors/suppliers are faced within their own peculiar specialties. He knows, for example, that the production or fabrication based on the shop drawings poses different problems for the structural steel, resteel, hollow metal, and many other types of subcontractors. The status of their shop drawings can greatly affect their capacity to produce and deliver products in the desired time and sequence.

The submittal/approval of shop drawings is one of the most important operations in the whole construction process, and not just in the production subprocess, because their status is critical to the overall progress of all the parties involved in the construction process, architects/engineers, general contractors, and subcontractors alike. It is a process where pragmatic thinking and initiative result in solid benefits to the production subprocess of the construction process.

Checking the Shop Drawings

Ideally, the operator must check the shop drawings for the following:

1. Contents.
2. Accuracy and adequacy.
3. Compliance with contract requirements.
4. Coordination of the details of work they contain with other work in the whole project.

In taking a realistic view of this construction operation, the reader must realize that this list represents the ideal for performing the steps in processing the shop drawing. To check all of these points thoroughly is a rather complex task, and it is an exceptional accomplishment if actually performed for every submittal on a project. In fact, the fourth step, coordination, is often simply impossible to perform completely, but since the contract demands, by implication, certain results from the shop drawing submittal/approval process, the operator should do his best to do what is practical and to do it correctly. In most cases, he realizes while checking the shop drawings that he only has time to accomplish only part of the ideal; he must determine the amount of time he can spend checking each submittal according to its relative importance.

Example. The operator cannot check resteel shop drawings in the same way he does a door schedule. Why? Because the resteel is needed immediately on the job for placement and the doors are not, and the door schedule needs to be more thoroughly checked than the resteel drawings. But the operator should realize that while the immediate need for the doors may not be that evident at the time, they (or at least their frames) are normally not that far behind the resteel in being needed on the job. In fact, on a wall-bearing masonry building they will be needed early in the construction. In addition, other more complicated submittal/approval requirements are involved in the production of doors (hardware and frames), which may make their early submittal for approval a priority.

Checking operations, by nature, involve a great deal of the operator's time. Therefore the operator, when preparing the schedule, must be able to gauge the time for checking submittals according to their importance to the overall production of the building. Checking resteel shop drawings cannot and should not take the same amount of checking time as that required for the furnishing of the doors on the job. This is true even though the need for certain items may precede the need for other items on the job. Apparent need is not always real need, and the operator must learn to carefully distinguish between the two.

The operator must carefully judge the amount of time he can devote to checking each of the shop drawings furnished under the provisions of work that constitute the job. These time problems are only half of the dilemma caused by the checking operations; the other half results from the operator having to convince the other parties involved in the submittal/approval process how important it is to submit critical items in the proper sequence and on time. This is sometimes a very difficult problem, and its resolution depends on the operator's ability to convince the subcontractors, the suppliers, and the architect/engineer of the urgency involved. This is not to infer that these other parties are ignorant of the process; it is just that most of the time they have to consider allotments of time and personnel within their own organizations. The capacity and organizational structure of an architect/engineer's firm, for example, may only allow for a certain sequence of approval operations, because of other work loads within their established organizational setup. At times it is extremely difficult if not impossible for them to vary their operations to meet the construction operator's demands, no matter how critical it may be.

The total operation of checking shop drawings must be highly organized in the construction operator's mind. Restrictions on time for the operation are essential and they must be based on realistic judgments of the importance of checking the drawings to the overall production of the building.

It is improbable that one will ever find a construction operator who really enjoys this operation; "checking shop drawings" is a phrase understood by many construction operators to mean repetitive, uninteresting, and boring work. Nevertheless, it is required by the submittal/approval operational process, and unpleasant as it may actually be, it must be done.

Coordination of Shop Drawings Within the Process

Coordinating shop drawings within the submittal/approval process is not the same as coordinating them for use. We have already stated that the coordination of shop drawings is an essential element of the checking operation (see step 4 in the list above).

The operator must also realize that the coordination of shop drawings continues beyond the submittal/approval process. They have to be used, and they must therefore be coordinated throughout the course of the project. This type of coordination is a distinct, separate operational task that must be learned by competent construction operators. Since it is a separate operation, it merits a fuller and more complete discussion than the scope of this book allows. Here the use of the term *coordination of the shop drawings* is limited to coordination within the submittal/approval process only.

Example.   Let us return to the metal door, metal frame, and hardware divisions of the specification and see what coordination operations are required.

Normally, hardware templates are required by the manufacturer of the hollow metal frames before he can even submit his shop drawings, much less start the production process of the doors. Templates cannot be supplied until a submitted hardware schedule has been approved. Some of the problems involved with sequencing the production in these particular trade divisions have been discussed earlier. Two additional problems having to do with the production of the whole work and that affect the submittals of these particular divisions must be examined. One problem occurs when the hardware is under an allowance and/or the owner has no intention of selecting the hardware until the job is in a more advanced state (even though the entire hardware schedule has been specified in the bidding and contract documents). This delay in selection will obviously delay the production subprocess. Another additional problem that can compound the above problem occurs when the masonry work into which the doors and frames are to be built is a subcontracted division of the work. This would lessen the expeditor's degree of control over the production operations, but he is still left with the task of supplying these materials to the job on time. This requires the expeditor to exert even greater efforts in his submittal/approval operation.

The first thing the operator must do in such a situation is to let everyone involved know exactly the consequences of any delay. In particular, he must insist on the owner selecting the hardware as early as possible, emphasizing the extreme difficulty and financial consequences of having to build the masonry walls without the door frames. The layman owner must be made to understand that the building of these walls depends on the approval of the hardware schedule, as unrelated an operation as it may seem to him. It must be made clear that it is his contract that demands the submittal/approval process, which in turn requires the shop drawings and their closest coordination within the shop drawing process in order to get these manufactured materials on the job in sequence and on time. Therefore the owner must act as fast as possible not only to expedite the job but to allow the general contractor to meet his obligations to his subcontractors as well.

There are many similar instances in the information coordination of this process that are more or less complex than the example given. Each instance requires that the operator know the problems a particular situation is liable to create, as well as how to mitigate any delaying effects and handle unsolvable problems of this nature. This operation of coordinating shop drawings at submittal time always requires solid and thorough analysis based on realistic construction thinking processes; also, the operator must be able to explain and convince others involved in the process that buildings must be erected in a logical (though not always apparent) progression.

11.7 THE TECHNICAL BROCHURE

Technical brochures are required submittals of the mechanical and electrical trade divisions of work in the contract; they are also used in many specialty areas of the architectural trade divisions. A brochure is a publication of technical data and descriptive information. The architect/engineer uses it for approval purposes to check the compliance of electrical and mechanical materials and systems and their installations with the contract documents. Mechanical and electrical contract drawings are usually schematic drawings: for example, a valve is not drawn in the shape of a valve but, instead, is designated by a symbol. The symbol is related to a schedule either contained in the drawings or in the specifications, wherein a trade name is specified for the item.

The submittal of the technical brochure furnishes the technical data for whatever item the subcontractor proposes to use on the job. The brochure may present information on the specified trade name product called for in the specification, or it may just as well present information on another acceptable "or equal" product. Even when the specific trade name item is to be used, the brochure is still required for record purposes and proof of compliance with the contract.

Substitutions for specified items are usually allowed in most bidding and contract documents under the "or equal" clause. The brochure submittal for a proposed substitution allows the architect/engineer to determine its "equality" from the technical data it furnishes.

The construction operator should understand that there can be variations in the extent of the demands for this kind of information. The approving authority on each individual project generally sets the requirements for the type and number of brochures that must be submitted by the contractor. Since these documents subsequently create obligations of contract the construction operator must be aware of these requirements when checking, referencing, and coordinating these types of submittals.

Field Measurements

This is now an appropriate point in the text to direct the reader's attention to a notation frequently encountered on the contract drawings and less frequently in the specifications. It says "for field measurement," "field measure," or something similar. When an item of work is so noted, the space into which it fits is to be built first and the item or items of work that are to be incorporated into the space are not to be built or manufactured until after the dimensions of the space are actually measured. This is commonly the case with such items of work as custom sash or toilet partitions. Custom glazing, for example, is always field measured.

This kind of notation changes both the shop drawings and the brochure procedure within the submittal/approval process. All of the ordinary procedures of the process take place, but they are applied to an already determined condition rather than being presented as a proposed means, method, or item type, as is the case with the ordinary shop drawing or brochure. Thus the operator can, on the one hand, be less concerned with coordination problems in his shop drawing and brochure checking operations when work is required to be field measured before fabrication and installation; but, on the other hand, he must make all parties responsible for building the dimensioned space aware that the field measurements must be maintained exactly. This is particularly important when, for example, only one of a series of typical window openings is built and measured. In order to safely allow fabrication based on only the one measurement, it is critical that the operator require all succeeding openings to be of exactly the same dimensions.

It is with the brochure item that field measurements frequently become a critical factor in clearly delegating coordination and building responsibilities to the proper party.

The Technical Brochure Process

A sufficient number of copies of the technical brochure are submitted to those approving them for use. The technical brochure process on a proposed item to be used on the job normally requires that all of the parties involved in the architect/ engineer's office have a copy of the brochure for their reference and checking. An architect/engineer can demand a brochure on each and every piece of material used on the job or he can demand very few; this will depend on the specific requirements of the individual architect/engineer on each job. After the architect/ engineer's approval the subcontractor must furnish sufficient numbers of copies of the brochures for distribution to the general contractor and his subcontractors and suppliers and to his own suppliers and subcontractors. It is common practice for the subcontractor to submit brochures to the general contractor in 8 to 12 copies, particularly in the cases of the mechanical and electrical products.

This process follows almost the same procedure as the shop drawing process as far as the flow and exchange of the information is concerned: both processes require the same recording, filing, and distribution procedures. However, there is one significant difference (based on the way each submittal is used) that the operator must bear in mind: usually, shop drawings are made to both extend and detail the work described in the contract drawings and specifications; but the brochure usually forces the contract working drawings to conform to the details of the submittal. The contract drawings, for example, will often simply designate a specialty item whose dimensions are not known until it has been selected and approved through the formal submittal/approval process. The selected item determines the actual space and measurement requirements in the contract documents, thus requiring the contract documents to conform to the submittal documents. This means that the approved submittal becomes the real contract document, which changes significantly the way in which the operator must view the documents, particularly in its reference and record aspects.

11.8 THE SAMPLE AND ITS PROCESS

The sample subprocess is somewhat different operationally than the other two subprocesses of the submittal/approval process. The sample is either an actual piece of material, a representation of a material, or a built-up combination of materials that are required to be installed in the building. Samples can take on many forms. The various samples required on an ordinary construction job today are furnished in traditional forms that have been developed to facilitate approval. Let us examine the accepted forms of required samples on four different materials.

1. *Paint.* The colors of paint to be used are submitted and approved from the manufacturer's color chip samples, which show ranges of color and tones.

2. *Masonry.* The brick and mortar manufacturers are specified. This type of brick sample is submitted as a packaged series of units of brick showing ranges of color, shades, and textures. Then, using a material, a built-up brick production run sample is made on the job, showing the brick in bond and coursing in the proper distributions of shade, color, and texture, with mortar joints in a finished

condition. The approval is made on the built-up sample, and this records approval of all the materials and the expected workmanship on the finished product.

3. *Acoustical Tile.*   The tile manufacturer and its type are specified. The contractor submits for approval pieces of the tile and trim proposed in triplicate.

4. *Sand.*   The type, grade, and consistency of the sand to be used in various parts of the job are specified. Sands of various types and grades, in sufficient quantities for examination and possibly testing, are submitted for approval. Generally, this type of sampling is done at the pit source of the sand.

There are all kinds of different samples required on a construction project. Certain samples may be required on one project but not on another, depending on the physical components of the project or whether the architect/engineer requires them. The operator can depend on there being several traditionally required samples on almost all building projects. Paint color chips, masonry materials, and carpeting fall under this category. Some samples are becoming required as the use of new items and systems increase. Colored concretes, mortars, and precast concrete facings fall under this category.

The sample is an extremely valuable tool for the construction operator to have at hand. The brick/block job sample, for example, sets the model for masonry work requirements on the entire structure. It eliminates, from the beginning of the work, most questions relative to the aesthetic results of the total construction, both during and after its completion; it also allows special precautions to be taken in the quality production of the work. Many times it brings to the attention of the job supervisor peculiar or unexpected difficulties with materials, particularly in coloration matches in work that uses component materials and in the maintenance and protection requirements until the work is turned over.

The sample is a reference and a record. Its submittal/approval process is generally similar in nature to the other two processes in its flow of exchange of information and in some cases (with paint chips, for example) it follows the exact patterns as for shop drawings and brochures. In other cases (for example, a masonry job sample) it is located in one place, observed, and approved or disapproved there, with formal correspondence on its status following.

The traditionally accepted method of processing single sample items for approval is to submit them in triplicate. The submitting party identifies and marks each sample, the marking being attached directly to the individual sample piece. The marking must give the following information:

1. The identification of both the submitting and approving parties.
2. The specification reference to the sample.
3. The name of the manufacturer and his type or catalog number.
4. The relevant dates of transmittals.
5. A place for the architect/engineer's stamp on the marking, or, if this is not feasible, a place for stamp on the piece itself.
6. The name of the project for which it is submitted.

The traditionally accepted method for processing a built-in-place sample is the following:

1. Gathering the specified and selected materials.
2. Assembling them.
3. Calling for the personal inspection by the approving party.
4. Following up with formal correspondence on its status of approval.

The sample, like the brochure, is an extremely valuable operational tool in areas where there is a need or desire to substitute other materials for contractually specified materials. When substitutions are allowed, the sample allows for a visual inspection of the physical variations between the specified product and the proposed substitute.

Samples submitted for testing purposes can be so varied in quantity and in nature that no specific formal process has been traditionally established that is applicable to all of them. Many times a certain test procedure must be specified for the special test sample which in turn establishes special requirements for its approval. Structural concrete is an example of such a sample; the construction operator must take special care that its formal approval is acknowledged in some form that makes it possible to use it for record and reference purposes.

11.9 OTHER SUBMITTALS

Besides shop drawings, brochures, and samples there are other important types of submittals where the submittal/approval operation is required directly in or is implied by the contract. Some of these are listed here:

1. The list of proposed subcontractors and suppliers must be furnished by the general contractor to the architect/engineer for approval.
2. The schedule of values, which is a detailed listing of the specific items of work and their costs, must also be submitted and approved. This item is used for billing and payment purposes and is discussed more thoroughly in Chapter 13.
3. The progress schedule is required by contract, but, at the present time, only for informational purposes and not for approval. Traditionally, it has been a document that required the approval of the architect/engineer, and this may still be the case in some contracts.
4. The details of certain operational systems or methods of construction are required for submittal and approval. Two specific examples would be the proposed design and system of sheeting and shoring, and dewatering and its layout when earth movement or water problems are present on the project. Usually these types of submittals are specified directly in the contract documents of the job. They serve the same purpose as the shop drawings, brochures, and samples, in that they are intended to coordinate work, but they also serve to solve some specific technical and engineering problems on the job and have the additional purpose of protecting people and property.

11.10 SUBSTITUTIONS AND THE SUBMITTAL/APPROVAL PROCESS

One of the most difficult contractual and operational problems the construction operator encounters in the ordinary construction project concerns the substitution of an item that is directly specified in the contract or bidding documents. The construction contract requires that the building be built exactly as specified. Theoretically, a contract without some provision to the contrary does not allow for any substitutions whenever some particular item, or items is specified.

Public policy in the public sector and good business sense in the private sector have traditionally required or allowed substitutions. The common "or equal" clauses used in the bidding and contract documents allow substitutions contractually. Not only are they allowed by the bid documents, but it is probable that several substitutions will be made between the initial bidding process and the final completion of the building. The "or equal" concept of substitutions calls for lengthy discussions on the bidding, contract, and contracting subprocesses of the construction process. We mention this concept here to remind the reader that the substitution concept also affects the submittal/approval process. This process gives efficient operational control over many operational and legal problems that are sometimes caused by the concept. It should be clear that the shop drawings, brochures, and sample processes provide additional protection to both the contracting parties in the area of substitutions, because of their detailed, organized built-in recording of all the transactions. For the construction operator and the architect/engineer the submittal/approval process easily allows for acceptable substitutions to be processed in a fast and thorough way.

11.11 DEVELOPING METHODS

If one were to realistically compare the amount of time the various operational divisions spend on operating with the submittal/approval process to the time spent on other construction operations, the contrasts are significant. Little time is spent on the process in estimating operations; job supervision and engineering operations are more involved, because considerable time must be spent in the coordination of shop drawings after approval; and the same is true to a lesser extent with samples.

Most of the time it is the expeditor who spends the greatest amount of time with the approval/submittal process, both before and after approval of the submittal. Sometimes during the course of a project this process will take up the expeditor's entire time; therefore this operator must develop means and methods within the established traditional procedures required in order to utilize time most efficiently.

Each operator must develop his own individual methods that work for him and permit him to do this work within a specific time frame. For example, the greatest bulk and concentration of the basic operations in this process are performed at the beginning of the project. The operator must seriously consider how the delivery of this or that material will advance job progress, but he must also consider how essential some particular bit of information might be to furnishing the total construction as well as how important it may seem at the moment. From these considerations flow others: What are the production times of various items of work? At what locations away from the job site are these items produced? Which production schedules of other jobs within the supplier's or subcontractor's operations might affect the production of an item for this or that job? From these and similar questions arise still more questions. The submittal/approval process demands a keen development of this question–answer kind of thinking, because of necessity for the process in efficiently constructing a building and because of its extraordinary demands on the operator's time.

A great variety of procedures adopted by construction professionals for use in the submittal–approval process have been accepted by the industry, but there still remains great potential for developing more efficient means and methods. In order to develop these individual means and methods the operator must direct his thinking towards the elemental features of the process.

We know, for example, that the submittal/approval process demands quickness of action and response, as well as thoroughness; we also know that the process develops, transfers, and records information; but above all we know that it makes severe demands on the operator's time, and therefore the careful organization of time through developing and using efficient procedures is essential.

The submittal/approval process often seems to be cumbersome and unreasonably time consuming; however, the construction operator should recognize that a building of any complexity cannot reasonably be constructed on time or at cost without it.

12

The Change Process

12.1 THE NEED FOR STUDY

The word *change* as used in this chapter means a change or the modification of the construction contract; the term *change in the work* means a change to the extent of work the contractor and owner mutually agreed upon under the original contract; and the formal contract document used for the contract change is called the *change order*. The operations required by the contract to process the change order are extensive and sometimes complex. The change process can take up a considerable amount of an operator's time. Depending on the type of project, the degree of completeness of the bidding and contract documents, and sometimes even the peculiarities of an individual owner, the change process can require very extensive construction operations.

Because of the need for construction operators to expend a great deal of time and because the process calls for concentrated efforts, operators must thoroughly understand what the change is, how it works, and how to operate with it. The expediting operator is the central controlling operator in this process.

In this chapter we discuss two general types of changes: the ordinary change and the constructive change. The processes required by both types are similar in form but significantly different in determining the way that the operator must initially approach them.

The ordinary change process involves the owner's invitation for an offer from the contractor to price suggested changes in the work; the owner, through his agent, the architect/engineer (who formalizes the procedures that finally result in the suggested changes becoming a modification to the original contract) initiates the change. The contractor quotes the proposed changes in the work and the owner accepts or rejects the proposal. When the owner accepts the contractor's proposal, he is bound to issue a formal change order. The original contract is consequently changed to the extent of the obligations and conditions required by the formal change order.

The constructive change is a change to the work that has not been expressly ordered by the owner, but it is a change nevertheless. It results from a contested interpretation of the scope of the original contract. Generally, the interpretation, implication, or deductions of the existence of a constructive change are based on some action taken by one of the parties. Once the owner or his agents, for example, do something (or will not do something) that implies a change and the contractor acts, there is every possibility that a constructive change has occurred. If there has been a constructive change, it is as real a change as an ordinary one and should therefore result in a formal change order (or change to the contract), using the same processes as an ordinary change.

The great difference in determining the approach that the operator takes in dealing with the two types of changes must be clearly understood: in the case of the ordinary change the owner deliberately determines the change in work and the contractor is given full notice and an opportunity to charge for the changed work; in the case of the constructive change it is always necessary for the contractor to "find" the change, define it, quantify it, and claim compensation for it.

In the modern construction process there has been a great increase in claims based on constructive changes by contractors, and this is mostly due to the inability of those who draw the contract and bidding documents to adequately describe the total work. This is not necessarily caused by incompetence, for often the time allowed by economic restrictions and the interest the owner has in using his building, is insufficient. If the owner must have a building in the shortest possible time, the abilities of the architect/engineer to fully design it, detail it, and adequately form the contract documents will be restricted. The results are that the contractor is faced with estimating, bidding on, and building from incomplete documents.

In order to protect themselves, contractors are forced to look for contractual and noncontractual ways in which to be compensated for necessary but uncalled for modifications to the work occurring during construction of the building. The fairly new constructive change doctrine that the legal processes presently follow offers a way of adequately and fairly compensating for extensions beyond the original contract that cannot be handled by the expressed provisions of the contract.

12.2 CHANGES AND TIME

It is necessary that the reader view time within two distinct contexts in the change process. First of all, time is a factor in processing the change itself and its restrictions are much the same in the change process as they are in other construction operations: time is short, time is critical to sequential progress, and so on. The operator must always think seriously, for example, about the available time he has to do the operations of processing the change.

But time in the change process must also be thought of as part of the change in work itself. Within this context it is an item of change that must be determined and quantified. A loss of time that a change in the work creates is something that must be compensated for as well as any physical change in the work. If a time change costs money, and its cause is the change in the work, then the contractor is entitled to charge the owner for its cost. The reader must not confuse the different uses of the word *time* in this chapter.

It is never easy for the operator to determine the extent of lost time caused by a change to the contract. If, for example, there is a delay of four days due to the late delivery of a material item involved in a change, then the four day period is obviously determinable. But an operator must always consider any delay this late delivery might have caused beyond the four days. Since it is as necessary to assign a cost to the time loss as it is for any other items of compensable change, the operator must be able to determine as closely as possible the total increase in time the change in work has caused. Since the total increase must always be based on indefinite suppositions, he must, after a reasonable analysis of the probabilities, come up with a total time increase on the basis of good sense and reason. To be perfectly honest, the total time increase normally allowed is usually simply the amount of time and its cost which the operator can convince the owner to grant and pay for.

The operator, in order to arrive at a convincing figure for delays, will be greatly aided in his task if he specifically directs his operational procedures to this factor of time in the change process. He must consider such things as keeping and maintaining excellent records of the loss of time at the time of the occurrence of the change in work not only in his own operations but also in the operations of his subcontractors. He must consider and be able to revise and adjust the schedule where the change has affected it. In most cases the operator must affirmatively mitigate the consequences of the time delay.

Let us examine more carefully the need to thoroughly analyze the total lost time.

Example. Shortly after the start of excavation for footings, the achitect/engineer decides to change the resteel in several footings. In this situation lost time immediately becomes a possible extra cost. The new resteel must be detailed, submitted, approved, fabricated, and delivered. It is simple to quantify time extensions for this change to this point; it is simply the time extension from the first cessation of work on the footings to the delivery of the new bars on the job.

This is not necessarily the only time that needs to be quantified and costed. If, for example, the excavator has removed his equipment from that piece of work and does not or is unable to return it to the work until sometime after the new delivery, a new question about the amount of compensable lost time is raised. If other succeeding operations are delayed by the inability to restart the delayed operation in reasonable order or sequence, there arise still further questions of time and possibly other affected parties enter into the picture. The construction operator should train himself to view lost time situations at hand from all possible angles when assessing possible compensable lost time in order to arrive at a reasonable compensation for its cost. If the operator remembers the simple rule that additional costs expended for changed work should be fairly paid for, he will quickly learn the means and methods of determining, recording, and keeping time in the change process.

12.3   THE ORDINARY CHANGE PROCESS

The total ordinary change process involves an extraordinary subprocess—the bulletin process. The operator must learn about how this process works, because its operations can require a great deal of work and time, depending on the number of bulletins issued on a job and the extent of the work involved in them.

The Bulletin

The bulletin is a formal listing of suggested items of work to be changed on a project; it is an invitation to the contractor to offer a price for the proposed changes to the owner, as well as a postcontract bid document issued for the purpose of pricing the suggested changes to the work for the owner's acceptance. The items of work are usually printed in regular (original contract) specification page form: they are listed and usually each change of work item is numbered separately. The bulletin's printed materials are accompanied (when required for informational purposes) by the prints of the appropriate original contract drawings (now designated bulletin drawings) that show the work to be changed. The bulletin items are directed to the attention of the contractor by marking the bulletin drawings. When and if a bulletin is approved and a change order issued, the bulletin drawing becomes the new contract document and the corresponding preceding issued drawings become void. Should succeeding bulletins change the new contract limitation drawings, then the latest issued bulletin drawing becomes the new contract drawing after the change order issued. Sometimes the bulletin items are described sufficiently in the printed matter so that the issue of new drawings is not required.

The Peculiar Nature of the Bulletin

The bulletin is the traditional instrument used in initiating the ordinary change process. It is used in the constructive change process as well, but there its use is approached somewhat differently.

As stated before, the construction operator must know exactly what the bulletin is and how it operates.

The bulletin is a very peculiar instrument when considered from both operational and legal standpoints; in many ways it is a burdensome and uncertain operational tool, both operationally and contractually. One peculiar feature of the bulletin is that the contractor is contractually bound in the general contract to quote bulletin work. This obligation is carried over and made a condition of the subcontract by the general contractor. But the bulletin is clearly an invitation to offer a price on the suggested items to work to be changed and thus it is, in effect, an invitation that cannot be refused. Because this invitation cannot be refused, the bulletin can be an extremely difficult and sometimes even unprofitable burden on the contractor, both operationally and financially. But it can also be a very profitable invitation.

For example, a bulletin having a great number of small-cost items of work that is issued at the end of a job is likely to be a losing proposition for a contractor, particularly if he is bound by contract to contractually fixed percentages of markup for changed work. On the other hand, a bulletin that contains a small number of large-cost items issued early in the job can be very profitable for the contractor.

Another feature of the bulletin that is most peculiar and creates many complex operational and legal problems is that on its face it usually says something similar to, "This bulletin is issued for quotation purposes only," and the statement then goes on to add quite definitely that "it is not intended to be nor is it to be construed as a change to the contract." This means that until the owner accepts all or part of the bulletin quotation, there will be no formal change in the original contract obligations and thus contractually the contractor must proceed with the original work as if there were not going to be any change at all. Obviously, just the issuance of a bulletin means that the owner has some changes in mind. With the issuance of the bulletin, and sometimes even before, the contractor knows that the owner is seriously contemplating changes in the work, yet he can do nothing legally and little operationally to implement the bulletin work until the change order is executed. This situation sometimes leads to impossible operational problems.

Example.   If a wall is already built or is being built and a bulletin is issued that excludes an item to install a new or additional receptacle in the wall, then the construction operator immediately has operational problems with the wall construction. Contractually he may not proceed with installing the receptacle until a change order is issued. But the change order process takes times, and if multiple subcontractors are involved, this takes even more time. Meanwhile the operator must face the question of proceeding with the work on the wall.

It is true that the time and costs required to go through the change process to the formal change order can theoretically be changed as part of the bulletin quotation and made part of the change order; but think of just the problems of schedule this situation creates. The realistic approach to this item of work would be to install it without delay; otherwise it might involve the movement of crews into another area or could cause other similar delay problems. If the operator opts to have the item of work installed before issue of the change order, then the installation is clearly not in accordance with the contract. Usually the great number of indeterminate consequences possible and the inability to properly cost the inherent delays involved make it necessary for the construction operator to operate deliberately outside the contract in such circumstances.

This is but one example of a small (yet typical) item of change that, when multiplied many times over on a heavily bulletinized job, can cause enormously complex operational problems. Problems of even greater complexity and uncertainty result when there are large bulletin items that involve the work of many trades.

Why is this paradoxical operational tool still used in the construction process? It would seem sensible to perform all critical change items within some sort of individual extemporaneous system of direct field order. However, architects/ engineers and owners are reluctant to deviate from the traditional bulletin instrument and its processes, which means the construction operator must deal with its contradictory and operationally troublesome process. Operationally, the bulletin process is the single most time-consuming operation in the entire change process; the operator must have great skill to perform its functions properly. It also forms a major operational task for the expediting operator in the contracting and contract subprocesses of the construction process and is a major area of trouble and concern for field supervisory operators in the production subprocess.

The Bulletin Process

The owner has the contractual option of making any reasonable changes to the work during the course of a project's construction. When he has decided to change enough items to make a significant change to the work, the architect/ engineer compiles them into a bulletin and issues a reasonable number of copies to the general contractor. There should be sufficient numbers of copies of the bulletin and its drawings for proper distribution to all the parties involved in building the items of work on the bulletin. Sometimes number of parties needing the bulletin can be very substantial.

With the bulletin in hand, the general contractors' expeditor does the following operations.

1. He notes the bulletin's receipt and records it into a systemized listing of all job bulletins. This listing records the necessary data on receipts and transmissions of the various bulletins received, as well as the status of the bulletin within the bulletin process.

2. He then analyzes the bulletin for its proper distribution for quotation and informational purposes. (We must make clear a significant point that applies to the entire change process in construction contracting. It is obvious that the parties involved in a construction project must be aware of changes made to the project's contract, but often they must also be made aware of proposed changes so that they may adjust their thinking and operations to the consequences of these potentialities, even if they are not directly affected or involved in the change.)

The operator must examine the bulletin documents very carefully for *all* pertinent subcontractor work requiring their quotations. These subcontractor quotations are required by the general contractor to quote its total price to the owner. This operation takes great skill at times. If he does not, for example, distribute the bulletin to a subcontractor who is required on the work and then quotes it without that subcontractor's quotation, he has a problem. When the owner accepts the bulletin either wholly or in part and issues a change order, the changed work is still automatically subject to all of the terms and conditions of the original contract, irrespective as to whether its cost was included in the general contractor's quotation.

When the general contractor's operator examines a bulletin for distribution purposes, he must think in a certain way, considering the nature and extent of operations that the entire bulletin item requires; he must think about time and cost considerations.

Example. If a suggested bulletin item places an additional receptable in a plaster wall, the operator must consider the physical completion status of the wall at the time the quotation is to be made. He must determine those operations that are completed at the time the bulletin is reviewed and quoted and the change is ordered, and those that are not. Depending on these determinations, the operator must think about how to direct the lathing, plastering, electrical, mechanical, painting, ceiling, and flooring subcontractors in making their quotations.

It would seem that if the wall has not yet been painted when the bulletin is received and ready to be quoted, the painting subcontractor's work can be stopped and he should consequently have no painting quotation to offer. But the bulletin does not allow the painter to price such a stoppage, yet he may be seriously affected costwise by time and other operational considerations. It may be significant to the painter that the work in a proposed change is to be performed either later or earlier on the project: it may cause internal problems with his crew scheduling or difficulties on other projects he has on hand.

There may also be many remote consequences of changing the work on the various trade divisions caused by the timing of the changes on a project. Such consequences (not anticipated at the time the bulletin is quoted) often become serious problems for the operator, particularly when their full extent is not known until after the issue of the change order. If one or another subcontractor runs into cost expenditures that were not anticipated at the time of the quotation of the bulletin, who pays these extra costs? Unless the general contractor's expeditor thinks carefully about these status–quotations–time problems at the beginning of the process, his company will pay, unless he can in some way convince the sub-contractor to do them for nothing.

3. The expediting operator distributes copies of the bulletin to the proper parties, including the estimating department in his own organization, for their quotations. This is done through a formal transmittal process that is much the same as the one used in the submittal/approval process discussed in Chapter 11. The time limitations for quotations specified by the general contract must be clearly understood by the subcontractors concerned. It is wise for the operator to explicitly impose corresponding limits on his subcontractors and suppliers in their respective subcontracts.

4. The expediting operator, faced with the time limitations imposed by the general contract on his quotation to the owner, must in turn expedite the sub-contractors' and suppliers' quotations.

5. The operator must set guidelines for the form of quotation he desires from the subcontractor or supplier (item for item, separated quotes, breakdowns, etc.).

6. After gathering all the required subcontractor quotations, the expeditor compiles them into a total and marks them up as allowed by the general con-tract. He also gathers the estimates of the work of his own forces contained in the bulletin and marks it up as allowed by his contract. Usually the contract allows one percentage of markup for the general contractor's work and another for the subcontractor's. The expeditor combines these two elements of the total price and makes his quotation to the owner.

The form of the quotation is governed by the requirements for either a simple lump sum quotation or an item by item quotation, which is usually specified by the architect/engineer and the owner at the beginning of the process. Certain owners and architects/engineers are quite specific as to how they want their bulletins quoted; usually they will require supporting breakdowns of prices and units of work for each item. The expeditor should be aware of this probable de-mand in order to guide his quotation and subquotation procedures and their form. The owner usually needs information in this form to allow him to make a reasonable choice of the items to accept or reject. At times it is to the contractor's advantage when the item by item form is used, because he is able to selectively quote high or low on the separate items as they affect the project's progress or financial picture.

7. The general contractor's operator sends his quotations to the architect/ engineer for their review, after which the architect/engineer makes his recom-mendations to the owner. The architect/engineer's recommendations are usually based on the reasonableness of the money amounts quoted. In some cases, extra communications are necessary between the architect/engineer and the quoting

general contractor to clarify points of misinterpretation of the cost of one item or another. Most architects/engineers will give a very significant consideration to the quotations for anything relating to time extensions and its costs.

8. The architect/engineer delivers the quotation with his comments to the owner. For internal operational budgetary reasons many owners like to review the bulletin quotation very closely themselves.

9. The owner can accept or reject the bulletin in any way he chooses, item by item or in its entirety. This option causes the operator considerable concern at times, because, for example, an item that he has purposely priced to be accepted might be rejected or vice versa. In addition, the owner usually has the option to accept or reject items and leave others in abeyance or subject to future negotiations.

10. When the owner accepts the quotation on certain items or the bulletin as a whole, he notifies the architect/engineer, who in turn notifies the contractor of the extent of the owner's acceptance. Contractually, this notification of acceptance should be done through the issue of the formal change order. However, the formal contract change is sometimes delayed for one reason or another. Usually, any indication that the owner will accept the bulletin quotation leads everyone to proceed full steam ahead in changing the work. The construction operator should be aware such action, without the issue of the formal change order, is done at the general contractor's and the subcontractor's own risks. We repeat again that *only* the formal change order changes or modifies the contract.

11. When the owner's change order is issued to the general contractor, the experienced expeditor will start, with good haste, the change order process in his various subcontracts. He sets up a definite, exact, and formal system within his subcontracting process and procedures for accomplishing this task (see Chapter 8).

When the bulletin quotation is made into a formal change order, the bulletin process ends and the formal change order process begins in the contracting subprocess of the construction process.

The reader should realize that the bulletin process, while it is a traditionally accepted method in the contract construction industry, is at times very cumbersome and frequently complex to fit into the continuing operations on a construction job. Like it or not, the industry has recognized this process as the accepted method of serving everyone's interests in changing and modifying the work. Developing and organizing methods of thinking about the nature of the bulletin and its required operational tasks is the most important prerequisite for the construction operator's successful operation in the change process.

A Major Problem in the Ordinary Change Process

For the convenience of the reader we included some discussions of the change order process within our explanation of the bulletin process. We must extend this explanation here simply because the reader must not confuse the bulletin in any way with a change order: the bulletin is just part of a sequence in changing or modifying the construction contract; only the change order changes the construction contract (this rule extends to the subcontract change process as well). This rule creates a major problem for the general contractor's expediting operator.

The new construction operator must realize that the possession of the formal executed change order is the only thing he can rely on completely to govern his actions in his other concurrent operations. For example, items of bulletin work should never be entered on the billing for payment (schedule of values) until the formal change is obtained.

In conducting operations under the restrictions of the change order process, the operator must give serious consideration to his other operations between the time that he knows about forthcoming change in the work and the time the formal change order is executed. This period can be very long at times and it presents difficult operational problems for the expeditor. He must determine which building operation can proceed safely without the formal change and which cannot, and he must determine this in relation to the legal implications of actually changing the work without changing the contract. To what extent, for example, does the owner take on obligations of contract by implication when a contractor proceeds to change work before the formal change order process has been completed?

All of these and similar problems are difficult, because they have no specific guaranteed pat solutions. The application of the law must always be adjusted to the particular circumstances of the case at hand. There is, however, some solid operational thinking that will minimize possible adverse legal consequences when deviations from the contract are necessary. The operator must go out of his way to keep on top of the bulletin process, always pushing it toward the change order. Subcontractors or suppliers who are late in their quotations must be reminded of their subcontractual obligations to furnish quotes that are complete, on time, and properly detailed. Since the general contractor's bulletin quotations to the owner demand the entire cost of the proposed changes, all quotations must be forthcoming in an organized and timely fashion.

When the operator directs his own forces or orders or allows his subcontractors to proceed with the work, he must make sure that this direction is well documented. The degree to which both existing and changed work is completed is of critical importance when and if the owner subsequently rejects an item in the bulletin. When concurrent or consecutive operations are involved in building a bulletin item, it is necessary to document when and where the participating parties have entered onto the scene. All of these facts should be carefully noted by the expediting operator, with on-the-scene verification and assistance from his field colleagues.

In dealing with the owner or his agent in this situation, the operator must make them aware that he is proceeding with the work. If this results in a negative response, then he has no recourse but to shut down the operation. Most sophisticated owners and almost all architects/engineers are usually well aware of the delays inherent in the change process, which may result in a vague response or none at all. The expediting operator must, however, force a definite response to avoid possible problems when operating outside the strict provisions of the contract.

Needless to say, the owner's subsequent rejection of work that is in the process of being changed calls for some inventive thinking on the part of the operator and he should certainly make the owner aware of the consequences. The owner, of course, has the strict interpretation of the contract to rely on in such a situation.

Failing to make the owner reconsider his rejection, the operator must take sterner measures. He must fully inform the owner of his intent to order a complete shutdown on any of the owner's future suggested changes unless a *field* change order is issued immediately. This suggestion should go a long way toward establishing more reasonable attitudes when future changes are proposed. We do not claim that it is common for an owner or his agents to deliberately cause the contractor trouble and expense by rejecting a suggested change that is proceeding without a change order. But many times this rejection takes place because of a real misunderstanding or different interpretations of the scope or meaning of the suggested change. No one is deliberately at fault in such a situation, but when the work has been done or is proceeding, it must be compensated for.

Acceptance, by implication, of work done during the ordinary change process can be a complex legal area. The operator should realize, however, that what he considers to be grounds for implied contract may be entirely outweighed by some other phase of the law which he has not even thought about. It is wise, therefore, for the construction operator to proceed very carefully with performing bulletin work before issue of the change order, particularly if he is relying heavily on the theory of implied contract as his basis for payment.

12.4   THE CONSTRUCTIVE CHANGE PROCESS

We have said before that the change process is the same for the constructive change as it is for the ordinary change, but only after issue of the bulletin listing the change. When the bulletin is issued, its items of constructive change are processed in the same way as for the change order. Therefore the major difference in operating with these two types of change processes results from the necessity of the general contractor's operator establishing that a constructive change actually occurred. Once this has been determined, the operator's objective must be to force the issue of a bulletin from the owner to cover the cost of the constructive change.

The ordinary change and its processes are initiated by the owner. The constructive change is caused by the owner, but its processes are intiated by the contractor. This means that in the constructive change situation the contractor's construction operators must know how to recognize constructive change (see earlier definitions). The construction operator must be able to recognize from a particular set of circumstances that a constructive change has occurred, which is by no means an easy task. Even more difficult is the task of fitting a constructive change situation into a successful claim that will result in the issue of a bulletin or a change order.

The Claim and The Constructive Change

In most cases a change order for a constructive change results from a claim for extra cost from the contractor against the owner. The claim is the process the constructive change process usually uses, at least in its initial stages. Establishing the existence of the constructive change is the prime step in the claim. If this step is handled correctly by the operator, the claim should ultimately result in a formal change order. The proper and correct handling of the claim process in its documentation and negotiations stages is what makes the claimed constructive change a formal change.

Establishing the Constructive Change

The operative terms in the definition of the constructive change that we are particularly interested in at this point are "that has not been expressly ordered" and "inferred, implied, or deduced on the basis of an action." Of the two expressions, the latter is the most difficult to understand, although the former also creates some complicated problems for the operator.

A prime example of the problems encountered with the "expressly ordered" term is whether the simple actions or inactions of the owner or his representatives constitute an expressed order as defined by the courts under the constructive change doctrine. If, for example, an architect's job inspector fails to act on something critical to a job's progress, to what extent is the owner responsible for a claim based on the inspector's inaction? The answer does not rest contractual on the relationship between the owner and architect. The term "inferred, implied, or deduced on the basis of an action" determines the criteria for establishing a constructive change. The inference, implication, or deduction arises out of the conduct of the parties, by which they do something required or allowed by the contract or fail to do something that is required. It can be deliberate conduct, mistaken conduct, or even unthinking conduct. The important thing is that the conduct has caused the change to happen.

The construction operator must always make a case upon which to lay his claim of change. He must be able to show a logical, causal relationship between the constructive change and the conduct of the parties within a particular situation, no matter how remote from each other they may have been, which can at times be very hard to do, particularly if there are intervening factors. Developing causal relationships between conduct and a basis for claiming a constructive change poses a very interesting but difficult problem in deductive reasoning. The claim for constructive change is a challenge to the operator; it requires the cold hard appraisal of the behaviors of individuals within a sequence of events, coupled with a solid application of the principles of the implied contract areas of the law. Very rarely are such things as the intent or motivation of the parties relevant in the constructive change process. This is one area of construction operation where the operator must view the situation impersonally and objectively to obtain the operational object of cost.

Proofs of Costs

One especially difficult operational task in most constructive change situations is that of proving actual costs, both as to whether additional cost was actually incurred and its extent. Additional costs must be capable of being proved. With the constructive change it is usually very hard to quantify the delays (which are generally the major item of cost) occasioned by the change. That is difficult enough in the ordinary change process. It is extremely difficult, for example, to know the exact cost consequences of delays in the constructive change situation, because the actual occurrence of the change can be obscure and sometimes very remote from its cause.

The owner will rarely pay claimed costs that are not quantified and proven and few courts or arbitrators are going to allow the payment of damages on some contractor's projection of the costs occasioned by a delay. The contractor's operator is going to have to show, in addition to the cause and effect of the change, the extent of the resultant delays and their reasonably exact costs. In order for him to do this properly he must detail and document the delays and their costs as close to the time of their occurrence as possible.

It is a waste of time and effort to establish a valid constructive change claim and to then have no way of collecting the costs caused by the change. The operator's objective in the constructive change process is to see that his company is fairly compensated for additional costs expended because of the change.

The Field Change Order

Sometimes processes and procedures are set up that give contractual authority to proceed with work while waiting for the change process to conclude. The document used for this purpose is called the field change order and it allows the contractor to proceed with the proposed change or part of it on the expressed written authority of the owner. This order is usually processed through the field representatives of the architect/engineer. It facilitates extemporaneous or emergency decisions of the owner or his agents to make changes in the work. It is ordinarily used only when there is an immediate need for modifications in the work and when waiting for the processing of the ordinary change would be clearly unreasonable. The subject matter of the field order is rarely extensive; often the modification is of such minor significance that the only reason for formalizing a change at all is to be nominally in accordance with the contract.

The owner is usually reluctant to utilize the field order change process, because the contractor can then, supposedly, charge exorbitant costs for changes. If this process is available on a project, it should be fully utilized by the construction operator and he should try to convince the owner of its validity and usefulness in the change process.

When changes are requested on the job, the operator should always at least request a field order change. Even if it is refused, the request indicates that the contractor wishes to proceed with the work on a strictly contractual basis.

12.5 THE STOP WORK ORDER

There are occasions when the contractor, because of a peculiar circumstance beyond his control, is forced to give the owner notice that he cannot proceed with the work. The contractor, by the terms of the general contract, cannot stop the work without being in breach of contract (except under some extremely specific contract provisions that allow stoppage). In any case, it is *never* wise for a contractor to stop work on his own initiative.

The contractor's operator must try to convince the owner to stop the work. It is usually difficult to get the owner to formally authorize the contractor to stop work, because the owner, by doing so, is actually changing the work.

The construction operator must develop means and methods for forcing the issue when he cannot reasonably proceed with his work. In the separate contracts system of construction contracting, this is an especially critical operational problem that each or all of the separate contractors may encounter on a project. If, for example, the work of one separate contractor actually stops the work of another, naturally the effected contractor should not be forced to bear the costs that result; he has no way to force the issue except through the owner. If there is no other reasonable alternative to the situation, he must get the owner to give him a stop order on his own work. This is necessary for the contractor to remain in compliance with contract, because he cannot stop on his own initiative without exposing himself to undesirable legal risks.

Different procedural methods are used to handle the stop order, but they are too varied to list here. The architect/engineer usually establishes the method if the stop work order process is used. The construction operator must learn to adapt his operations to whatever procedure is available to him on a particular job.

12.6 PERSPECTIVE

It is an extremely rare occasion when a project is not changed in some way during the course of its construction. Therefore the operator should view changes in the work as such common occurrences in the ordinary course of the construction business that handling their process becomes second nature to him.

There are many things the operator learns through experience that help him put the change and its operational processes in their proper perspective. Admittedly, some changes create complex problems, and sometimes these problems are impossible to solve in a wholly satisfactory way. But the operator can take heart in one respect—he will find that almost all of the other important construction operations have integrated the change into their processes and procedures, and this fact alone eliminates some of the frustrations of the change process. For example, the change always affects the schedule, but the scheduling and its processes are specifically designed to be adaptable to the change process. The change also affects payment operations, the closing out process, and, in fact, almost all of the operations that we discuss in this book.

The operator must learn to act efficiently and quickly in operations involving change. His thinking must be thorough and analytical, but, most importantly, the operator must develop an ability to determine the total effects of changes on the job. This is the single most important area in the change process in which to develop his thinking skills.

The required processes and procedures of changing the work must be learned in an organized, systematic way, but only after the operator thoroughly understands the fundamental nature of the contractual aspects of the change.

13

The Payment Process

Remember that time is money.

Benjamin Franklin

13.1 THE CONTRACT AND PAYMENTS

The product of the construction process determines the contractual conditions of paying for it. It is both the physical size and cost of the building, coupled with the way in which money is used and managed today, that determines the contractual processes of payment.

The payment in the construction process is an exchange of money for completed work. The amount of payment at any point in time indicates the status of completion of a building. Theoretically, the percentage of payment paid is the same as the percentage of work completed. The construction operator knows that this premise does not hold true very often; aside from the possibility of there being miscalculations in payment billings, there are always areas of work completed between pay periods that are not paid for on time, resulting in a lapse between completion and payment.

Customarily, the flow of payments between the parties on a project is as follows:

1. The payment is made by the owner to the general contractor, less the retention.
2. The general contractor pays the subcontractor, less the same percentage of retention.
3. The general contractor pays the material supplier, with or without retention.
4. The general contractor pays the labor he has hired directly.
5. The subcontractors and suppliers pay their labor forces and suppliers and their sub-subcontractors.

The owner does not directly pay every laborer and materialman on the job, because he contracts with the general contractor to handle this task for him. In the contract construction arrangements used today the general (or separate prime) contractor shifts this burden to his subcontractors and suppliers (except for those labor services and materials he employs directly). Within the operational organization of the general contractor it is the expediting operator who performs the bulk of this operation. It is the general contractor's expeditor who is responsible for seeing that each and every person or entity who furnishes labor, materials, or services on the project gets paid their due on time.

The general construction contract, through its payment provisions in its general conditions, is the instrument the industry uses to make sure that payments are made. The contract obligates the owner and contractor (and hence his subcontractors and suppliers) to pay what is due on time, and this obligation is reinforced by the precedence of common and statutory law. There are strict laws that are stringently enforced to guarantee the payment for labor and materials expended in a building.

Since both business and legal consideration put such an emphasis on guaranteeing payment, all construction operators in a construction organization must realize the importance of the payment and its processes. We reemphasize that payments between all parties in the construction process must be on time and for the amount due.

The construction contract lays the groundwork from which the various parties operating within the subprocesses and phases of the construction process form their own individualized budgeting, payment, billing, receipt, and disbursement processes. From the general contractor's standpoint the processes that must have specifically set-up procedures, means, and methods are those involving the receipt and disbursement of money paid to it by the owner. The development of all kinds and types of specialized operations is required for the payment process.

In the area of payments the typical construction contract as it is formed and used today is not quite satisfactory in implementing its intent. The contract is fundamentally weak in its affidavit and proof of payment provisions. In practice, these conditions of contract are not practical and in some cases they are blatantly ignored by construction operators. A strict interpretation of the contract's payment provisions obligates the prepayment of subcontractors by the general contractor before payments from the owner. In addition, the contract usually demands a statement from the general contractor that he has paid a proportional part of his previous billing to his subcontractors and suppliers, irrespective of the fact that the payment provisions of the subcontracts are different from those in the general contract. As a matter of practice, however, these provisions simply do not fit the way in which the great majority of contractors conduct their business, even though they are supposed to be operating under the specific terms of the construction contract. This presents a troublesome "fiction" in the area of the payment processes in the construction process, because these strict contract provisions for payments are deliberately avoided by contractors and are not enforced by the great majority of owners.

13.2 TYPES OF PAYMENTS

There are two types of payments required by the ordinary construction general contracts and subcontracts:

1. *Periodic payments.* These payments are made from an agreed upon "schedule of payment," and their amounts should be in direct proportion to the amount of work completed up to a point in time. Most often the designated time period is a month, and thus the term *monthly payment* is synonymous with *periodic payment.*

2. *Final payment.* This is the last payment of money, and with the other payments it should add up to the final contract price (including modifications) agreed upon by the parties.

The periodic and final payments made by the owner flow downward through the general contractor to the subcontractors and suppliers.

Retention

All payments except for the final one should withhold from the payee that construction contract peculiarity, the retention. The retention is an amount of money withheld from all periodic payments (except the final payment) amounting to some percentage of the total periodic billing. A billing of a contractor for $10,000, for example, would result in the actual payment from the owner of $9,000 if there were a 10% retention provision in the contract. The percentage of retention is usually 10% of the periodic billing for work completed. Sometimes the contract may have a provision allowing a reduction of the percentage of monies withheld in retention as the work progresses satisfactorily toward completion.

Retention is a peculiar feature of the contracting subprocesses of the construction process. Its purpose is supposedly to protect the owner from the nonperformance or improper performance of the contractor, and it creates trouble and problems for the contractor from both a business and a financial standpoint. It would seem to be an unnecessary burden to the contractor under the ordinary interpretation of contract law, even more so when there are contract bonds required from the contractor that give the owner the protection similar to the retention; yet it has been a traditional contract requirement and operators must adjust to its being in the contract and operate accordingly. A full understanding of the retention and its effects should show the operator the importance of always expediting completions on the job and of the final payment and job close-out processes.

The general contracting expediting operator should keep in mind the following example when dealing with retention in the payment process.

Example.   If there is a markup of 5% on a $1,000,000 job for overhead and profit, then anticipated overhead and profit is $50,000. Remember that this $50,000 includes a part of the money the contractor must have to pay his overhead; what remains is profit. If retention has been mentioned at 10% throughout the progress of the job, this means that at final payment time $100,000 has been earned by the contractor that has not been paid to him. It should be evident, from an examination of this money amount withheld, that the operator should find the most rapid way possible to collect the final payment. This should be the main goal and purpose of his operational efforts at final payment time, because the retention amounts to twice the anticipated entire overhead and profit for the project.

The retention is withheld by the general contractor in the same proportions from its subcontractors and most of its suppliers in their periodic payments. Sometimes general contractors find it beneficial from a business standpoint to pay certain suppliers in full on their periodic billings, but even when they are highly selective in this practice, it usually creates financial problems for themselves.

The whole concept of the retention and its effects within the construction process should be thoroughly studied and understood by all construction operators, mainly because retention directly and adversely affects the conduct of the construction business.

13.3 THE PERIODIC PAYMENT PROCESS

The Schedule of Values

Periodic payments made by the owner to the general contractor are based on a schedule of values, which is generally composed by the contractor's expediting operator at the beginning of the project. The schedule of values is a formal document that is drawn and submitted (and resubmitted if necessary) by the general contractor's expeditor to the architect/engineer and owner until there is mutual agreement on its contents. Once an agreement is reached, the schedule of values is established as the basis for all payments until the end of the project. In most cases the schedule of values used by the contracting parties is a traditional set-form that has been found to operate efficiently; in other cases it is custom formed for the particular project, usually in accordance with the wishes and direction of the owner.

The schedule of values is simply a listing of the items of work, with various amounts of the total contract price assigned to each of them. The form of the schedule is adapted to the cumulative nature of the periodic payment process. Basically, then, the form is as shown in Figure 19.

The form has columns in which to list the percentages of completion of work and money figures. Percentages of completion are used in the process to shorten the amount of time necessary for the periodic certification of completed work.

Job					Payment No. _____ of _____		
Job Numbers					Date :		
Item	Scheduled Amount Allowed	% of Item Completed to date	% of Item Completed this P.P.	% of Item Left to Complete	Amount to be Paid this P.P.	Total Amount Paid to Date Incl. this P.P.	Amount Left to be Paid
Excavation	100,000	50	10	40	10,000	60,000	40,000
Re-steel (mat.)	50,000	20	10	70	5,000	15,000	35,000
etc.							
At Completion	2,000,000 Contract Price	all items 100%	—	all items 0%	—	2,000,000	0

Figure 19. A basic schedule of values form.

The listing of items of work in the schedule of values can be as detailed as the contractor and owner agree on. Obviously, the greater the detail, the greater amount of time is required of the expediting operator to handle and process the schedules of values. The use of greater detail, however, has several advantages to the operator: for example, it serves to guard against overbilling and forces the expediting operator to know more details about the actual production of the job. (Here the term *schedule of values* is synonymous with *schedule of payments*.)

Once the form of the schedule of values is determined and its listings made, it must be approved by the owner through the architect/engineer. Once approval is given, the owner and the general contractor are bound to accept its listing of work and values throughout the project unless they mutually agree to revise the listing. It is very rare for the listed items in the schedule of payments to be substantially changed or modified once it has been approved and used through one periodic payment.

The schedule of values serves the final payment process as well as the periodic payment process. The final tabulations of values for final payment in the various columns of the form are shown in Figure 19.

To summarize the operations involved with the schedule of values, we first assume that the general contractor's expeditor handles them. The process is as follows:

1. A set-form is selected by mutual agreement of the parties or the expediting operator composes it in compliance with the owner's instructions.
2. The expeditor lists the items of work as he sees fit.
3. The expeditor lists the money amount for each item of work as he sees fit.
4. The expeditor submits the schedule of values to the architect for approval.
5. The architect, after review and notations, submits it to the owner for his approval.
6. Depending on the owner's response, the schedule of values is revised until its content is mutually satisfactory to the owner and contractor. Usually the revision process is done informally until the schedule is finally agreed upon.
7. The architect returns the completed approved schedule of values to the contractor for his use.
8. The schedule of values for the job is then set once and for all.

Provisions for payments for materials stored on site and for the changes in the work at various stages of the job must be considered by the parties to the contract in the payment process. It is advantageous for the schedule of values form to have space to list these items.

When the schedule is approved, the general contractor's expediting operator takes operational control over the use of this document in the payment process. He has the form printed in its approved format for use throughout the remainder of the job.

The General Method of Billing

The methods the general contractor uses to bill the owner vary under different construction contracts, but generally they are as follows.

1. The contractor, after receiving his subcontractor's and supplier's bills and invoices, bills the owner by a set time that is specified in the contract in order to collect his actual payment the next month. The general contractor must have his bill prepared by, say, the twenty-fifth of the month in order to be paid by the tenth of the following month. This means that the general contractor's operator must set time limits for the receipt of the subcontractor's and supplier's invoices and billings so that he may correlate and enter this information into the general contractor's billing to the owner. Operationally, he sets time limits for his subcontractor's billings through his subcontracting operation (i.e., by including the limits in a subcontract provision). The correlation of these billings and their entry into the general contractor's billings are accomplished by the operator setting up operational procedures for both himself and the financial department of his organization.

2. The expeditor sends the billing to the architect/engineer or his field representative. The architect/engineer reviews the schedule at the job site and either approves it, modifies it, or rejects it outright. This review allows for the visual inspection of completions if they are disputed. If there are changes or modifications, the architect/engineer then generally comes to an informal agreement on completions with the contractor.

3. Once the schedule of values and billing problems have been resolved between the architect/engineer and the general contractor, the architect/engineer formally certifies to the owner that the contractor's billing is correct. This is done formally through a request for payment submitted by the contractor through the architect/engineer to the owner.

4. Once the billing is certified for payment, the owner, with few exceptions, is obligated to pay the contractor his bill.

Two Major Objectives in Periodic Payment Operations

The construction operator must develop specific operational procedures in both the periodic and final payment processes, because of the extreme importance of exchanging payments for work completed in full and on time in the construction process. Except for some differences in its required formalities and its meeting certain special contractual requirements, the final payment process follows the same course as the periodic payments. The final payment is, in fact, the last periodic payment on the job.

Generally, the expeditor operating with the periodic payment has two major objectives to meet: he must make sure that the payments received from the owner and disbursed to his subcontractors are adequate and on time. Making adequate payments calls for a certain kind of operational thinking and the development of specific procedures.

Adequacy of Payments

From the time the billing process begins until the time the owner actually pays, there are only about 15 days in which to do all the work involved. Within this time the general contractor's entire operational organization is held responsible for the receipt of payment (including its financial operators). The organization's operators are required to ration and set aside time for these operations; the procedures, means, and methods that they develop to do all these operations must be efficient. Establishing communication within the contractor's own organization is especially important. The computing and estimating of work completed to date on the projects are operations that the field supervision forces usually perform. The expeditor usually develops the specific method of billing that is used from the start to the finish of the job. Quantities are taken and recorded by the job engineer and the superintendent and the expeditor usually determines how completed items of work are translated into a percentage of completion when they are only a component of an item of the schedule of values. The correct assignments and interorganizational coordination of all these operations insure meeting the objective of being adequately paid.

Timeliness of Payments

The absolute requirement of timeliness throughout the payment process demands perseverance on the part of the operator, as well as rigid and strict operational thinking. Once procedures are established that solve the time problems in the payment process, the operator can allow no deviations. If, for example, a subcontractor is given a time to bill and he is late, he must be made to suffer the consequences of his carelessness. As harsh as it may seem to the reader, the operator should either delay that subcontractor's payment or, in the case of repeated carelessness, not bill for them and consequently not pay them at all. This action will no doubt prevent such carelessness from recurring.

If a subcontractor is obviously overbilling in his invoices and billings, the expeditor does not have time for extensive arguments or discussions with the subcontractor; he must react in one way or the other to such practices, because of time considerations. He must insist that the subcontractor cease overbilling once and for all.

The time allowed in the construction process for the operations of payment is very restricted. The construction operator must set up for each individual job procedures the payment that will fit into these time restrictions.

He must closely consider the owner's attitudes toward paying on time. Problems might arise because of the owner's cash position, his reluctance to move invested money, budgetary problems in his own operations, or perhaps just because of the type of owner he is. Who or what type the owner is often has a significant bearing on the operator's ability to process payments.

Example. A manufacturing company can be an owner; a school district or a municipal corporation can also be an owner. These two types of owners can be a problem, but for different reasons, when dealing with the timeliness of payments. It would seem that a manufacturing company should be in a perfect position to pay in full and on time, but the construction of a new building project may cause serious problems with their overall cash flow position, thus affecting payment for construction. The reader may only guess who suffers when a choice has to be made as to where available cash goes. This type of owner has many opportunities to delay payments or the payment process, and he can become very hesitant in paying the contractor.

On the other hand, a public body operating through, say, a government council presents other problems for the operator. One problem would be when the council, by law, meets for business only at certain specified times. If these times do not coincide with the periodic payment process, then serious delays are likely to be encountered in the payments. Operationally, the expediting operator has no choice but to adjust the payment process to this situation. The fact that the council cannot act on its approval of a contractor's payment request does not lessen the contractor's need for payment or his responsibility to pay his subcontractors and suppliers within a reasonable time after they have performed their work.

This last thought raises a specific point that the reader should consider.

The general contractor's expediting operator must always consider the adverse consequences of the owner's failure to pay in full and on time (no matter what the reason), because this failure can have serious consequences on the general contractor's ability to operate effectively with the people to whom he owes money. A subcontractor who has not been paid very rapidly becomes a most uncooperative person to deal with, not only on the affected job but on all work it is doing with the general contractor.

The construction operator must also consider the architect/engineer's position in the operation of payments. It is the architect/engineer who certifies to the owner the correctness of the contractor's billing. The expeditor must establish and insist upon set times for these people to perform their functions in the process. These times should be reasonably adjusted to the time limits established in the contract.

The operator must give special consideration to the attitudes of the architect/engineer about the urgency of seeing that the contractor is paid on time. There can be great variations in attitudes and expectations between various architects/engineers about the payment process, which can greatly affect the time it takes to process payments.

Example.   One architect/engineer's office will be extremely careful about certifying the quantities, percentages, and billings in every periodic payment; another office will be very careful about these same things, but only when certifying the billing for the final payment. Some will require extravagant proof of completion percentages and others will question very closely the validity of subcontractor billings and invoices; some will demand peculiar proofs for payments for materials stored on site and still others will be extremely cautious on the billings for changes to the work. All these demands have their varying effects on the time it takes to process payments.

This example should indicate to the reader the broad range of possible difficulties that can be encountered when dealing with architects/engineers' offices in the payment process. Usually the operator begins to learn these things during the first billing period on a job, and he continues to learn about them as the job progresses. As he learns about the different problems that can be encountered, he must adjust his operations to them.

The construction operator must view the adequacy and timeliness of his own payment and billing operations realistically. If he expects proper action on the part of the owner and the architect/engineer in the payment process, he must also be efficient in his own operations. Being the initiator of the process, the operator is required to set an example, and there are many considerations he must give to the efficiency of his operations.

Example.   If percentages of completions are to be determined from cumulatively recorded information based on the direct observation of the completed work, then the completions must be correctly reported and be observable. No matter whether it is the job engineer, superintendent, or the expeditor himself who makes the quantity survey of the completed work, this operation must be done in a logical and systematic way. The procedure for doing this task should be standardized and be operationally easy to perform and the reports and records of the quantities of completions should be easily understandable.

Listing percentages of completed work that are exaggerated or based on wishful thinking will usually fail to pass the close review of the architect/engineer. Besides the consequence of this failure, consideration must be given to the time-consuming task of redoing the schedule of values and the billings. Percentages asked for initially on the schedule of values must be realistic. It is the operator's duty to make his own estimate of the work complete and accurate and to see that those of his subcontractors are the same. For example, the expeditor must take strong corrective action against subcontractors and suppliers who deliberately or accidentally overbill or who are careless in their invoicing and billing habits.

Material Stored on Site

An initial clear-cut statement of the policy for billing for materials stored on site must be made at the beginning of the payment process and understood by all the parties involved. This statement must include the means and methods of quantification for stored as opposed to installed materials. The exact method of billing and furnishing backup information must be established. An inordinate amount of time can be spent on these phases of the billing and payment process if the operator does not have certain guidelines clarified at the beginning of the payment process. A well-recognized general rule to apply, for example, is that billings made for materials stored on site must be proven by invoices. If this requirement is made known at the beginning of the job, then invoices must be provided. Estimates of quantities of stored materials will not suffice.

A Suggested Method of Processing Periodic Payments

A well-known method for processing periodic payments that is used in the construction process is as follows.

1.   The expeditor makes out the schedule of values according to his own estimates of the percentages of work completed.

2. He takes this to the job site and meets with the job superintendent and the architect/engineer's representative.

3. They all review the percentages of completion, attempting at the same time to settle any discrepancies between the architect/engineer's and the contractor's representatives. Most differences in views on percentages of completion figures are settled at this meeting, but if they are not, one of the three following courses can be taken by the operator:

    a. He can bill as he views the percentages.

    b. He can compromise.

    c. He can accept the view of the architect/engineer's representative.

He must consider taking courses (b) and (c) with caution, however, because he is not in the process of collecting money only for his company but also for many subcontractors and suppliers.

4. Assuming that any difference in estimates of percentages of completion are worked out in the job meeting, the contractor's operator revises his billing on the schedules of values agreed upon and requests payment. The request for payment is a formal document ordinarily furnished by the architect/engineer or owner that suits the owner's requirements.

5. The formal request for payment accompanied by the schedule of values, the contractor's billing, and all supporting requirements is sent to the architect/engineer for his certification for payment. The *certificate for payment* is another important formal payment document. It is the architect/engineer's document, and its form and content are determined by him; it embodies the general contractor's request for payment. It certifies to the owner that the architect/engineer agrees to its contents and that the amount listed therein is correct and should be paid.

6. Theoretically, on a certification, payment is due and owing in accordance with the terms of the ordinary construction contract and accepted industry practice. There have been many legal questions connected with this point, but unfortunately they cannot be fully examined and discussed here. It suffices to say that a basic objection by the owner to the contents of a certificate for payment, regardless of his agent's certification, constitutes a serious problem for the contractor's operator. Whether the owner's refusal to pay is legal or not is immaterial, because the fact still remains that the contractor cannot stand still for an appreciable delay in the payment process. Should this problem arise, it calls for the expeditor to make an immediate effort to meet with the owner and settle any objection face to face. Even if the contractor stands in the correct legal position, it is of very little practical value or significance when the owner refuses to pay on the certification of the architect/engineer. This situation creates enormous problems in the periodic payment process, but it is not nearly as serious as when it occurs in the final payment process. In the case of final payment, the owner has his building and is most likely using it. Under these conditions the contractor and owner are in totally different positions than they were in the periodic payment situation.

7. With certification and all required supporting data in his hands, the owner pays the contractor. The payment most often passes directly from the owner to the contractor.

All of these periodic payment operations must take place in a very short time. Therefore, operational procedures must be exactly timed, with very little allowance for unforeseen difficulties. The operator must make sure that all necessary operations involving all contractually required data are in order. This data must be gathered and delivered as expeditiously as possible. We repeat here that *all other construction operations in the contractor's office and field must take second place to the payment operation when its time comes.*

Supplying Supporting Requirements

All billings and payment requests (final and periodic) contractually require the general contractor to supply supporting data including the following:

1. The sworn statements on prior payments to subcontractors and suppliers.
2. Full or partial waivers of liens.
3. "Other data" (which, unfortunately for the general contractor and its expediting operator, can mean almost anything).

These documents are required by contract. (They are discussed further in Chapter 10, which examines the closing-out process of a job). Here we only wish to discuss the sworn statement on prior payments.

The sworn statement is a formal document in which the contractor swears to the fact that he has paid his legitimate debts from funds paid to him by the owner for past billings on the individual project at hand. This sworn statement is another "fiction" in the construction process; it is a rather meaningless document in the contracting subprocess, because it does not function at all in the manner for which it is intended.

Practically speaking, the actual payment of money to the subcontractors and suppliers does not take place until the money is actually in the hands of the general contractor, who is bound by contract to distribute it. The contract requires him to swear, as a condition of his next payment, that he has distributed the proceeds of the prior payment. The trouble begins when the general contractor, for legitimate reasons, withholds from his subcontractors and suppliers the monies paid to him by the owner in previous billings. Legitimate withholding can be based on reasonable back charges to a subcontractor from the general contractor or on the general contractor's real concern about the acceptability of the work of a subcontractor or supplier. The general contractor cannot pay his subcontractors and suppliers the full amount of their billings, because to do so would put himself in an untenable position; yet in order to receive the next periodic payment, the general contractor's operator must swear that he has paid these people in the same proportion as the owner has paid the contractor in the prior periodic payment, in other words, the operator must swear to something that he knows is not true.

While the general contractor's operator can make exceptions on his sworn statement for these withholdings, it is practically impossible to assign amounts to these exceptions, particularly when they are matters of serious dispute with the subcontractors or suppliers and involve substantial sums.

The sworn statement, purportedly used as a method of assuring the owner that the contractor has made prior payments to his subcontractors and suppliers, is really only a fallacious compliance with the contract; it is certainly not proof of the payments nor is it realistically thought to be by most construction operators. Unfortunately, the only thing that it really does is plant a seed of uneasiness in the mind of the operator who must personally swear (with reservations, it is hoped) to something that is false. The construction process and its operators deserve something better in this area of operations.

13.4 THE OPERATION OF FINAL PAYMENT

We want to be brief about this operation here. Final payment, besides being a payment operation, is also a close-out operation. Most of its operational and contractual requirements are discussed in Chapter 10. Final payments require furnishing "as built" drawings, operating instructions, final waivers, and so forth. Its supplemental requirement of processing the punch list for close out are best left to the discussion of Chapter 10.

Here we wish to examine only two aspects of the final payment process for the purpose of understanding their effects on the way the operator approaches its operations: first, we must examine the relative difference in the positions of the parties at the time of final payment as opposed to their positions during the periodic payments; and second, we must examine the final payment with respect to substantial and final completion.

At about the time the final payment becomes due, there has been a significant change in the bargaining positions of the parties. At this time the owner is about to occupy or has occupied the building. The contractor, however, does not yet have his total money. Unfortunately, sometimes this change in position will be indicated by significant changes in the attitudes of some owners as to what is owed to the contractor and what is not. These changes in attitudes are reflected in the dissatisfaction with this or that item of the construction and extended punch lists. The owner simply wants to get as much as he can from the maintenance and repair of owner-caused damage on the new construction.

The construction operator can combat these attitudes by handling them in special ways. A general rule for successful operations is to force the owner to commit himself to accept the building or parts of it at some specific point in time; for example, the owner must accept the responsibility for an area of the building when he occupies it. This is done operationally through meetings and by correspondence. The construction operator must never allow an owner to take unfair advantage of the situation by forcing the contractor to do extra work on the job as a condition of final payment.

At some time prior to substantial completion clearly defined guidelines must be established for the use of the building by the owner's and contractor's forces. The available areas of work, their accessibility, and the working times in occupied parts of the building must be determined and scheduled. Once determined, the responsibilities of each party must be upheld; this extends to everything from cleanup and dust protection to the transference of insurance obligations.

Once these responsibilities are established, it is easier for the operator to make a clear delineation of the requirements of completion. The gray areas of overlapping responsibilities are cleared up and this expedites final completion and payment.

There are other areas of interest to the operator in the final payment process operations. We briefly note them here.

1. Careful consideration must be given to the status of unresolved claims and disputes as they affect the final payment processes.

2. A careful analysis must be made of the effects of refusals by subcontractors and suppliers to furnish final waivers of lien, and methods must be established to cope with this problem.

3. The operator must have at his fingertips the exact amounts of money that are owed to his firm and that he owes to others. It is particularly important for him to have fully processed (if only on paper) all change orders to the contract received from the owner and all those of his subcontractors and suppliers as far ahead of final payment as possible.

4. There must be a clear understanding on disbursements and withholding of monies to subcontractors and suppliers between the expeditor and his own financial department before final payments are made to them.

5. The operator must realize that his superiors expect final payment within a certain amount of time after the owner occupies the building. He must be able to give them a reasonably accurate date of receipt of the final payment.

13.5   USING THE PAYMENT PROCESS OPERATIONALLY

The Owner's Use of the Process

Nothing can make a contractor more nervous than the owner threatening or even indicating that a portion of a periodic or final payment is not going to be forthcoming for one reason or another. Contractually, of course, this is a serious threat, but the experienced construction operator knows (particularly during the periodic payment process) that it may be an indication of the owner's displeasure with the contractor's operations. This should provide an incentive for construction operators to shape up their operations in such a way as to not unduly aggravate the owner. On the other hand, if a building is being built correctly and completely, the owner should be made to understand that he has no right to withhold or refuse payments.

From the beginning of the job to its end, the contractor's expediting operator must make the owner and his agent aware that any kind of action in withholding or delaying payments, in addition to being unfair to his company, will become a matter of personal concern and dissatisfaction. The operator can best get this point across by always conducting his part of the payment process with exactness and having an uncompromising attitude about the adequacy and timeliness of the owner's payment policies. The owner must be made to realize that the operator knows what he is doing in the process, knows the project, knows the worth of each completion, and intends to collect money for it on that basis.

The General Contractor's Hammer

The payment can be used as an operational tool with erring or reluctant subcontractors and suppliers. When a subcontractor or supplier, for one reason or another, gives an outright refusal to perform, something must be done by the general contractor. We attempt to give pragmatic approaches to operations in this book. We do not hesitate to say therefore, that threatening to withhold payment or actually doing so is in many cases the only practical way of getting a delinquent subcontractor or supplier to perform. This can be done quite legitimately and safely by the general contractor's operator by refusing to bill the subcontractor's work to the owner. But even when he does include the subcontractor's billing in his billing to the owner, there is still a justification at times to withhold payment from subcontractors who consistently refuse to perform, until they improve their performance on the job.

Withholding payment is not a tool that can be used without great discretion and good judgment; to use it haphazardly would be a sure indication of either outright incompetence or dishonesty on the part of the general contractor and its operator. When using this operational tool, the expediting operator must realize that he walks a fine line. He knows from his experience with owners the seriousness of not being paid for work done. A great deal of analysis on the individual subcontractor or supplier's business and financial positions must be done by the operator before he uses the payment hammer. Ordinarily, it is legally much easier and safer to use this tool with a smaller rather than a larger subcontractor.

The operator must also realize that withholding monies from subcontractors and suppliers is very dangerous from a business standpoint. Many times a supplier or subcontractor can be a company much larger than the operator's. This does not mean that big companies cannot be wrong when they refuse to perform or deliver; it simply means that the construction operator should give more thought to the risks involved when dealing with a large company.

There is really no more effective way for the general contractor to force performance than by withholding payment. It hits straight and hard at the subcontractor's or supplier's most vital interests, but it must be used very carefully and with great discretion. It must be used fairly and usually only as a last resort.

Owners should not object to the general contractor using this tool fairly; they should realize that the relative amount of protection they have against the nonperformance of the general contractor is not the same as the amount the general contractor has against the nonperformance of his subcontractors. The performance bond, for example, is one such protective measure that is very often required in the general contract but not in the subcontract. This gives the general contractor some reason for using the payment hammer with his subcontractors. Under some circumstances there is no practical alternative to withholding payments, even when there is a cancellation clause in the subcontract.

13.6 THE IMPORTANCE OF TIME IN PAYMENT OPERATIONS

We cannot close this chapter without again warning the reader about the importance of time in the payment process. Payments must be made, and on time. *No other construction operation can displace processing the payments at those times allotted to them during the performance of the contract.* One need only consider the consequences of delayed payments on the entire scheme of the construction process, its subprocesses, and its phases.

Payments owed to the labor and materialmen on a construction project come ultimately from the owner down through the various channels of the payment process; they go from the owner to the general contractor, and from there to the subcontractor and materialmen. This forms a chain. Any disruption in the schedule of payments causes all kinds of difficult financial and mangerial problems for the general contractor, subcontractors, and suppliers. Late payments make it difficult to meet payrolls (which *must* be met) and to distribute monies to remote parties, they cause strained relationships between all the components of the forces on the project, and they result in late performances and reluctant service on the job by resentful subcontractors and suppliers. Late payments place unwarranted burdens on the accounting and financial departments of all the contractors and their work forces, which sometimes has serious additional cost consequences.

Every construction operation has some peculiarity in its functions. In the payment process the construction operator must consider time in a special way: payments must be adequate but, just as importantly, they must be on time.

14

The Bonding and Insurance Processes

14.1 THE INSTRUMENTS

Operating efficiently with insurance and contract bonds is a fundamental requirement within the subprocess of contracting in the construction process. This chapter explains how construction industry operators are concerned with these two instruments; no attempt is made to give more than the very broadest explanation of the substantive nature of these documents. A knowledge of their legal purposes, contents, and effects is presumed or to have been acquired elsewhere.

When there is a requirement for these instruments in any construction contract, the construction operator must deal with them both initially and throughout the entire life of the contract. He must learn how to operate with them in a general way; he must know how to obtain them, how to execute them, and how to develop practical means, methods and procedures to implement and integrate them within the subprocess of contracting.

The following is a broad and brief definition of each instrument:

Insurance is a contract whereby the insurer (insurance company), in return for a fixed payment (premium), guarantees the insured (contractor, owner, third parties) that a certain sum will be paid for a specified loss or liability caused by a specified contingency (risk). The insurance contract takes the form of an insurance policy.

A bond is an agreement under which one party (e.g., bonding company) will pay, within stated limits, for the financial loss caused to the obligee (the owner) by a "principal's" (contractor's) failure to either perform the contract (i.e., the performance bond) or pay for labor or material expended in the contract (i.e., the labor and material bond).

The bond that guarantees that the general contract bidder will come to contract under its proposal is called a bid bond, and the performance and the labor and material bonds are called contract bonds. The bid bond furnishes bid security and the contract bonds furnish contract security.

Each insurance policy and each type of bond instrument creates the need for certain operational processes, which in turn call for the operator to develop certain procedures, means, and methods for properly and expeditiously handling and working with them.

Each instrument is a contract, but each contract is separate and distinct in its purposes and effects on the parties within the construction contract. At the same time each of these instruments, when required by a construction contract, is limited and restricted by the terms of the instrument and the operation of law. An insurance policy, for example, would limit its effectiveness to occurrences on a single project operating under a specified construction contract. The performance bond for a construction contract furnishes its guarantees of payments and performance only on that one contract.

14.2 THE BID BOND PROCESS AND OPERATIONS

The first step in the bid bond process requires the construction operator to obtain the bond from the surety company. It is usually the case that the performance, labor, and materials bonds (contract bonds) will be obtained from the same bonding company that furnishes the bid bond to the contractor. It is the exception rather than the rule when a company that issues the bid bond will not furnish the contract bonds. It is the bid bond that, for all practical purposes, determines the ability of the contractor to furnish the contract bonds after award of contract. Therefore securing the bid bond is an operation that is essential for the contracting firm and its responsible operators to learn and perform properly.

The bid bond must be secured before bid time, and the sooner the better. The bid bond and its required copies must accompany the bid proposal on a project that requires security for the bid: it forms a part of the proposal. In most cases the bid proposal is considered to be incomplete if the bid bond does not accompany it or if it is improperly or incompletely executed. The physical instrument must be obtained, properly executed, and it must accompany the proposal.

Bid Security

Bid security takes on two forms: a bond or a certified check. The amount of the bid security is generally specified in the bidding documents, and it is normally required to be 5% of the proposed price in the bid, whatever that may turn out to be. In some instances, the bid security is specified as a set money amount rather than a bond.

The bid bond is a bid security whose requirements are generally stated quite clearly in the instructions to bidders. The instructions to bidders may set the requirements of another recognized form of bid security—the cash security, which usually is in the form of a certified check. The reader should take note of this, because operating with a certified check is not at all the same thing as operating with the bid bond.

Ordinarily, the bid security in either bond or cash form on a $1,000,000 proposed price would amount to the sum of $50,000, and if a certified check is used, it must be executed for $50,000. It is obvious that a cash transaction takes place when using the certified check method of bid security; there is no such transaction involved in the bid bond form. This requires the construction operator to view and operate with the bid bond much differently than with the certified check security. The following examples illustrate this point.

Example 1. The owner might specify in the instructions to bidders his option of holding the bid security for the proposals of the three lowest bidders until he selects a contractor and awards a contract. In most cases the time he may hold the bid security is not specifically stated in the instruction to bidders. The practical effects of the owner holding the bid bond as opposed to his holding a certified check are significantly different for all of the bidder's, but particularly for the three lowest bidders. Any security in the form of cash is tied up until the owner awards a contract.

Example 2. On bid day the security must accompany the bid to the place of delivery. Since the bid security is based on a percentage of the proposed price, many difficult operational problems can arise when using the certified check as opposed to the bid bond. Not knowing the amount of the bid in most cases until the very end of bid time makes it very difficult to physically obtain the certified check for the correct amount. It requires the operator to develop specific operational procedures to accomplish this task.

On the other hand, in some cases the bid bond needs to be determined prior to bid day so that the bonding company can issue a bid bond to the contractor relative to his bonding capacity. At times much confusion and difficulty are involved in obtaining the bid bond, caused directly by this fact. This is a particularly serious problem when the actual proposal price may far exceed the architect/engineer's prebid estimate of the work and the bidder is close to the limits of his bonding capacity. The construction operator must develop procedures, ways, and means of dealing with problems like these.

A full awareness of the processing times for both these types of securities is essential. The contracting firm first has to determine which form of security to use if the type of security is optional with the contractor. Once this determination is made, the operator is required to operate according to the demands of whatever type is used. It is safe to say that the advantages of using the bid bond as bid security far outweigh those of using the certified check or cash deposit.

14.3 THE PERFORMANCE BOND PROCESS AND OPERATIONS

Construction operators do not operate with the performance bond; rather, they operate *around* it. Specifically, there are three major operational procedures for the operator to develop that are involved in the contract bond process, and they are the following:

1. The development of a close and continuing relationship with the local bonding agent. This relationship must have as its basis both the agent's and the operator's trust and confidence in each other's competence and integrity.
2. The development of accounting and record-keeping procedures that allow for the fast and complete exchange of information between the bonding company and the contractor.
3. The development of guidelines and procedures for the total company personnel whenever the possibility of default is indicated and most certainly when default is inevitable. These guidelines for the operations of individual operators are generally very restrictive in nature. The habit of talking carelessly, for example, which is always undesirable, becomes even less desirable where a bond default situation is at hand. Construction operators must watch their own actions as well as those of others in the organization. The reasons behind a default or its causes are always important in determining the legal and financial responsibilities involved. The construction operator should be especially careful about what he says and does

whenever extraneous factors or circumstances may have forced a default situation.

The construction operator should realize that the performance bond by itself is never the cause of the problems that might arise out of it. It is a series of circumstances that leads to a default that cause the major problems. Operations should be conducted with an awareness of the provisions contained in a bond in order to prevent a default. Once the bond is defaulted, little can be done to help the contractor.

The default is the last thing that either the bonding company, its agents, the contractor, or the owner wants to happen. Unfortunately, the functions of the bond take over, and then very little can be done to avoid the disaster. It is ironic that those who have the responsibility (generally the architect/engineer) for recommending to the owner that a bond be called in default have the least direct stake in the situation; but even these professionals recognize the seriousness and unproductiveness of the bond default in relation to the entire construction process.

The "Cost to Complete" Operation in the Bonding Process

The major area of information that a bonding company and its agents will be interested in (in the bond process) is the financial status of all of a bonded contractor's work. This requires the contractor to furnish the detailed completion and financial status of each individual job to the bonding company, allowing it to make judgments on such things as bonding capacities and the contracting firm's money-making abilities relative to granting any increases in bonding capacity. This information allows the bonding company to study the risk factors.

On the contractor's side the bonding process is handled in most cases by the top executives in the construction firm. In order for them to act properly, they must have accurate and complete information available that will permit them to deal intelligently with a bonding company in order to promote the interests of the company.

When dealing with a contractor on bonding matters the bonding company is interested not only in the status and condition of the firm's projects on hand but also in other factors, for example, information about past performance on projects. In cases where the bonding company is considering increases in bonding capacity or the issue of new bid bonds, it will want to know the status of all jobs at the time that any and all of the bid bonds are requested. When dealing in or near a default situation on a performance bond, it will want to know the very detailed status of the particular job in trouble at that time. Since a current knowledge of these matters is vital to both parties' interests, the operational organization of the contractor must be able to furnish usable and accurate information. The construction operator can be assured that if he has not developed methods and procedures for supplying the necessary information, he will have to do so, and sometimes very rapidly. This is not the type of operation where means and methods can be developed haphazardly or extemporaneously. It is therefore best for the operators to be aware of this potential need and to act accordingly by developing applicable procedures and methods beforehand.

When operating around the contract bond, the construction operator must develop his documentation and communication skills in specific ways. The operator who is most involved in operations with bonds is the expeditor and one of his major tasks is the documentation and communication operations used to determine the "cost to complete" the work.

So as not to confuse the reader, let us make it clear that the term *cost to complete* used here is based on the same concept of *cost to complete* required in other construction accounting procedures and operations.

The cost to complete operations has many applications in different subprocesses within the construction process, and not only in the bonding process. However, it is necessary for the operator to understand that both performing and applying the cost to complete operation in bonding matters are significantly different from the same tasks in other construction operations.

The difference is that in the bonding process the cost to complete the work must be made available at any time it might be needed. The operator in bonding matters is not usually given the same length of time for this operation that, for example, is allowed for him to compute "costs to complete" for a financial statement. The determination of the status quo of job expenditures to a certain date relative to actual completion is sometimes demanded by the bonding company with very little or no notice.

The basic acceptable procedures for determining the cost to complete are the same for all the uses of this information. Fundamentally, cost to complete is an accurate computation of work actually done relative to monies expended on it to the same date, along with an estimated computation of work left to be done in relation to the estimated monies to be expended in completing the work. All costs to complete are reviewed by the operator in relation to the monies left to be paid on the work to which the contractor is entitled under the contract.

It is the operation of estimating the cost of work to be completed that requires the most skill and knowledge. The cost to complete is in many respects a "mini-estimate": it is not done under the same kinds and types of pressures encountered with the original estimate on bid day, but it still employs the same fundamental concepts of estimating units of work and the same guessing at the cost of whatever unknowns remain, and it is still a projection of work to be done within a time frame. Because less time is allowed for completions, it sometimes becomes more difficult to perform this operation than to do the original estimate.

There are, however, some features of this estimate that make it easier to do than the original. Actual quantities of work left to be done can be accurately determined because there are fewer items to quantify and, most importantly, the actual costs of production on the particular job are a reality. In addition, while there are still some pressures involved, they are not the same as those encountered in the bid process.

When determining costs to complete, the operator generally uses the following information:

1. The original or rewritten estimate.
2. All records of actual costs available to him.

From these documents he forms the estimate of the cost of what is left to complete on the job. For work left to be completed by his own forces, he compares the cumulative quantity of work actually done to the work quantified in the original or rewritten estimate, item for item. When proportioning the costs to complete, however, he must correct any quantities that are obviously wrong in the original estimate.

For example, if the original estimate quantified wall form at 100,000 square feet and it is obvious that there are actually 110,000 square feet to be done on the job, then the overage must be accurately reflected in the projected cost to complete. The additional 10,000 square feet can be reflected in a cost to complete as a money amount based on either multiplying the number of square feet by the unit price used in the original estimate or the cost reports of actual unit prices. The construction operator must be able to judge which of the two is the realistic unit price to use in projecting cost to complete; for example, the actual unit prices are not *always* either correct or valid to use in this operation. In forming these judgments, however, the operator must consider the possibility of having to explain in detail to the bonding company why he did not use actual unit costs instead of estimated unit costs. The operator goes through each and every item of the work of his own forces in the same way, using the same thinking process.

When dealing with subcontractors, the operator faces other problems. Theoretically, the operator can determine a subcontractor's costs to complete from the totals of each subcontract, the amounts already paid, and those left to be paid. This would mean that he must only consider the known payments of money to subcontractors, regardless of the amount of work each one has done. Using this method, he need not worry about a subcontractor's overbilling; but to accomplish the true purpose of using the cost to complete in the bonding process, it is of course necessary that the operator consider and accurately report actual status information on the subcontractor as well as on the work of his own forces. If he has a good idea that there is a subcontractor in trouble on the job or one that has overbilled, he must realistically appraise these conditions and report such findings in the cost to complete. In this operation it is vital, not only to the bonding company's interests but also to the long-term interests of the operator's own firm, to think along realistic lines; a fact can only be ignored or hidden for so long. While there might be a sincere hope that a different situation exists, the only way to deal with this particular operation is to see things as they are, not as one would like them to be.

Bonding companies are extremely interested in the contractor's projections and assessments of the actual financial condition of subcontractors on a job. A major subcontractor who has been overpaid or underpaid by the general contractor in relation to work actually done gives the bonding company cause for alarm. It is best for the operator to realize this when he makes projections and assessments about his subcontractors in the cost to complete operations in the bonding process. The operator will be most embarrassed when, upon being forced to give a future cost to complete to the bonding company, he realizes that assessments in his prior costs to complete were the result of wishful thinking rather than reality. The bonding company will come to one of two conclusions about the operator and his company: they are either incompetent or, in extreme cases, dishonest. Neither conclusion will benefit the construction organization in its future relations with the bonding company.

14.4   THE LABOR AND MATERIAL BOND PROCESS AND OPERATIONS

The labor and material type of bond protects more than just the owner's interests; it protects the suppliers and the laborers on the job.

There are a considerable number of interrelating legal problems connected with this bond and its processes. The legal problems cannot be given the space they require in this text, so here we only state very general principles for operating when problems on this bond are at issue.

It is pertinent to emphasize some particular points about the contract payment bond at this point. In order to efficiently operate around this bond and its potential problems, the operator must thoroughly understand the complete processes of payments of all types that are either made under the terms of the construction contract or that might be required by law. Many times it is necessary to know, for example, the exact channels through which payments are to be made either by contract, by statute, or both.

The operator must understand basic contractual relationships. For example, he must know to whom his company is responsible and to whom it is not, because restrictions on liability contained in the bond may be affected by the concept of privity of contract.

In addition, the contractor (and its operators) must be aware of the legal responsibilities in relationships with and between subcontractors and materialmen, not only as a principal in the contract payment bond but as a legal entity who must operate under the law in general. What, for example, are the effects of the bond protection and responsibilities relative to its position under a state's lien laws? There are a multitude of this and similar kinds of potential questions for the contractor (and its operators) to be aware of, even if they do not know the answers. This awareness allows the contractor to be cautious in his operational actions. For example, the operator's procedures in payments to subcontractors and materialmen might have to be based strictly on questions of common law peculiar to the labor and material bond and its processes.

The payment bond process is probably the most misunderstood process in the whole bonding and insurance process. Fortunately, it is not a frequent source of problems, probably because the ordinary owner, architect/engineer, contractor, subcontractor, or supplier simply does not understand it and, knowing this, wishes to avoid its complications. When problems do occur, however, they can be among the most legally complex in the entire construction process.

In order to develop his operations for the labor and material bond process the operator must at least be aware of the following:

1. The exact interests protected by the bond in their proper order of priority as creditors.

2. Applicable supplemental statutory laws and something of their legal procedures that fall outside of the common law of surety itself. For example, an awareness of the laws, and their procedures, that protect the rights of creditors is especially important. Even though there is extensive application of creditors rights law in the payment bond process the law of creditors' rights is a separate and distinct area of legal study and application be-

yond surety law. Is demanding study of this area of the law a big order for the operator to fill? It certainly is, and it grows in proportion to how competent the operator wishes to be when operating around the payment bond. Any degree of knowledge of the subject matter is better than none. It is necessary that the construction operator vigorously take on the task of learning about the legal nature of the labor and materials bond. He will have little opportunity to rely on prior experience when operating around this type of bond.

14.5 BONDING SUBCONTRACTORS

Bonds, when they are furnished by subcontractors, are of the same general type and form as the general contract bond; only the parties involved are different and there are some differences in the operation of the law between the subcontractor bond and ordinary contract bonds. Essentially, subcontractor bonds guarantee the performance and payments of the subcontractor, as principal, to the general contractor as obligee.

There is some question as to the economic worth of the subcontract bonds when it is possible to balance their purchase against using methods of selecting only good, qualified, and capable subcontractors.

As in the case of any contract bond, the worst situation that can occur is a defaul of a subcontractor.

The contractor must judge the value of requiring bonds from subcontractors when the overall cost of a default cannot be reasonably justified by the cost amount of the subcontract. To require a bond from a mechanical subcontractor with a $5,000,000 subcontract might be good, solid business practice, but to require one from a toilet partition subcontractor on the same job whose contract amount is $700 would be unrealistic. In between these two extremes, however, is an extensive area of subcontractor bonding where the general contractor must judge the economic benefits of requiring their bonds.

Even in the case of the mechanical contractor, it is still possible that problems in the interrelation of the work of this subcontractor into the total work might make the default situation so costly that it would not be economically justifiable to bond him, even when his contract amount is substantial. It is possible for the limitations on liabilities contained in the subcontract bond to be greatly exceeded by the total costs involved in a default situation, and thus a bond only partially protects the general contractor and owner. Therefore the operator must not think that it is always better to have a subcontract bond on large contract amounts than to have no protection at all. The situation is not so simple: such a default will still have very serious and harmful effects on the project. The operator must be able to assess and then weigh the advantages against the disadvantages of bond protection as opposed to the realities of a default when thinking about and operating with subcontract bonds. The best answer offered for any subcontract bond question is making sure the subcontractors are able to perform the work, regardless of the amount of the subcontract.

It is not likely that the general contractor's bonding company will object to the general contractor requiring subcontract bonds. Many times the bonding company will strongly suggest or insist upon subcontract bonds for major subcontractors. This occurs in many instances where the issue of a bid bond for the general contract bid might be at stake, particularly when the general contractor is very close to its bond capacity. On the other hand, it is unlikely that a bonding company will increase bonding capacity on the contractor's promise to make it a matter of practice to require bonds from its subcontractors.

14.6 THE DEFAULT

When a contractor defaults on a bid bond, one type of situation exists; when he defaults on a contract bond, a different situation exists. One important feature of any type of bond is its particular limitations of liability of the surety. Terms such as *face amounts* or *limited liability* become important for the construction operator to understand in order to operate correctly with bonds.

Example. The default on a bid bond results in the following situation at the time of default.

1. The limitation of liability (or that amount for which the bonding company is responsible on a default) has been established at 5% of the proposed base-bid. This results in establishing liability of the surety. If the
   Base bid price   = $1,000,000
   Then the
   Surety's liability = $ 50,000
   This $50,000 is not, however, a face amount in the same sense as the face amount of a performance contract bond.

2. Upon failure or refusal of the contractor to come to contract within the specified time after the owner's acceptance of the proposal, the bond, at the option of the owner, is in default.

3. If the owner calls a default and chooses to go to the lowest responsible bidder, then what is the situation? The amount payable to the owner by the bonding company is the difference between the bids of the defaulting bidder and the next selected low bidder that comes to contract with the owner.

Example.

1. A's bid is $1,000,000—bid security is $50,000. B's bid is $1,025,000. If the owner accepts B's bid, the bonding company is liable for $25,000 to the owner.

2. A's bid is $1,000,000—bid security is $50,000. B's bid is $1,100,000. If the owner accepts B's bid, the bonding company is liable for $50,000 and no more to the owner.

3. A's bid is $1,100,000.
   B's bid is $1,050,000.
   C's bid is $1,125,000.
   If A or B refuses to come to contract with the owner and he accepts C's bid, A's bonding company is liable for $50,000 and B's bonding company is liable for $52,500.

4. If the owner does not choose to select another bidder but chooses instead to rebid the project, then A's bonding company is responsible for paying the cost to the owner of having to go through this process, but in no event is it liable to pay more than $50,000.

Continuing Liability

The reason for illustrating the default on the bid bond to the reader is to emphasize the importance of the operator understanding the limits on the liability of the bonding company for any type of bond. The construction operator must understand a fundamental fact here: the bond, while being a type of insurance, does not operate in the same way as an insurance contract. While the insurance company pays on an occurrence, a significant resulting liability does not necessarily follow from the contractor to the insurance company because of the occurrence. However, with the performance and labor and material bonds there most certainly is a continuing liability running from the contractor to the bonding company. The bonding company pays for the default on a performance bond, but it has legal recourse to recover all it paid on the default (if anything is available) from the defaulting contractor. The construction operator must realize this in operating with construction bonds.

In the case of the bid bond, it gives him pause, for example, when faced with the choice of refusing to sign a contract for a project if he suspects that his bid is too low; he must ask if in fact it is too low. He must analyze his bid in order to balance its lowness with the cost of the bond default to the contractor. In fact, in many cases a contractor will decide to take the job even at loss in order to prevent default on the bond because of his continuing liability to the surety.

Avoidance of Default

Construction thinking in the bond process must be directed toward avoiding default. Even if this entails a great deal of cost and risk, it is necessary for the operator to balance the effects of default. This is true not only for practical economic considerations but also for business considerations of reputation and goodwill.

It is rare for a construction firm to survive default on even one job when significant money values are involved; the figures are simply too large and overwhelming. Ordinarily, a default on contract bonds is the first and last symptom of a sickness that is eventually going to kill the construction firm. If this is true, then it is obvious that all construction operations involved with the bond process must be directed toward avoiding default.

14.7 INSURANCE AND OPERATIONS

This area of the contracting subprocess is so broad that only two specific points of interest relative to construction operations are discussed here. The first specific point pertains to the subject of "proof of insurance"; the second pertains to the "occurrence." They are discussed here because handling then involves specific operational procedures.

Proof of Insurance

The construction operator must fix firmly in his mind the following rule and *always* follow it:

> All those who are responsible in any way in the entire construction process must never allow work to *commence* on a construction project without the *physical proof of insurance* being in the *possession* of those contractually or legally entitled to them.

One of the fastest ways for a construction firm to fail is to have a major uninsured accident that causes multiple injuries or deaths on a project. The consequences of such an occurrence are simply disastrous. But it is not only the major accident that should be the serious concern of the construction operators; supposedly minor accidents can result in a major injury, which can have very serious financial consequences for the construction firm.

All construction operators must make sure, within their areas of responsibility, that absolutely no one is allowed to do work on a construction project unless proof of insurance is physically in the general contracting company's possession. A subcontractor's assurance over the telephone that he is insured is *not* what is meant by physical possession.

The job superintendent must never allow people on his job when he does not personally know whether their employer is covered by insurance. Operationally, the expeditor must emphasize and reemphasize to subcontractors this absolute necessity and make it a strict requirement stated in writing in the subcontract document itself. He must not, however, stop at this point; one of his major operational functions is to pursue, follow up, and police this subcontract requirement. For the expeditor, then, proof of insurance requires not only initial attention but also constant attention to the currency of the subcontractor's and his own firm's insurance coverages.

Operationally, it is a good idea for operators to set up a formal procedure for accomplishing this part of the insurance process. One such procedure is to develop a system of communication between the expeditor and the job personnel for recording whether physical proof is in the possession of the company. A special tab showing proof of possession, attached by the expeditor to the field's copy of the subcontract, should be part of the procedure. This does not mean that the job superintendent is not bound to question the expeditor about the subcontractor's proof of insurance; he must go beyond such checks to make sure the general rule is followed exactly.

Proof of currency of coverage is also essential. The expeditor must develop a system that is easily adaptable to clerical use when he develops a system for obtaining and updating this information from subcontractors. Because of the varying expiration dates of the different insurances and because of the numbers of subcontractors that may be involved on a project, this is an essential operation. Its procedures must be set up by the expediting operator so that all operators, including even the clerical help, can do it efficiently and completely.

The Occurrence

The operator must perform various operations of a precautionary nature that center around the word *occurrence* in the insurance contract. These operations are necessary both before and after an occurrence, and they are rather simple.

Accidents are a common occurrence on the construction job, and they can happen to persons or property. The operator, when developing operations around the insurance policy, is most concerned about recording the facts, both before and after the occurrence. This calls for a close analysis of what to record and when to record it.

An accident requires the operator to record the fact of the happening. He can do this in certain ways, using different operational procedures. He can take statements and record them; he can clear up ambiguities when numbers of conflicting statements have been made. He must realize that his best record of what happened will usually be based on firsthand observations.

The results of an occurrence, however, present a different set of operational procedures for the operator to develop and use. Procedures must establish a permanent record of facts to the greatest extent possible. Needless to say, the photograph is an extremely important aid in any procedure that requires recording facts at a particular point in time. It is very simple for the operator to establish the operational practice of photographing the results or effects of an accident. It is somewhat more difficult, but not impossible, to even establish cause through this medium, by its skillful use.

The recording of relevant facts before an occurrence is not an easy thing to accomplish. Formal notice of an unsafe condition accompanied by evidence (photographs) of that condition to a party making or causing the unsafe condition would be an example of recording before-the-fact data that could be of extreme importance after the fact of the occurrence. The ability to record facts like these, however, calls for close and continuous observation and analysis of the construction project by all parties responsible.

In operating around the word *occurrence* in the insurance policy, the operator must use all the tools available to him. In addition, he must be fully aware of the emotional, mental, and physical conditions of parties who have been involved in or witnessed an occurrence. The objective gathering of real facts, as opposed to hearsay evidence based on wrong impressions or emotionally related suppositions of witnesses, should be the goal of the operator in this operation.

14.8 DEVELOPING METHODS

In the bonding and insurance processes it is necessary for the construction operator to develop his thinking around what these instruments really are (a contract) and how they function. There is little opportunity to deviate from the formal legal requirements of the instruments. It is the nature of the bond, the insurance policy, and the way they work that forces the need for only supplemental procedures that make these formal requirements work to his benefit as much as possible. The operator may take consolation in the fact that once he does develop his own supplemental operations and their procedures, they are the operations least likely to be changed in any construction process he may be involved with. Once he learns about bonds and what they require operationally, he can depend on his good, solid fundamental knowledge to serve him throughout his career.

It may seem to the reader that in this chapter we emphasize the need for operational thinking that is protective in nature. This seems logical, since we said at the beginning that this chapter is devoted to developing and using procedures to work around the bond and insurance contracts. But, in addition to the need to promote practices that are protective in nature, it is also true that some of the operations discussed must be based on preventive thinking as well. It is essential for the operator to understand that the default on a bond, for example, must be prevented if possible. If default cannot be avoided, *then* the protection of his company's interests becomes his paramount objective.

The development of procedures and methods of operation must therefore be directed primarily toward avoiding any situation where these instruments and their effects, because of the operation of law, are taken out of the operator's control. In this area of the construction process, the operator has only a minor influence on the legal processes required by the instruments themselves.

15

The Claims Process

Beware
of entrance to a quarrel; but being in,
bear't that the opposed may beware of thee.

William Shakespeare

15.1   GENERAL COMMENTS

Definition

Contract documents used in the construction process today use the word *claim* or *claims* in different provisions within their conditions of the contract. Webster's dictionary says that a claim is a demand for something rightfully or allegedly due—an assertion of one's right to something. In the construction contract this word is used in the same general sense. The significant addition to this definition in construction contracts is that the "something due" results from some obligation arising outside of the terms of the contract. The concept of claim is inserted in the construction contract in order to permit the equitable adjustment of disputes over the inclusion of extra or additional work actually performed that was not specifically called for in the contract.

Attitudes

The construction operator must develop two special attitudes toward the construction contract claim in order to successfully function in claim operations. These attitudes result from the following general rule, which applies to most contract construction.

The conscientious operator must consider that only that work included in the contract is to be furnished to the owner—no more and no less. This policy is a good legitimate business and professional practice. Some very difficult attitudinal problems arise for the operator when he is faced with a situation that demands doing more work than the contract calls for. Most of these problems arise because of the human tendency to try to avoid disputes or adverse situations. In addition, there are times when an operator becomes careless or lazy about pursuing compensable items of extra or additional work that his company is legally and morally entitled to. There are times, too, when good business practice demands trade-offs or outright contributions of this type of work to the other party.

The successful construction company must not permit its operators to give extra or additional work to the owner (except under some special circumstances), unless the company is reimbursed for the work; it would not be good business practice, and ordinarily the economics of the construction process in today's industry will not allow it either. The operator must assume that the other parties privy to the contract are reasonable and honest, but at the same time he must take the attitude that his demands for extra work are equitable and justified too.

290

The claim provisions made within the typical construction contract for the claim are mutually agreed on, and it is a duty of the construction operator to take advantage of these provisions. He need not feel unprofessional, much less guilty, about using the claim provisions to the best interests of his company. (It should ease the mind of the construction operator to realize that there are special courts that deal exclusively with contract claims; they exist solely for the purpose of the adjudication and settlement of claims.)

It is in the nature of the claims process to create adverse situations and positions. This can result in some very unpleasant situations relative to personal relationships and professional associations between the operator and his industry associates. No one likes these situations; nevertheless, the operator must perform with competence, perseverance, and thoroughness with whomever he is dealing with, no matter how unpleasant the situation in which he finds himself.

A most important attitude that must be developed in the claims process is one of caution. The construction operator must always bear in mind that while claims situations will not arise in every construction job, every job has the potential for claims. This is an essential point.

Let us clarify a point here before we proceed with our discussion of attitudes so as not to give the reader a wrong impression about the use of the claim in the construction process. We do not suggest or promote using the claim as a standard business practice.

No competent construction operator representing an honest and reliable construction firm should specifically look for claims as a matter of business practice. This will not be beneficial to the firm, particularly to its long-term success. But the construction operator must be pragmatic when dealing with claims. When an adverse situation does arise, the operator must be prepared to handle an operation that is unfamiliar to him. A claim is not an everyday operation; he must think and act cautiously.

Example. One of the vital steps in the claims process is the earliest possible documentation of facts and events surrounding an adverse situation. But how does the operator recognize when a simple dispute becomes a basis for a claim? (Early recognition of a claim situation aids the operator in all the succeeding operations required by the claims process.)

The operator must start documentation operations as soon as he suspects the potential for a claim, just as a precaution. At times even the slightest hint will start him on documentation operations, which will eventually give him a great advantage in the claims process.

The adverse nature of a claim causes many of the unpleasant features of the process. There is, for example, a simple basic contest inherent in the claims situation where one party has to try to collect money from another when this other party does not honestly believe he owes it. Even when both parties are acting with sincere and honest motives, the claims situation can still sometimes lead to unpleasantness, bitterness, and mistrust. Taking objective and impersonal attitudes would seem to be the answer to such difficulties, but business is conducted by ordinary people who often find it hard to see the justice in an opponents' position.

To the greatest extent possible emotional and unreasonable attitudes must be replaced by good, solid reasoning presented with the greatest number of supporting facts available. The principle task of the construction operator (up until litigation) is to convince the other party of the sincerity and validity of the claim, with the idea of settling the dispute between the parties without the need for outside judgments and decisions.

We claim that the best practical results are obtained for both parties when claims are settled in their early stages. The introduction of third party adjudicators in the arbitration process or in court should always be considered to be a last resort. Too many times this last resort becomes the only alternative because of personality conflicts and unreasonable behavior on the part of the parties in the early stages of the process. The construction operator must bear firmly in mind that as the claim progresses, time and money are wasted. The construction operator should develop and refine helpful attitudes that will give him an advantage in the claims process.

Disputed Work—Confusions

Disputed work is generally extra work in some form or fashion. The courts have defined extra work as "work arising outside of and entirely independent of the contract and not required in its performance." This is a simple definition that the construction operator can relate to in assessing a claim situation. But a common error of many construction operators is to confuse the terms *additional work* and *extra work,* they are not the same thing. The courts, following the directive of the industry itself, define additional work as "that necessarily required in the performance of the contract, not intentionally omitted from the contract, and evidently necessary to the completion of the work."

This distinction, however vital it may be in making a legal decision, does not eliminate additional work from being the subject of a claim; but it does alert the construction operator that when additional work is involved in his claim, he must take a different approach in his operational procedures than if he were dealing with extra work. He must, for example, give more thought to the supplemental and attendant costs of additional work than he does in the extra work situation. Additional work (which may or may not be compensable) may involve additional expenditures that were clearly outside the contemplation of the parties at the time of contract. It then follows that the claim's operator must emphasize the facts about the *contemplation* of the parties, depending upon which type of work is involved, because each type calls for his using different thinking approaches and procedures in his handling of these facts within the claims process.

Terms such as *delay, compensable delay, excusable delay, change, constructive change,* and so forth describe extremely complex subjects in the area of construction operations. Therefore the competent construction operator must have some general idea of what they mean in order to facilitate his overall operational capacity. But in the claims process he must know exactly what they mean, not only as they apply to operations but, in addition, what problems their causes and effects create in the law. Experience will teach the operator about these things. Unfortunately, a new operator may be faced with a claim situation or be involved in the claims process as part of his responsibilities at the very beginning of his career. The new operator must immediately learn what these terms mean, because the steps involved in the claims process demand operational actions and, more importantly, operational reactions to the adverse situation.

The Operator and the Legal Process

In the claims process there is a strong probability that legal assistance will be required. This is not to say that this is a foregone conclusion. Quite to the contrary, the main purpose of this chapter is to hopefully lessen the need for attorneys in the construction process, because we believe that the more competent the construction operator becomes in this process, the less he will require their help. After all, construction operators regard themselves as "contractors," do they not?

It is not suggested here that the construction operator should be as competent as the lawyer in contract law. The reader must realize that while contractors unfortunately think of themselves as laymen in the law (as contradictory to their self-impression as this may seem), they seem to forget that the attorney is a layman in the construction process and its procedures, technology, practice, and operations. In claims situations (or any adverse situation involving the construction process) a "guessing game" is often played by the operator and attorney as to who is to educate who in their respective professions. Bluntly speaking, the contractor should know more than any ordinary attorney about contract formation, contract conditions, and their implementation and proper performance.

The technical and procedural aspects of the legal process are, of course, the domain of the attorney, and this is as it should be. But it is hard to imagine that an attorney advising a contractor client in the legal aspects of the claims process would not welcome the aid and assistance of a construction operator who can provide him with relevant and substantive data to work with. With these materials the attorney has a better chance of winning the claim; without them he can have trouble winning, not through any fault of his own but because of the inadequacy, incompleteness, irrelevancy, and in some cases total lack of material to support the claim he must press.

Unreasonable or disorganized actions taken by the operator in gathering data or in communication procedures before the attorney's arrival on the scene can be equally disastrous to a claim. The attorney, for example, cannot be expected to know the detailed technical aspects of the construction process, yet the fact is that in order to win a claim he must be familiar with the technical aspects of the claim at hand. The construction organization is therefore forced to educate the attorney at its own expense. This can be very costly to the contractor claimant if the technical aspects of the claim have not been organized in such a way as to make it easily understandable to the attorney.

It is always necessary for the construction operator to attempt to minimize legal expenses in the claims process. In order to do so, at some point in a real or suspected adverse situation he must develop and direct his operations specifically to the claims process. This may cause a major and extensive change in the direction of his operational thinking, but it will facilitate and shorten the attorney's time on the scene.

What kind of directional change is necessary and when should it take place?

The directional change should be toward protecting specific interests and it should take place as soon as it becomes evident to the operator that the substance of the claim cannot be settled by resorting to the terms of the contract.

Determining the exact moment when a disputed item will result in a claim (thus raising the potential need for legal assistance) is always a matter of the greatest uncertainty; nevertheless, the construction operator must do so at some point. As soon as this is done, his operations automatically require him to direct his entire operational thinking to satisfying the requirements of an attorney in his area of expertise. A good guiding premise for the construction operator to rely on in determining when to change direction is, "the earlier an attorney can be called in and furnished with as relevant and valid data and documentation as possible, the more able he will be to adapt his operations to both the procedural and substantive legal issues of the claim."

The Contractor Versus the Owner

This operational process would be much simpler for the operator to perform were it not for the adversary relationships that necessarily exist between the parties to the dispute. No sensible contractor likes to be engaged in dispute with owners or their architects/engineers. The great majority of owners and agents are honest and fair people who sincerely desire to pay fairly for what they contract. Most owners are laymen and novices, as far as construction and the construction contract are concerned, but who nonetheless possess good common sense and have a keen sense of fairness. The claim situation usually matches the professional against the amateur, and this can often cause a feeling of discomfort for the professional who has a sensitivity about fair dealings in his operations. This, from a personal standpoint, is the hardest type of owner to claim against.

A few owners are more sophisticated, and some are extremely competent when it comes to the claims situation and its processes; they have learned through practice. In dealing with this kind of owner, the construction operator can operate much more objectively in the pursuit of his claim. He does not have to feel that he is taking unfair advantage of the situation, since he is operating against professionals who are more than capable of taking care of their interests. The construction operator must then take special care of his interests with these contestants. He should not, however, think that dealing with the sophisticated owner necessarily presents a more difficult operational task. In many respects it is much easier to operate with this kind of owner, mainly because he knows the rules of the game, its methods, and the operations that regulate the contest.

The Contractor and the Architect/Engineer

With certain owners the construction operator must temper his attitudes considerably because he looks to repeat business when he deals with them, but not to the same extent that he looks to working again with the owner's architect/engineer. There is a greater chance that a contractor will do business more frequently with the same architect/engineer than with the same owner, and the construction operator must therefore develop somewhat different attitudes when dealing with these professionals. Many times in the claims situation the contractor must, from an economic standpoint, initiate the claim, regardless of any adverse consequences that might arise between himself and the owner's architect/engineer. The claimant should carefully weigh the alternatives and exercise good business judgment when making its decision to claim, because such action will likely affect its relations with the agent on other current work or future work.

Architects/engineers are in a unique position in this operational process. In most cases it is they who have formed the bidding and contract documents out of which the claim has arisen. This results in making them repsonsible in some respects for the claim itself when it occurs; for example, an item of work may not have been included in the documents by mistake or otherwise. The architect/engineer is thus forced into a defensive position. Depending upon the degree to which his interests are at stake, there is every possibility that the agent will be an obstacle in any attempt to settle the claim, even when the owner may be willing to make an equitable adjustment. The construction operator's main task is to convince the architect/engineer of the real deficiency in the documents and the integrity of the claim. It is always wise to try to settle substantive issues of claim, if possible, with the architect/engineer before the owner is presented with the formal claim.

The Contract and the Claim

A claim cannot be settled by or through the provisions of the contract. The claim provisions in the contract make the claim an extracontractual matter. The ordinary construction contract really only allows for the possibility of claims and, except for specifying certain required procedural steps of notice and time, does nothing to regulate or even guide its settlement. Therefore the resolution of the claim depends on the development of extracontractual procedures by the parties in dispute for use in the process. Since the initiator of the claim is usually the contractor, its expediting operator is the person who starts the process on its way.

15.2 THE THREE STAGES OF THE CLAIMS PROCESS

There are three distinct stages in the claim process:

1. The documentation stage.
2. The negotiations stage.
3. The litigation stage.

Each stage requires establishing definite patterns for the operator's performance.

Example. The operator from the beginning stages of the claim will start to operate according to a definite set pattern. Negotiations must be patterned to the same set of negotiators meeting time after time. Litigation requires set patterns for the dissemination of information between the contractor's attorney and the operator.

Each stage requires organization.

Example.   Documentation must be listed in the order of its writing and importance to the claim situation. Negotiations must be systematized with agenda, and meeting procedures formalized. Litigation requires the organization of technical data to the extent that the attorney needs it to win.

Each stage also requires the adaptability of thinking processes in carrying out the total claims process.

Example.   The contents of all prior documentation will vary widely from subsequent documentation. The detailing, tone, extent, and content of correspondence will become extremely varied as the claims process progresses. Negotiations require wide adaptability to extemporaneous dealings between the parties, because the atmosphere or tone of this phase of the claims process can change so rapidly. The litigation stage usually calls for the transposition of stress from previously strong technical points in the informal negotiation stage to legally important points depending on the attorney's impression of the court's attitude or the arbitrator's demands. The reader can see, therefore, the wide stage on which this process can take place. Each role played in each act calls for many adaptations of the thinking processes of the operator.

The Documentation Stage

This is the initial stage in the claims process. At this point all parties involved in the construction process begin to be concerned with the dispute. The construction operator begins to consider a much wider dissemination of information than he may have thought necessary in the ordinary course of his operations. Both field and office members of the operational organization are alerted that a dispute exists or is imminent. This allows them to start coordinating their efforts toward protecting the interests of their organization. They should be fully alerted when a dispute starts to become a real adverse situation, even when at the time it might be of a minor nature. The construction operators who are generally first aware of the dispute are the expeditor and the field supervisory forces. It is up to them to flag the entire operational organization to the possibility of a claim.

When a dispute takes a real turn toward a claim, documentation procedures and processes also take a turn. Documentation relative to the matter in dispute must be carefully analyzed both as to its effects on the adverse situation and also as it affects the company business as a whole.

In the claims situation documentation must be formed according to certain rules; it must form evidence and be the following:

1. Exact in detail where detail is required.
2. Made and kept with a strict concern for chronological order.
3. In writing, supplemented by graphic or photographic displays when practical.
4. Clear and concise in its contents.
5. Attested when necessary.

6. In copy for distribution purposes when required.
7. Totally focused on its objective, which is the satisfaction of the claim.

Documentation must be kept current with the events that occur within an adverse situation. To try to recollect detailed fact situations after the fact is a difficult if not impossible task at times. As the period after the occurrence of the facts increases, so does the difficulty of adequately documenting them. This means that the facts should be documented formally as soon after events occur as possible.

Formal documentation means just that; it means, for example, proper and complete correspondence in letter form and not merely an assemblage of memorandums or notes, which might be only meaningful to the writer. It means the presentation of facts (as understood by the operator) to the other party for comment or rebuttal as close to the time of their occurrence as is possible. The documentation process is to be directed toward the single purpose of formally establishing a record of facts that a reasonable person at some future time will accept as being accurate.

There are some notable exceptions to insisting upon formality in documentation for a claim, including the job superintendent's job reports (or log) and the contents of the minutes of job meetings. These two writings have higher credibility as evidence if they are made informally (as they usually are), because of their spontaneous nature, and this spontaneity must be maintained in making these two types of documents. But this does not mean that other documents supporting or noting them should be informal. Quite to the contrary, objections to the contents of either type must be in the most exact form possible. In addition, formal documentation supporting the logs and minutes must be done thoroughly and as expeditiously as possible.

It is a good idea, as a general rule, for the construction operator in the claims process to eliminate or minimize the use of memorandums and notations as a substitute for formal documentation. However, there is one notable exception to this: when extemporaneous verbal conversation has a bearing on the dispute. It is a good idea to immediately verify all verbal conversations by a memo signed by the parties to the conversation. But even within this exceptional case memorandums should be converted as quickly as possible to formal documentation.

The construction operator should give special attention next to the "double-edged sword" nature of the documentation stage of the claims process. Ideally, documentation should establish a true record of events and facts, but this can work both to the advantage and disadvantage of the operator in the succeeding claim stages of negotiations and litigation. The operator must adapt his thinking to this fact so that he is never thrown off guard by the story told by the total facts. To prevent this from happening, the operator must physically put the total documentation into a sequence for examination and interpretation. This can be done by what is known as "bed sheeting."

A "bed sheet" is simply a graphic display of excerpts from the total documentation in their order of occurrence. The form and content of the bed sheet vary according to the needs of the individual claim. Essentially, this document should present those parts of the total documentation that are pertinent to the claim itself in as total and convenient a graphic form as possible. These ordered excerpts of documentation thus form a separate claim document.

The excerpts should form a complete picture of the claim. In fact, the basis of the claim and its source should stand out clearly in this one document if it is made correctly. Only those facts that establish and prove the claim are included in the bed sheet. This document is extremely helpful in keeping the presentation of facts straight in the negotiation and litigation stages of the process.

The claimant's operator should be aware that the other party can bed sheet as well; it is possible to present quite different pictures of the claim by using different presentations of the total documentation. He must also be aware that his opponent can establish counterclaims from the claimant's document. Nevertheless, this type of document facilitates the claims process: it makes a picture and can be used throughout all stages of the claims process.

The bed sheet should include pertinent excerpts from the following:

1. The superintendent's reports and log.
2. All correspondence.
3. The minutes of the job meetings.
4. Published reports of the owner's agents (i.e., architect's job log and reports).
5. Supplementary documentation such as subcontractor correspondence, consultant reports, test reports, inspection reports, and so forth.

These excerpts, of course, must be supported by actual complete documents. Special documents like photographs however, can be inserted as references where required. An easy method for bed sheeting is as follows: use the backs of old drawings; cut and paste the excerpts in their proper sequence down the sheet, keeping separate columns for the separate types of documents excerpted; and when the composite is complete, print it to make a new document.

The bed sheet can be a very extensive and time-consuming document to compose in major claims, but the reader should compare its advantages to the physical and cognitive difficulties involved in sorting, sequencing, and remembering the application of the several hundred individual documents that are likely to be pertinent in presenting a clear picture of a major claim; he should also consider the relative ease with which attorneys, arbitrators, courts, and juries can use the bed sheet.

The Negotiations Stage

The construction operator should always aim for an equitable resolution of a claim before the negotiations stage. The prime reason for this is because while the operator has some real control over the processing of the claim before the negotiations stage, his control diminishes as this stage progresses. He will start to lose control over the situation particularly when the involvement and participation of attorneys increase. In the litigation stage the operator becomes at best a consultant to the. attorney, with little or no control over the claim processing or resolution. Many times his role is reduced to one of merely accepting or rejecting a negotiated settlement that the attorneys adjudge to be satisfactory or possible.

The negotiations stage is the intermediate stage in the claims process. It therefore mixes some aspects of the documentation and litigation stages. It is not the objective of the operator in this stage to merely establish facts, as in the documentation stage, nor is it desirable to emphasize the application of law to the facts. The latter is the main objective of the litigation stage. In the negotiations stage the operator must attempt to achieve a settlement, without destroying the validity of the claim should a settlement not be reached.

This stage must establish positions wherein movement toward settlement becomes possible. The documentation stage only starts establishing positions and the litigation stage sets the positions finally; between the two is the negotiations stage. The operator must realize this important fact and take positions that are adaptable to change in the negotiations stage. His adversary must also recognize this fact. There is usually only the finality of winning or losing in the litigation stage. In litigation it is not nearly as easy to change one's position as in negotiations. Litigation is to be avoided if at all possible.

Compromising

In all probability, the contestants will have to approach negotiations with an attitude of compromise. If this were not so, then there would be no reason for including this stage in the process at all. Since a great number of claims are settled in negotiations, and negotiations by their nature demand compromise, the adversaries must direct their thinking to what this stage requires in the way of compromise to reach a successful conclusion. The development of thinking processes for this stage of the claims process should be directed toward a compromised settlement. The contestants must learn how to compromise realistically and to determine methods of negotiation by which they can gain as much for their respective interests as possible. They must avoid putting their opponents in a position where they must resort to litigation.

Understanding the Adversary Relationship

The construction operator must realize that he is an adversary, because his objective is to get something from his opponent who does not want to give it. But he must also realize that his opponent, by participating in negotiations, is aware that he must give something up, or else the issue would be settled and there would be no alternative but litigation. If the operator can get the negotiations stage started, he can have some confidence that he will gain *something* from his opponent if they are successful. Both opponents must regard each other as being capable of protecting their respective interests, and both must operate accordingly. A contest begins in the negotiation stage of the claims process, and the attitude of the construction operator here should be "let the best-prepared man win." The business ethic becomes the predominant consideration at this point. The setting of this process is competitive, and what is considered to be fair and equitable by the business ethic relative to competition should govern the thinking of the operator. This is an extremely difficult attitude for the conscientious operator to assume in some situations, particularly where he knows that his opponent is operating under a strong technical or business disadvantage. Nevertheless, the construction operator must be impersonal and competitive in his purpose (but not in his attitude of compromise). The negotiations stage must be approached on the basis that it has by far the most personal contact between apparently honest and sincere opponents than any other stage of the claims process. The operator's attitude of compromise and his competitive purpose must be skillfully integrated in this stage of the process.

Personality

The next thing the operator must consider in developing his thinking processes is to adapt his personality to this stage of the process. The process of negotiating a claim requires the participants to adapt to each other's personalities and dispositions. But some general adaptations of personality to the process itself are also required: cool behavior, clear communicative skills, and patience, for example, are clearly helpful; quick anger, boorish behavior, and bad manners hinder the operator. Successful negotiators always gauge their actions and behavior in negotiating sessions to the participating personalities.

Acting

The construction operator should be keenly interested in developing his acting skills for use in the negotiations stage of the process. In most adversary proceedings it is necessary to playact at times. The reader should not regard this statement as being facetious. Playacting is often an essential part of the negotiating process; to use it to portray indignant righteousness, offended honor, and the dramatic but controlled expression of emotion is perfectly acceptable in this process. Remember, negotiations are a contest. Each side has the right to playact to gain advantage over the other; it is up to each party to develop an acting skill that is suited to his or her personality.

The proof of the value of deliberate playacting is that it works. For example, in negotiations operators will often combine as a team for the specific purpose of presenting themselves to their opponents as extremes in personalities or dispositions: one will play the part of the hard, uncompromising, straight-down-the-line individual, while the other will be easygoing, flexible, and good-natured. Each role has its purpose in the negotiation stage of the claims process. The art of acting allows greater flexibility in the art of compromising.

Timing the Start of Negotiations

There are certain times when negotiations can be more successfully conducted and, more importantly, concluded than at others. The construction operator must recognize this and consider it carefully in his negotiating strategy.

Example.   When is the best time to start the negotiation stage of the process from the claimant's point of view? The most appropriate time is when he feels that a face to face meeting will be the most productive in satisfying his claim. It is not likely, for example, that the owner will have more money available toward the end of a job than at its beginning. This means that the claimant's strategy should be to start negotiations as early in the course of the job as possible. But he must be careful, because, ordinarily, the full costs of the claimed extra work can usually be more accurately determined as the job reaches its completion, and it is better to negotiate when the costs are determined fully rather than to project or guess at the amounts involved.

The start of negotiations may be determined by other influences. Business considerations of cash flow, for example, can force the decision on the claimant; cash flow problems can determine the timing of the entire claims process.

There is another matter of strategic timing to consider in the claims process: the operator must determine when to introduce other participants into the process. When, for example, is it appropriate to involve attorneys and subcontractors in negotiations? Introducing them either prematurely or too late in the negotiation stages can cause serious difficulties in the total claim process. Aside from cost considerations, the introduction of legal counsel, for example, into the process may result in a complete change in the attitudes and behavioral patterns of the negotiators. As for subcontractors, they may have extremely strong self-interests, which could delay or obstruct the proceedings.

The Process of Negotiating Claims

The claimant initiates the negotiation by simply asking the other party to participate. If the other party responds affirmatively, some sort of negotiation procedure is established. An initial procedure might be set up as follows:

1. A time and a place are set for an initial meeting.
2. Participants in the initial process are agreed to in advance.
3. One of the parties will make an agenda.
4. The meeting is held.
5. Succeeding meetings are scheduled and held until a settlement is reached or the parties decide to litigate.

If the party claimed against demands that the formal requirements of the contract be followed exactly, then usually a request for a change order for the claimed work should be made by the claimant, which, upon refusal, formally sets the stage for negotiations.

There are usually three substages in the negotiations stage of the claims process, the last two of which are forced when a satisfactory settlement cannot be achieved:

1. The preliminary substage.
2. The intermediate substage.
3. The final substage before litigation.

Each substage has a purpose and its own particular requirements of thinking for the construction operator. Each one allows for the possibility of settlement. Within each substage there are indications of the possibilities of success in negotiating to a settlement, thus eliminating the need to resort to litigation. The construction operator should never count on settlement in any one of the three substages, but it is just as true that he should never discount the possibility. Many claims are settled in the first session of the first substage; others are not settled until the final substage. Settlements can happen at any time.

The Preliminary Substage

The preliminary substage of negotiation generally involves no more than becoming better acquainted with the opponents who are directly related by contract to the adverse situation. There are sometimes great differences in even well-acquainted peoples' attitudes toward each other in the claims situation as compared to their behavior under ordinary conditions. This substage gets the parties together for conversation. At this point there is no room for threats or intimidating behavior. No more is necessary here than the parties' informal efforts to persuade each other of the validity and sincerity of their respective positions. It is not wise to try to obviously gain advantage over the other party or to bring up irrelevant matters at this substage. There is no need to bend the truth here or mention legal counsel or suit. It is essential that behaviors be cool, calm, and collected.

Paper (documentation) impresses an opponent, particularly where he is short on paper himself. Preparation and thoroughness are also impressive. The operator must realize that although much information might have passed between the parties during the documentation stage, the face to face presentation and filling in of facts, opinions, and viewpoints are just beginning at this stage. The objective of this stage is to convince the opponent that it is to his advantage to carefully examine his own position.

In the negotiation stage of the claims process the question always arises as to how much and at what time information is to be divulged that is not already contained in the documentation. No set answer can be given, because an individual negotiation takes on its own character, depending upon its contestants and the nature of the claim. It is safe to say, however, that the preliminary stages of negotiations are somewhat of a discovery process. Many times the first substage of negotiations brings out previously unrevealed facts that have a serious or even conclusive bearing on the substance of a contestant's position.

In addition, the dispositions, opinions, and viewpoints of the parties are almost always much more thoroughly revealed in the preliminary stages of negotiations than in the documentation stage. By careful watching and listening to opponents' statements, attitudes, feelings, and even facial expressions, their deeply held convictions can sometimes be detected with great accuracy. Again, as in so many other construction operations, the ability to observe and listen is a prerequisite to gaining experience in negotiations.

Here we must discuss, for a moment, a particular point about timing: there is always the problem in timing the introduction of money matters into the process.

A main objective of the claimant's operator is to see that the negotiations are kept going to the point of outright refusal to bargain further. Nothing, for example, will undermine negotiations faster than the untimely introduction of maximum money amounts. This is so because the party claimed against must pay something and in most cases he does not believe that this should be the case. The mention of a significant amount of money has a tendency to back the opposing party to a wall.

It is wise for a claimant to be as vague as possible in the preliminary stage about demands of actual dollars. It is very unlikely that the full cost of a claim involving large amounts of money can be determined exactly at this time, particularly in large or complex claims that extensively involve complex time considerations. If the issue is forced, the claimant will tend to inflate his claim. It will never do to present unreasonably inflated claims, because damages must be actual and capable of being proven. It is more than likely that money amounts presented at the wrong time will scare the opponent, causing him either to reject negotiations or to delay them to his advantage.

However, at some point in the negotiations stage the money amounts at stake must be presented and discussed. The operator must learn to determine the best time to do this and at the same time prevent a breakdown of negotiations. The subject of money in a claim cannot be avoided. Sometimes it is good to introduce it early, and sometimes it is better to introduce it late. The presentation of real money amounts calls for the operator's persuasive skills and maneuvers, because as the negotiations progress he will usually increase the unease of his opponent over the amount of money demanded. In revealing significant amounts of money, it is especially difficult for the operator to maneuver when his opponent is pressing him for firm figures that he does not want to reveal.

Inflation of Claim. Next we must examine and discuss the inflation of a claim. Early revelation of the claim amounts of time or money almost always forces a claimant to inflate them because of the uncertainty of these amounts early in the claims process. The total effect of a time extension, for example, can be a very complex substantive issue to determine correctly if it must be done before it is fully determinable. The issue can only be fully defined after all its effects are known. In addition, it is usually difficult to fit the extension of time factor to other factors of cost in the claim, and even more difficult to establish accounting procedures to price the effects of the time extension. Therefore, when time extension is a major issue in a claim, it may furnish a legitimate reason for inflation of claim if, say, the opponent forces the claimant to make a comittment on time or a revelation of money amounts too early in the negotiation.

But there is very little if any justification for claim inflation, when the *substance* of the claim is inflated. This is no longer a matter involving just money amounts; it is a matter of falsely adding contractual obligations through the claims process by relying on expedient interpretations of the contract that the claimant knows were beyond what was actually contemplated at contract time. Where there is no actual damage or cost incurred, no damage should be claimed, even though the language of the contract could possibly be interpreted to allow compensation for it. An extension of time and attendant costs, for example, for the changing of reinforcing bars in certain footings can be legitimately claimed if the change actually modified the contract by delaying the project. But if at the same time a claimant changed his sequence to avoid those affected footings and did not, in fact, lose time and money, then the extension of time can be claimed, but it should not be charged for.

The construction operator should think very carefully about inflating the substance of a claim. If it is revealed in negotiations that such an inflation has taken place, it will tend to seriously, if not fatally, affect the atmosphere of the negotiations. Distrustful attitudes on the part of the opponent followed by indecision and unwillingness to bargain are the likely consequences. The operator should assume that from then on each detail of the substance of the claim will be closely inspected and interpreted by the opponent. When a claim is inflated this way and it is found out, it forces the opponent to go into a protective stance and to demand detail. Getting into too much detail is a disaster for the negotiating process at any substage but is especially damaging in this substage because its success depends so greatly on keeping the negotiations general in nature. In addition, claim inflation is more than likely to be discovered at litigation time, and when it is discovered, it will not help the claimant.

The Intermediate Substage

This substage obviously comes about because the preliminary stage has not resulted in a settlement of the claim. The operator must therefore change his tactical approach to the negotiations. The tactics of the new substage of negotiations are governed by the realities of the new situation.

Both adversary parties at this point know much more about the claim substances, the facts involved, the attitudes of their opponents, and are better armed with their own counterarguments to the validity of the positions of each other. They have learned as much to their advantage in the first stage as their opponent has learned to his. The claimant's operator must therefore realize that, while he might have gained advantage from the first stage, he has also revealed much information to his opponent.

In this substage the operator must introduce the "devil's advocate" concept into the examination of his claim. Too often an operator will become so involved in the claim, both mentally and emotionally, that he will avoid or completely miss the possible validity of the other party's position. There are two points that must be thoroughly examined by the operator when using the "devil's advocate" approach. The first requires an *objective* examination any contradictions or confusions in his interpretations of the fact situations, the accuracy of his recording of time sequences of actual happenings, and his honest appraisal of any adverse substantive issues involved in the claim.

An examination of the second point is the more important, because it is at this substage that the substantive law is usually introduced into the claim. The factual basis for a claim can be perfectly exhibited by documentation, but if some unthought of point of law becomes operative to the issues presented, then part or all of the claim can be fatally damaged. If, for example, there is a legitimate claim for time extension but notice was not given in accordance to the contract requirements, this could be fatal to that claim. At the very least the operator must be aware that these and similar possibilities exist and might ultimately result in the destruction of the claim. He must adjust his approach to the claim when this type of substantive legal issue comes up within the setting of this substage of negotiations.

Sometimes during this substage it is important that the operator reveal to his opponent the potential damage and the actual consequences the claim has had on his company. The operator must use skill to avoid threatening his opponent, but at this stage there must be a push to start the opponent moving toward settlement; many times this approach works. One can mention, for example, the difficulties being encountered by the field construction organization in placing or maintaining manpower on the project owing to the uncertainties caused by the claim situation. Depending on their effects, these and similar statements of difficulties can be mentioned to the opponent. If they are true statements, it is certainly legitimate to use them in the negotiations process.

The construction operator should realize, as the sessions in this substage progress, that negotiations are getting more delicate to handle because by this time both the exposed and settled positions of the parties are starting to harden. It is important at this stage that the operator rely on and emphasize hard facts that are beneficial to his position. By this time most of the relevant facts are known to the participants and so it becomes a matter of constantly emphasizing the facts that support the claim.

This is the critical substage in the negotiation process. This is the point in the process where there is the best chance of settling through negotiation or where there can be a total breakdown. This is so because at this point the contestants know the facts and have definite impressions of each other's strengths and weaknesses. The value of conducting this substage correctly is that it is here where admissions and acknowledgments are made on at least some major points in the dispute. One prime purpose of this substage is often to still try to convince the other party of the validity of the claims and to also make him aware of the consequences of not moving toward a mutually acceptable settlement. Another purpose is to set the stage for the final scene of the negotiations process, which is the scene that determines either settlement or litigation.

The Final Substage of Negotiations

Sometimes this substage is not necessary because settlement has been reached in one of the other preceding substages, but it is most likely where *failure* of the negotiations process takes place. One main objective of the claimant's operations should be to prevent the failure of the negotiating process. The next step—litigation—is never desirable; it is time consuming and very costly. These two factors alone should motivate the construction operator to exert every skill and effort in trying to settle by negotiations, even to the point of compromising and making major concessions that result in the loss of a portion of the claim or claim amount, in order to avoid litigation. It becomes a matter of balancing the economics of the options existing at this substage in the claims process.

At this point the construction operator must approach the negotiations generally in the same way as he does in the intermediate stage, except that he should increase the emphasis on the major and substantial parts of the claim. The operator must think of other factors in this stage as well.

The negotiation process, when considered as a whole through the claim process, becomes more general in nature as it progresses. The final substage usually requires broad discussions and arguments on the general issues of the claim, both item by item and in its entirety. It is at this point that substantive issues of claim (with dollars attached) are seemingly cast about carelessly in an attempt to reach settlement. Here the operator must look carefully at the total amount of money he is offered in a settlement offer, regardless of where the money might be formally assigned to specific parts of the claim. It makes little difference as to which issues of claim are allowed or disallowed if the money amount awarded by a settlement is satisfactory.

The final substage of the negotiations require the operator to be extremely cautious as to what he says and does. Remember that the next step is litigation. While the issue of whether this claim or that is allowed might become immaterial in the negotiations, the operator can be assured that the proceedings of the negotiations become a matter of record (at least informally) in the litigation process. Therefore, from a practical point of view, it is never wise to admit to specific points of the substance of the claim at this point, unless the opponent formally commits to settlement; rather, it is a much better tactic to maintain the substance of the various claims and to negotiate on the amount owing on each of them in order to reach an acceptable total settlement in money. If an acceptable total amount can be agreed on, then the point of maintaining the substance of the individual claim becomes moot. This tactic, of course, really depends on the operator's correct analysis of the possibility of settlement with an acceptable amount of money in the negotiations. His willingness to compromise on the substantive issues at this point depends on his belief that there is a real possibility of a just settlement.

Legal Assistance

It is difficult to propose a general rule as to when attorneys should directly participate in the negotiations process. The introduction of the attorney by one contestant will usually cause the introduction of an attorney for the other side. Sometimes this puts a damper on negotiations. It is safe to say, however, that the attorney must usually enter the practice at some time in the advanced substages of the negotiation stage, or at least be available in the background for certain purposes. This should usually be no later than the intermediate substage.

The attorney should be the ideal person to take over the role of the "devil's advocate." This is necessary if for no other reason than to protect the operator from prejudging his legal positions should negotiations fail. This need for protection usually requires the attorney's presence, if not participation, in the actual negotiations themselves.

The attorney, when he does come on the scene, must begin to give direct instructions to the claimant in the application of the law to the claim. He must caution and direct the operator about the realities of the law substantive in the claim but, more importantly, he must inform the claimant about the exact legal procedures required to be followed should there be a failure in negotiations; he can also give valuable advice as to the implementation of a successful legal settlement, particularly when third parties' interests may be potentially involved. Incidentally, throughout the entire claims process the construction operator should be aware of the potential problems that can occur when third parties have cause to intervene, and do so particularly if they intervene after the claim has been settled between the original contestants. It is necessary for him to protect all the interests of all the parties in the claims that are directly affected (and even those parties that might become involved through the operation of the law).

There are certain cases where special circumstances can require the last part of the final stage of negotiations to be conducted exclusively between the attorneys of the contestants. These negotiations may consist of a series of meetings, which are often valuable, because objective views can be expressed strictly on the operation of the law to the substantive legal issues of the claim, without confusing or mixing the technical and legal aspects of the claim. The significant advantage of this kind of meeting is that it throws fresh parties with more objective views of the claim together to negotiate. At the very least, these meetings give the opposing attorneys a chance to view their client's legal positions realistically through relatively unbiased and impersonal eyes. The dangers in using this negotiation strategy is that the attorney might not fully recognize the application of some vital technical aspects of the claim or fully understand the absolute minimums of the amounts of settlement that will be acceptable to his client before he makes concessions or commitments.

In every negotiation on claims there is a point at which hard decisions have to be made. No rule can be stated for determining when this point has arrived in the negotiations; it is simply a judgment the parties must make at a particular time based on the options available to them. The negotiations process is constantly in motion, even though it has its definite stages of development. The construction operator must realize above all that a successful settlement is possible at any point in any substage of the process, as is complete failure.

The Litigation Process

For purposes of this chapter arbitration and litigation are grouped together as being the last resort in the claim process. We realize that there are significant differences in the concepts, procedures, and rules of both actions, but they are both formal proceedings that are time consuming and costly. They both require the time and effort of the construction operator, and they both involve outside parties to judge on the merits and substances of the claims and counterclaims.

It is at this stage of the claims process that the construction operator ceases to be the main operational player. He is still an important character, but because of the demands of the claims process at this point, he becomes subordinate to the attorney. It is true that the operator maintains a more important role in the arbitration process as opposed to litigation, but even in arbitration it is usually unwise for the contractor to enter the scene without full disclosure of his positions and goals to his attorney and giving full authority to his attorney to determine the issues of substantive law involved.

Roles are reversed in the litigation phase of the claims process: the construction operator becomes the consultant and the attorney becomes the operator. Legal procedures require the attorney to become the main operator in the litigation phase, because he knows how to operate within the legal process. The construction claims process has at this phase been surpassed by the legal process.

In order to litigate, a decision to do so has to be made by the claimant contracting firm. In making its decision, the firm must very carefully balance many considerations against others. There is no doubt, for instance, that considerations of the time required in the litigation process are always a matter of major concern. It is well known that court dockets today are crowded, and the necessary prelitigation procedural processes of discovery and deposition usually required in a claims litigation can add to the time it takes to litigate a dispute. In fact, these prelitigation processes can be added as a legal tactical maneuver. It also costs money to hire all the extraneous people (stenographers, experts, reporters, etc.) who usually become involved in the legal process. These kinds of costs must be paid by the claimant, because legal expenses are not included in the estimate of a construction project. While these expenses can be charged to the opponent if he loses, they will be a direct overhead cost if the litigation is lost by the claimant. We recognize that attorneys work under differing fee structures, but even in contingent fee situations supplementary costs to the claimant can be considerable.

Throughout this chapter we have emphasized that litigation should be avoided if at all possible. The monetary costs are one main reason for this, but the construction operator must also realize that litigation costs not only in terms of money but also in terms of more subtle adverse effects on the business considerations of goodwill toward his company and his own professional reputation. The litigation stage can be long and hard; it requires the parties involved to withstand many outside pressures that can affect their thinking and their perseverance. Nevertheless, if the claimant has a genuine claim that is properly documented and of significant value, and the claim cannot be settled otherwise, then litigation is the only good business alternative left.

There is only one single goal for the parties in litigation—to be a winner. In claims litigation the stakes are usually too high at this point to lose. Once into the litigation process the claimant should consider himself to be in it for good. No holds should be barred in this phase, excepting, of course, those prohibited by law. The law provides the methods of litigation and the rules that govern them. The construction operator, once in, must do his best to win. The chances of maintaining goodwill between the opponents at this point is no longer a significant factor. Quiet, courteous, but, most importantly, impersonal behavior toward the opponent is the practical demeanor to assume in the litigation phase of the claims process. Nothing is to be gained by behaving otherwise.

An arbitrator can always allow for negotiations within this process, as can the court. It is desirable, in most cases, if the court can "influence" the parties to reopen the negotiations, but negotiations conducted under the order of a third party are different than those engaged in voluntarily. The desirability of agreeing to reopen and continuing negotiations when requested to do so by the court or arbitrator is that the claim is brought back to the parties for settlement, thus enabling those who are most informed to come to an equitable agreement. Such agreements are usually welcomed by the courts. Judges, for example, are usually very competent in their professional field, but they are laymen when it comes to construction matters. This is somewhat less of a problem with arbitrators.

The reader may ask why the objective opinions or judgments of laymen are not a valid basis for deciding the merits of a claim. It is simply because objective judgment is not always the fairest one particularly when highly technical aspects are involved in a dispute. Justice demands a reasonable understanding of *all* the technical aspects involved in a construction dispute, as well as some reasonable (but not all-inclusive) understanding of the legal aspects of the case by the party or parties judging the merits of a construction claim in suit. The parties to the claim are most likely the parties who should best understand the overall content and nature of a construction claim.

15.3 THE OBJECTIVES OF CLAIMS STUDY

The objective of this chapter is much the same as all the others, that is, to try to convince the construction operator to develop his thinking processes toward accomplishing his objectives by means of a particular operation. Claims, by their nature, require unpleasant and tedious tasks. They call for special attitudes and efforts from the operator that are quite unlike those required in other areas of the construction process. A successful settlement of claims calls for behavior that is uncharacteristic of a sensitive person, as well as rapid and varied adaptations of one's personality. There is an appropriate time in the process, for example, to be highly personal, and there is a time to become impersonal. Sometimes it is extremely difficult for the ordinary person to adapt his personal behavior to the operation.

The natural personal behavior of an individual in any type of endeavor is usually the best behavior for producing the optimum desired results. The claims process demands more: the construction operator is forced by circumstances and the nature of this process to change character often and to act in different roles.

In the negotiating stage the construction operator must think carefully about the individual characteristics of his opponents, and usually he must do so without the benefit of having prior acquaintance with them. He must attempt to fully understand their positions and attitudes about the claim as it passes through its stages.

Many times the operator will be forced to be in conflict with the personalities of his opponents. He must try to understand their dispositions and personal attitudes about the claim, as opposed to their official attitudes. He must also watch for nuances in attitudes and behavior. For example, an opponent's sympathetic remark about the operator's position may be evaluated by the operator as a distinction between this opponent's official position and his personal opinion; such a remark may be taken as reflecting a conciliatory attitude upon which the operator can build and expand to his advantage. On the other hand, a derogatory or sarcastic remark or attitude can be viewed by the operator as an indication of a weak spot in his position that might require reanalysis or avoidance altogether. A simple remark can at times mean a break or concession in a position; it depends on the circumstances that surround the remark. Judgments made on these kinds of nuances should usually be based on a very close analysis of the circumstances under which they were made; they should be given their proper weight and should be based on strict objectivity. Usually, the more information the operator can gain about his opponents, the greater his advantage in the claims process.

The construction operator must have the ability to realistically and objectively assess the value of pursuing a claim before he opts for the litigation process. He must, for example, be able to assess the following:

1. The time that is inherent in the process and such things as the company's cash flow or ability to absorb any additional financial burdens caused by pursuing the process.
2. The value of the goodwill and future business of his opponents, since it is most unlikely that the claim will enhance these relationships.

3. The value of his own time spent in this operation and that of spending it on other necessary tasks.

4. The main risk of losing the claim or being forced by the circumstances of the process to settle below the minimum acceptable compensation.

5. Such supplemental considerations as the differential on the interest paid on monies borrowed to support the pursuit of the claim and the apparent inability to fully recover these costs in the claim.

No claim should be instituted on principle alone. Every claim must have solid economic reasons for its initiation and processing. The basis for claims should never be deliberately sought from either the circumstances of a project or the contract documents. A company that relies on making claims an ordinary, acceptable, and deliberate course of business practice is not likely to succeed. The cause of a claim in the claimant's mind should be self-evident from facts that demand an equitable adjustment.

The most important psychological aid the construction operator should develop is a frame of mind that can reject and withstand the pressure caused by uncertainties in the claims process. This type of pressure is unique to the process; it is unlike the pressure exerted on the estimator by his operations, mainly because of its duration. The claims process can be a very long one and the uncertainty of winning or losing can become a real problem for the construction operator. He must conquer this uncertainty through perseverance and determination.

Temporary setbacks, discouragements, and disappointments must never prevail over the objective of winning. The construction operator who perseveres in his claim need not be ashamed of his involvement in the process: its proper performance should be a matter of pride to himself, and should he win, one of considerable satisfaction and value to his company.

16

Documentation and Communication Procedures

Careful with fire is good advice,
we know.
Careful with words is ten times
doubly so.

Anonymous

16.1 DEFINITIONS

Documentation is the supplying of documents or supporting references. A document is anything written or printed that is relied upon to record or prove something.

Communication is a giving or a giving and receiving of information. In construction operations methods of communications generally take form within a system.

16.2 GENERAL COMMENTS ON METHODS

Documentation and communication are used extensively throughout the entire construction process. Both are vitally important operations on which the various operators must spend significant amounts of time. In most cases the processes and procedures of these operations are developed by the construction operators to suit their personalized ways of writing or speaking, together with the customary procedures and processes required by their company, industry practice, or the contract.

Ordinarily, uniform documentation and communication methods are established by a construction company for efficiency in operations. Uniformity of method allows the documents to be understood and used by all members of the firm.

Example. A set company documentation procedure would be the established way for a job superintendent to use a set-form to report on his job. A set communication procedure would be the established sequence or system of taking telephone subbids on bid day and the methods used to include them in the bid price.

A personalized documentation procedure would be an expeditor's own listing methods for recording shop drawings or bulletin status. A personalized communication procedure would be an individual's choice of using the telephone, telegraph, or written correspondence when a set communication procedure has not been established by the company.

There are certain types of operational documentation that are both set-form and personalized procedures of recording and communicating information.

Example. Keeping a job superintendent's log is an operation whereby the super-intendent documents certain facts to record happenings on his job. Besides documenting for the superintendent's own benefit, this recording also serves the purpose of commuicating information to others. It is individualized in that it is an original, extemporaneous composition, and yet it is a communication procedure that is expected by the company organization from its supervisory operators.

Documentation and communication are both construction operational practices that must be learned thoroughly by all the various operators in the construction process. Contractors, subcontractors, and suppliers, architects/engineers, and owners must all be proficient in these fundamental construction operations.

16.3   WRITING

While there are many other forms of documentation, the one most commonly used in the construction process is writing. We know that writing must convey thoughts. Since a composition is going to be interpreted when it is read, it must always be governed by two important principles:

1. The writer's ideas must be expressed clearly so that they are adaptable to interpretation by the reader.
2. The writing (which is evidence of professional competency) must be proper in language form and usage.

Various types of documentation will be discussed later in this chapter. Each type develops its own character in its form, but no matter what the form, it must be capable of being understood.

Writing for Understanding and Interpretation

Each writing (and there are many) made on the contract requires its peculiar statements relative to the contract provision that mandates the writing. All written statements must satisfy the requirements of the contract provision. For example, a letter requesting an extension of time to the contract might take the form as shown in Figure 20. The contract provision that requires this kind of writing is usually known as "delays and extensions of time," and it requires written notice of the occurrence of the cause for the extension, as well as a written request that the extension be made by a certain stated time limit. The contract provision does not tell one how to write the letter; it just states that a writing is required. The construction operator must utilize his thinking processes to adequately satisfy this provision and all others like it in the contract.

In the letter (Figure 20) the reader should note that it is:

1. Dated, thus establishing the time of the notice.
2. Stated to be in "accordance with," thus giving the contractual reference and the reason why it is written.
3. Properly addressed and clearly written so that it is a good, concise communication and a document of record.

```
┌─────────────────────────────┐
│         ABC  INC.           │
│    General  Contractors     │
│        108 Main St.         │
│    Columbus, Ohio   54112   │
└─────────────────────────────┘
```

```
Smith and Smith Associates
Architects and Engineers
700 N. Main Street
Cleveland, Ohio 59842

Att: Mr. W. C. Smith

Re - Midtown Office Complex - Your Job 7816

Gentlemen:

Your changes to the reinforcing steel in certain footings has caused a
delay on the reference job.

This is to give you notice in accordance with Article 8, Section 8.3.2,
of the AIA-A201 General Conditions of the Contract, that we wish to claim
an extension of time for this delay.

We shall furnish you the costs occasioned by this delay when they can be
fully determined.

Sincerely,



John Doe
Expeditor

cc. __
    __
    __
```

Figure 20.   A letter requesting an extension of time.

It follows from this example that the same thought processes which went into the composition of this letter can be equally useful in composing documentation for every contract provision that requires a writing.

Thinking about requirements of time, reference to the contract, and language form should be useful in any type of construction documentation procedure. By thinking this way, the construction operator can develop an individualized method of composing all different kinds of specific types of writings. He forms an outline of the information he is required to provide when writing a certain type of document; this promotes efficiency of operation.

Writing With Discretion

The construction operator must write with discretion. This calls for good judgment and pragmatic thinking approaches based on the facts of the situation at hand. Let us pose some questions that illustrate the range of discretionary thinking that should be exercised by the operator when composing a document:

1. How many facts should be given to the reader? All or some?
2. How much personal feeling or opinion should pervade the ideas presented in the letter at the particular time it is written?
3. What tone should the letter take? Should it be demanding, stern, or pleasant?
4. How much detail should be given about each stated fact? For example, should the operator state exact or general amounts of money? Should he state exactly the potential effects of a particular fact or situation, or should he leave them to the reader's imagination?
5. How should the letter be written in relation to the expected answer? Should any answer be expected at all? Is the letter being written to provoke or force a certain answer?

It is possible that all of these questions (and others) have to be considered by the construction operator when composing a specific document. Many of them are automatically answered as the writer develops his writing skills and categorizes specific types of documents that have to do with contract matters and other specific construction operations. Thus the letter requesting an extension of time, the cover letter, the letter requesting approvals, and so forth all take on different but definite patterns of formation.

16.4 CONTRACT REQUIREMENTS

The construction contract itself is a document, and within the contract are the conditions of contract. These conditions provide the means and methods of implementing the performance of the contract. The performance of the contract depends, in a considerable number of instances, on meeting certain requirements for writings that are set up within the provisions. These requirements can be either directly spelled out, implied from the provisions, or just obviously necessary to facilitate the performance of the contract. Here are some examples of contract requirements for writings:

Direct. A letter requesting extension of time.
Implied. A backup letter for an oral request for information.
Facilitating. A cover letter sent with a payment request.

If one carefully examines the typical conditions of contract, it will be seen that there are many expressed demands for written notice. There are also many provisions that clearly imply documentation. Written documents that facilitate implementing the performance, even though they are not expressly required or implied, are found to be necessary because they allow the construction operator to expedite his operations more efficiently. The construction operator becomes familiar with the respective needs for these types of documents through experience.

Form

While the contract demands a document, it does not always specify its form or exact content. A payment request, for example, must usually be documented in set-form, but rarely, if ever, is a written notice. Affidavits and waivers are ordinarily required in set-form, but a letter requesting an approval rarely is. Therefore it is necessary for the construction operator to find out where set-forms are to be used and where they are not. Procedures for using the set-forms become familiar to the operator within a short time. Other procedures (where a set-form is neither specified nor required) require the operator's own personalized methods of forming his documentation to meet the requirements of the contract provision.

Content

Obviously, the provisions of the contract conditions require the contents of the documentation to be consistent with implementing the performance of the contract. While the content of writing on the contract should be consistent with the legal requirements of the provisions, the reader must not overemphasize the importance of composing a content required for meeting legal objectives in relation to the practical need for composing a content required for the implementation of contract performance purposes. Indeed, when the operator is composing strictly for the purpose of implementing the physical performance of a contract, he must deliberately avoid unwarranted references to legal matters. Documentation composed to meet contract requirements must be primarily directed toward expediting building the project. Certainly all documentation has its potential legal effects, and this must be considered in its formation, but to document strictly with legal considerations and purposes in mind is not a desirable tack to take in expediting the construction of a building. In fact, legally oriented documentation can be very detrimental to the timely and orderly progress of the construction. There is usually a specific point in time when it becomes necessary to give primary attention to legal considerations in documentation operations. Legal considerations should never be the initial or only basis for documentation on the conditions of contract. Initially, documentation on the contract is to be composed primarily with the idea that when it is done properly, it will get the project done faster and most efficiently. When composing documentation on the contract, the operator should attempt to move people and events toward the timely performance of the contract.

16.5   BUSINESS REQUIREMENTS

At times certain types of documentation and communication are required that are strictly related to the business aspects of the construction process. They are not related directly to the contract or to its performance. Examples of this type would be correspondence promoting sales or the operator's company. Correspondence on financing, insurance, and bonding matters are other examples. These documents require the same general thinking processes as for other construction documentation, in that they must be clear, concise, in proper form, and so forth, but they have a different objective: they are primarily directed at promotion, and their main purpose is to gain something. Here we are talking generally about documentation for business purposes. The documentation can be a letter of introduction to a client or architect, a brochure on the company and its personnel, or a letter to an auditor. It can be written for any number of reasons and in any form. Generally speaking, this type of document should be informative, thought provoking, and interesting. We say "interesting," because its purpose should be to attract and keep the attention of the reader. There is nothing wrong with adding color and tone to this type of letter.

Thinking about legal considerations when composing this type of letter can be important if the documents are directly related to, for example, an insurance claim or a report on the facts of an accident. Usually, however, documents involving complicated legal matters, while they may be written by the operator, should at least be edited by an attorney.

16.6 LEGAL CONSIDERATIONS

There are two main reasons why learning about the law and its application is important in the construction process:

1. The proper interpretation of the law to everyday practice can benefit the documentation and communications process and all of the parties engaged in it.
2. Some of the legal pitfalls in the process can be avoided.

We are concerned here primarily with the second reason. It is important that construction operators develop a "sense" about the legal aspects of their documentation and communication operations.

It is safe to say that construction operations of any sort can always involve and raise questions of legal consequence. Should we say further, then, that construction operators should learn all about the law or at least all about contract law? This would be neither practical nor desirable. The various fields of law are too vast to allow this complete study outside of the legal discipline. The specific application of the law to construction operations, however, is such an important factor in their implementation that it cannot be altogether avoided or ignored. When examining the whole range of construction operations, the law's application is of particular importance in documentation and communication operations.

Ideally, legal considerations should never be a prime concern of the construction operator until an actual adverse situation occurs on the project. Unfortunately, the possibility of such situations arising in the construction process forces the operator to be continuously concerned about the legal aspects of his operations.

We have said before that it is not proper to construction thinking to automatically assume the probability of an adverse situation occurring on a project. It is absolutely essential, however, that the construction operator recognize the potential of such a happening, mainly because of the nature of the construction contract and its requirements of performance and also because of the many facets and different interests involved in the construction process. The problems created by this potential must be realistically faced by the construction operator, even though he thinks of himself as a constructor and not as a lawyer.

Recognizing this potential, the construction operator is required to conduct his documentation operations accordingly. This is necessary both in the formation and application phases of the documentation and communication procedures. The reader must be aware that the time he is required to spend on legal considerations in the construction process is increasing currently, and it is not likely to decrease in the foreseeable future.

It would be a mistake for the construction operator to think that his approach to his work must be based on feelings of mistrust rather than trust. It would also be a mistake for him to approach his work with the idea of outwitting someone else by a legal maneuver. The objective of the construction operator should be to construct quality buildings on time for cost. Unfortunately, however, construction operators and their operational procedures must keep pace with the current trend in the process toward legal confrontation and litigation.

This being the case, construction operators must absolutely develop their thinking processes to give proper attention to the legal aspects of construction. Construction operators must, especially when forming and applying documentation, give appropriate attention and priority to legal considerations and their consequences as the situation warrants.

The need to increase attention to the legal considerations of documentation grows as a dispute or adverse situation worsens. More time must be spent by the operator in determining the effects of his documentation on the potential legal consequences caused by the adverse situation. Greater care, precision, and discretion in documenting and communicating are called for as the situation gets worse.

Usually, in an unresolvable adverse situation a point is reached where legal protection becomes the only reason for documentation. It is at this point that the construction operator must switch from normal operational procedures to another deliberately set course of procedures. He can assume at this point that the main objective of his documentation is directed solely toward protecting his company's interests rather than accomplishing the building of the project. It is extremely important to remember and apply the following rules to documentation operations and their procedures:

1. They must be done on time.
2. They must be dated with an attestation when required.
3. They should be addressed to a general recipient (with attention noted) rather than to an individual.
4. They should be concise and to the point.
5. They should be considered and formed as if they were a part of an evidential record of the particular dispute or situation.
6. Responses should be very carefully checked for incorrect facts, dates, and allegations. Immediate formal protests must be made to any and all such errors.
7. If a response to documentation is required, it must in all cases be provided, if only by a letter noting that it has been received and reviewed.

8. They must be sent to all interested parties other than the addressee in copy, if this is required.

Documentation required for adverse situations should be processed keeping the following thoughts in mind:

1. It must be formed, delivered, and recorded as if it were evidence.
2. It must be properly organized and filed so that practically anyone in the company organization can use it for reference.
3. It must be composed in such a way that an attorney can understand and use it efficiently.
4. It must be composed with a chronological order in mind, and it must be processed in the same way.
5. It must be followed through, and it should not be started unless it can be followed through.

In developing documentation in adverse situations, some general principles for developing its content should be observed by the construction operator:

1. It must be directed to the issues.
2. It must be based on fact not fantasy.
3. It must be pertinent to the situation.
4. It must be directed to support a position, but not unreasonably so.

In developing a documentation procedure for forming a legally oriented record, it is important for the operator to note the following:

1. Different kinds of documentation may have to be collated into clear and understandable composite records.
2. The record will ordinarily adhere to the chronological order of its compositions.
3. The record will be developed so that the most important documents are given greater prominence than the lesser ones.
4. The record will be developed to form the simplest possible narrative of the dispute or situation.

16.7 EVIDENCE

We cannot leave our general discussion on documentation and its processes without considering the fact that it constitutes evidence. This is a particularly important point to bear in mind when documenting for an adverse situation. Documentation, whether it is intentionally directed to such a purpose or not, often serves to prove or establish something or other. The construction operator must assure himself that the documentation he composes will act to his advantage in an adverse situation.

The construction operator should not be expected to know the law of evidence like a lawyer, but he can and should know some elemental principles of this phase of the law so that he can use them as guides in his documentation operations. In this way this part of his work will be easier for him to compose and will better serve the company's interests.

We list for the reader some very rudimentary rules of the law of evidence that are applicable to documentation operations:

1. A copy of a document is not of the same evidential value as the original document.
2. A document must ordinarily be signed in order to legally identify its writer; it must be witnessed in some cases and attested in others.
3. A written statement of fact is generally better evidence than an oral statement.
4. Statements of fact or opinions, however communicated, should be as directly recorded from their sources as possible; otherwise, implications of hearsay can become an issue.
5. A photograph can be a statement of facts.
6. Spontaneous written or oral statements have great evidential value, because spontaneous expressions are often regarded as demonstrations of true feelings and motives as they bear on the fact situation.

All of these rules are very general in nature and most of them are unrelated, but they all result directly from the law of evidence. All of them, if they are weighed and applied carefully, will serve the construction operator in documentation operations. There are certainly exceptions to these rules, and they can be quite complicated legally. When writing or speaking, these general rules should serve as a reliable guide for the construction operator in his documentation procedures.

16.8 TYPES OF DOCUMENTATION

There are several types of specific documentation that are used by the construction operator and the purpose of the document governs the form and content of each type. Common purposes for documentation include the following:

1. To give notice of something.
2. To report on something.
3. To give an opinion or viewpoint on something.
4. To question something.
5. To request something.
6. To record something.

Each purpose calls for a specific approach to documentation. The following are some examples of various types of documents:

1. Documents of notice—notices to proceed, notices of cancellations, etc.
2. Documents of request—payment requests, approval requests, etc.
3. Documents of record—test reports, photographs, minutes, etc.

A specific type of construction document demands that the operator compose the document so that it is easy to use and understand. Ordinarily each type of document demands a different substantive content directed at a specific purpose. It is never advisable, to combine different types of documentation into one document.

Correspondence

Correspondence is specifically defined here as communication through the exchange of business letters. In construction operations correspondence is expanded to include the telegram and the memorandum; it also includes the exchange of other written data, whether or not it is covered by a letter. Some examples of other kinds of correspondence are payment requests, shop drawing, transmittals and test reports.

It is the business letter and its exchange that we wish to briefly examine here. It must be clearly understood by the reader that a letter, once delivered, becomes and remains a record as long as it exists. Therefore even the ordinary business letter should be written seriously and responsibly. The letter does not disappear, as does oral communication; it can serve the writer, or it can come back to haunt him. The writing should be the following:

1. Correct in language usage.
2. Correctly formed.
3. Concise and clear in language.
4. Properly developed in thought and content.
5. Exact in expressing the writer's intent.
6. Properly addressed and attested.

The memorandum is also correspondence, but the construction operator ordinarily uses it differently than the letter. There is one major consideration that the operator must give to its use: the memorandum should never be used as a substitute for the full, formal letter. Even when there is little time, the operator must guard against using the memorandum without considering the greater value of the usually more detailed letter. While the memorandum is acceptable documentation in certain situations, the construction operator must make sure he does not overadapt its use. There are appropriate times and situations where it should definitely not be used. A useful guide for restricting its use in construction operations is for the operator to think of it only as being a one-way communication.

Reports

Reports can take all different forms in construction operations. Reporting means giving information about something seen or investigated. The purpose of a specific type of report may require it to be given at certain time intervals or it may be that the need for the report is dependent only on some occurrence. The following are some types of reports of various kinds commonly used in construction operations:

1. The superintendent's report.
2. The test report.
3. The cost report.
4. The inventory report.

These and other types of reports often have supplements or will consist of a series of documents. For example, a final consulting report will possibly be based on a series of preliminary reports. Reports can be made for all kinds of different purposes: some contain just facts, some contain opinions and viewpoints and some purposely avoid including them, and some contain analyses and conclusions and some do not; many reports have traditionally established forms, others must be original in form; and some reports contain only general information and some are extremely detailed.

Once the form of a report is established, it fairly well determines the required content of the report.

Example.   An established job superintendent's report form usually asks for the notation of certain observed data. The weather conditions, the number of trades, the number of tradesmen, and so forth are required data. This established form of report is purposely made to obtain *specific* information that is necessary on *all* projects. It allows for originality only in its "general comments" section. It is efficient and easy to use.

When contrasting the job superintendent's report to the superintendent's own job log, we see that the job log is a different type of report. In this document we might very well find exactly the same information as contained in the job superintendent's report, but in the log the superintendent records it for different purposes or reasons and in a different way. The job log is a report, but it is also the personal expression of an individual. It is written as much for his own benefit as for anyone else's, whereas the formal superintendent's report is primarily written for the benefit of others. There can be significant differences in the way in which an objective observer of the two reports interprets each one of them.

Contrasting these two types of reports illustrates the great variety of purposes that a report can have when made in different forms.

A test report is another special type of report: it reports test results. Many times test reports are done in a series, and the entire series of reports attests to the validity of the item tested. Thus the concrete test report can consist of a series of reports on the different time interval breaks of the test cylinders, with the last report on the cylinder being the 28-day break. All of the information contained in the interval reports is valuable and usable, but it is the last report that furnishes evidence of the structural validity of the concrete.

In order to make certain reports valuable and useful documentation the con-

Code	Quantity Estimated total	Unit Cost Estimated	Allowances Estimated total	Quantity This pay period	Expense This pay period actual	Unit Cost This pay period

ABC INC.

Job:

Quantity Total to date	Expense Total to date	Unit Cost To date actual	Total Gain To date	Total Loss To date

Figure 21. A labor cost report form.

struction operator must consider his main purpose in forming the report. Let us briefly examine the labor cost report in this respect.

This report contains information that is vital to construction operations and their main objective of cost. It allows the construction operator to know exactly what his job labor cost is at different points in time. To do this the report must be kept current and it must be cumulative, complete, and accurate. It must be adequately broken down and itemized and it must show estimated cost allowances for expenditures of labor, as well as the actual expenditure of labor. Let us examine a form that performs in the manner we have just described (see Figure 21).

The report in Figure 21 is formed as follows: The estimated labor on a project is broken down to a set-form cost breakdown by coding the labor cost headings. These headings are the following:

1. Temporary buildings and facilities.
2. Supervision.
3. Excavation.
4. Demolition.
5. Formwork.
6. Concrete placement.
7. Concrete finish.
8. Masonry.

9. Metals.
10. Miscellaneous.
11. Carpentry.

These coded items are further broken down and itemized as follows:

5. Formwork
    5A. Spread footing forms.
    5B. Wall footing forms.
    5C. Pier forms.
    5D. Wall forms.
        5D.1. Circular wall forms.
        5D.2. Short wall forms.
    5E. Slab forms.
    5F. Beam forms.
    5G. Column forms.
8. Masonry.
    8A. Exterior face brick.
    8B. Interior face brick.
    8C. Backup.
        8C.1. 12″.
        8C.2. 8″.
        8C.3. 6″.
    8D. Interior masonry partitions.
        8D.1. 12″.

These form headings are used by the field forces on the project in assigning the man-hours expended to the individual coded items. These costs are recorded daily and weekly and compiled and reported to the company operational organization.

The actual labor cost of each item of work on the project is known no later than the time it takes to report it.

This simple cost recording system gives a wealth of important information to the construction operator. Almost all of the operational organization has use for this information.

Let us examine how each division might use this report. The general superintendent uses it for the following purposes:

1. To immediately identify areas of bad cost on individual jobs.
2. To make judgments on using premium time on certain jobs in relation to the total labor cost picture on all jobs.
3. To keep a record on the cost performance of his subordinate supervisors.

Expediting uses it for the following:

1. To estimate the cost to complete at any point in time.
2. As an aid in periodic payment requests.

Estimating uses it for the following:

1. As historical data for future estimating purposes.
2. To analyze the cause of variations between actual unit prices and estimated unit prices.

Its most important purpose, however, is in its use by the job superintendent because the report furnishes the basis for all future job operations.

This report, its system, and the way in which it works are without doubt the main educational tools used for learning the main cost objective of construction operations.

Photographs

Photographs are documents. A photograph is simply a statement of fact. Construction photographs are taken for all different kinds of purposes, including the following:

1. To record job progress.
2. To record the results of a job accident or structural failure.
3. To give proof of a particular physical condition of the project at a particular time.

In any case, the photograph provides superb documentation for construction operational purposes. If the procedure for taking the pictures and their attestation is proper, there is, in fact, no better way of recording actual physical facts and conditions for future use. The photograph is especially valuable because of the ease with which it can be understood and interpreted by both construction operators and laymen alike.

The photograph is easy and quick to make. It can be adapted to recording facts and even fact situations of widely varying natures. It can record today what is covered up tomorrow—or record that which will be hidden in an hour. This is an extremely important attribute for any document to possess in construction operations, because of the staged phases of building the construction project.

The following are a few of the important things that an ordinary photograph of a construction project can record:

1. The physical state of the components of the project as well as its entirety.
2. The weather and its effects on the site.
3. Areas of work that are clear or clustered.

A series of photographs taken within a time interval can show the following:

1. The production progress within the time interval.
2. Interferences in the sequence of operations.
3. Proof of actual conditions that are not apparent at the present time but that did exist in the past.

Photographs, when interpreted reasonably and honestly, will tell the truth of the matter. Obviously, photographs taken at daily intervals will give a more complete and truer picture of conditions or situations than will those taken a week apart. If taken at reasonable time intervals, photographs will not lie.

The use of photographs to prove job progress has long been used in the construction process. Using pictures to show cave-ins, collapses, fires, or the results of some other catastrophe has been common practice for a long time. These photographs, however, ordinarily show facts after an occurrence; they rarely show the occurrence itself.

It is entirely possible that the construction operator or his company will require pictures for record purposes as a matter of ordinary operational practice. This photographic record is simply a device that provides a continuing picture of the project. It is an excellent documental procedure.

There are certain criteria, however, for composing the photographic record both as a valuable working document and as a document of proof. Objectivity/subjectivity interpretations can always present problems in written documents between the correspondents, but the nature of the photographic record, as both an operational tool and as proof, ordinarily precludes the subjective view of facts. When the problem of objectivity arises with photographs or photographic records, it is usually due to the manner in which they are made. To be as objective as possible photographs taken for the record must be the following:

1. Taken on time and in a time sequence.
2. Taken in reasonable numbers and in proper view.
3. Taken at a set time of the day if possible.
4. Taken with care to distinguish between the need for broadness of view and/or detail.
5. Identified and attested as required.
6. Taken with good photographic technique and composition.

The ideal photographic record would be one that allows a layman or a court to use simple common sense in interpreting the recorded physical facts and conditions, with a minimum of accompanying technical explanation. The photographic record should be composed so that a minimum of editing for meaning is necessary. It should speak fairly well for itself.

The construction operator should view the use of the photograph and the record in certain ways. He should, for example, be aware that the skill and knowledge necessary for directing the taking of one or a series of photographs for a composed record are essentially the same as those used in directing the project itself. In addition, he must be aware that pictures will show the facts whether or not these facts will serve his interests in adverse situations. They are statements of fact and may be used and applied in any way and whenever facts are required.

Minutes

Minutes of job meetings (which are an integral part of the construction project) are reports of the meeting. These minutes are published documents and require a documentation process that involves the following:

1. The taking of the minutes.
2. The editing of the minutes.
3. The publishing of the minutes.
4. Correspondence and discussion on the contents of the minutes.

It might be suggested here that to simplify this documentation process the minutes should be taped. Many people engaged in the construction process strenuously object to being taped in conversation for various personal reasons. But aside from this, conversations in a meeting that is taped will usually be more hesitant and restrained. The meeting held for construction purposes should be as informal and unrestrained in discussion as possible. Meetings conducted in an informal atmosphere tend to be more valuable in expediting job progress.

The contractor, subcontractor, and owners and their architect/engineers' representatives ordinarily participate in the construction job meeting. Depending on the procedures established by the contract and the project's operational requirements, a representative of one of these parties chairs the meeting. The chair is usually responsible for recording the minutes of the meeting. The chairperson takes notes of the contents of the meeting, edits and transcribes them to readable form later, publishes them, and distributes them to interested parties.

These documents should be handled in a specific way. After publication they should be carefully reviewed by all of the parties concerned as to their factual content. Any inconsistencies in understanding what happened or what was said at the actual meeting should be reported back immediately to the publisher. To allow minutes to go unread and unnoted by the construction operator is a very serious mistake. See Chapter 18 for a further discussion of minutes.

The minutes are extremely important as reference materials in adverse situations, because they express the problems, views, complaints, intentions, contentions, ideas, and decisions of all the parties to the meeting at the point in time at which it was held. The minutes help recall facts and the positions of the parties at the time.

The construction operator must therefore develop a system for minute making, when he has this responsibility, as well as the ability to adapt the minutes to their best use.

16.9 THE PROPER USE OF LANGUAGE

There should be no need to dwell on the necessity of the proper use of language in documentation; we will therefore make only one central point on its importance in the documentation process.

When one writes a construction document, one does it for the benefit of others. It is put down on paper, and it remains there. It does not go away and it does not change. It may be seen by many other people and used by persons other than the addressee for one purpose or another. It is absolutely necessary for the construction operator to write well in all of his construction operational functions. If he has not learned how to write, he will be forced to by the demands of the construction process. He will find that he cannot assume that the responsibility for writing properly and well rests with someone else. One bad experience resulting from such an assumption will make him realize how important it is to write correctly.

Writing, to be useful and valuable, must express the thoughts and intentions of the writer, but, above all, writing must be understood by the party who reviews it. The proper use of language makes the former easier and the latter possible. Writing must be correct in spelling, grammar, punctuation, and structure. When done correctly, writing is almost certain to be understood by its recipient. For the construction operator to ignore or deny the practical importance of proper language usage in documentation and its processes and procedures is unrealistic, unproductive, and illogical.

If construction operators desire to be regarded as professionals, they must express themselves as such. Fortunately, or unfortunately, the correct use of language in writing will do more to show the degree of true professionalism of a construction operator than any other of his common operational functions.

16.10 TIMELINESS

One other feature of construction documentation and its processes and procedures that must be mentioned specifically is the need for timeliness. Almost all documents, processes, and procedures in construction operations are governed in some way by time restrictions. When making construction documentation the operator must realize that time is a critical factor in its formation, transmission, receipt, and response. Sometimes there are rigid and absolute contractual demands for documentation within certain time limits. Even when there is no expressed time demand at all, one can still be assured that time is of vital importance. The requirement for timeliness in the documentation operation calls for the construction operator's careful regard for the time limits imposed by the contract and the nature of the construction process.

We do not know of a single documentation process or procedure that is not somehow restricted by time. Sometimes the contract sets very arbitrary time limits for documentation and communication operations. Sometimes construction operations demand time limits on documentation because of impending or subsequent operations. There are all kinds of varying time demands placed on the construction operator by the construction process. He should be aware of these demands and develop his thinking processes to comply with them.

16.11 DEVELOPING METHODS

A construction operator will find that the required documentation processes and procedures will not happen by themselves; they must be developed and formed by the operator himself. We conclude this chapter by offering some general suggestions that should help in his development of effective methods in these operations:

1. Consider the relationship of documentation and communication operations to the entire construction process, to all of its participants, and to all of its subprocesses and phases.

2. Consider the relationship of the contract to the documentation. This is a special consideration.

3. Consider its potential effects on business relationships.

4. Consider the relationship of the law to the documentation.

5. Consider what type of documentation is required by the particular situation or construction operation at hand.

6. Consider time—always.

It is through documentation and communication that construction and its operations are accomplished. Therefore proper operational processes and procedures for documentation and communication are as necessary to good construction operations as are the proper physical and manual building construction processes and procedures. When analyzed from a purely practical standpoint, they are probably a great deal more so.

Remember always that there is no other occupational tool that the construction operator will use more during the course of his entire career than the pen or the pencil.

17

The Bid Day Procedure

If you can make one heap of all your winnings,
And risk it on one turn of pitch-and-toss,
And lose, and start again at your beginnings,
And never breathe a word about your loss:

Rudyard Kipling

17.1   THE TIME OF THE BID

At the beginning of this book we stated that its contents are based on a model construction organization. One of the criteria for this organization was that it bid primarily by the competitive bid method in both the private and public sectors. A significant point about the competitive bid method of construction contracting is that it establishes a specific bid time or some time limit for the submission of the contractor's bid proposal to the owner. It also establishes a place for the receipt of bids by the owner, which can have significant time implications, as we will see later in this chapter.

The bid time is the exact time established in the bid documents up to which the owner will accept the proposals (or offers) of the various general contractors bidding on the work. A specified bid time is set for a specific clock time (2 PM EST, 8 PM CDT, etc.) on a specific day. Theoretically, all bidders are bound to furnish their proposals at a specified location at or before that time. If the contractor's proposal is not furnished by bid time, it is not supposed to be considered as a valid bid by the owner. Since any adverse consequences to the owner of recognizing late bids are minimal, there are instances where the owner chooses to ignore or evade this bid condition. It would be a mistake, however, for the construction operator to consider that instances of accepting late bids are common, because they are not.

It is essential that the operator develop his thinking processes to being on time in the bidding process. Ordinarily, it is a drastic mistake to submit a late proposal. The operator should realize that whatever the time specified as bid time (whether it be 2 PM or 12 midnight of a certain day), it is the time by which he must have the bid in the owner's possession. It is an extremely bad habit for an operator to start ignoring or making compromises in his thinking about the requirement of submitting the bid on time. To do so engenders sloppy and careless operational behavior that is not acceptable in the estimating phase of the construction process. Two o'clock PM EST, February 24, 1979, should mean not one second afterward to the construction operator. It would be poor competitive practice for those contractors who do meet this requirement not to vigorously protest the owner's acceptance of a competitor's late bid arrival.

The construction operator should realize certain things about specified bid times, depending on whether the job bid is for private or public work. Public work customarily requires that the bids be on time, and because this work also requires the public opening and reading of bids, all competitors are furnished the opportunity to observe the arrival of late proposals. Protests are easily registered with the parties opening the bids in public works bidding. In private work, where there may or may not be the requirement for a public bid opening, there is little chance of a late bid arrival being observed by the competing contractors. There is much more latitude for allowances of late arrivals in private work than in public work. The restrictions on the practice of accepting late bids are usually less severe, depending on who the owner is, in the private sector. On the one hand, some private owners are very strict about not accepting late bids, while, on the other hand, some take the attitude that their own specified bid time can really mean that a bid will be accepted at any time.

For the purposes of this discussion of bid day procedures, we start with the premise that the specified bid time is to be strictly adhered to. Therefore the "bid day," as we use the term here, consists of the time from the opening of the general contractor's office on bid day to the specified time (bid time) of the receipt of bids. Every operational procedure of bid day must be developed within that working time in order to get the proposal in by bid time.

For the purposes of illustration let us make the specified bid time 2 PM EST, February 25, 1979, Room 333 in the local city/county building. The time and place are thus set for all bidders. The construction estimating operation should examine certain aspects of this specified time and place. For example, he should relate local time to the standard time given. The estimator should immediately, on beginning his operations, gauge the time required for all estimating processes to fit into the time allowed, thus giving him the opportunity to fit the capacity of his organization to the tasks. He should carefully note the place of the receipt of bid in relation to the location of his own office, but, most of all, he should evaluate all the difficulties that might be encountered on bid day, because with the set bid time he has a limited number of hours on that day within which he must do the work. But in making judgments about such things relative to the time available on the bid day alone, he is severely restricted by the set-time demand. Consequently, a 10 AM bid time on bid day becomes much more of a problem for him than an 8 PM bid time. Each set time has its own advantages and disadvantages. A 10 AM bid time would, for example, be obviously more restricted in operational time, but, on the other hand, it forces the estimator and the subbidders to have more of their work done prior to bid day. An 8 PM bid time allows more total time for completing the estimate but also extends bid time some three hours beyond the normal business hours of his subbidders. Communication with them is much more difficult to maintain under this circumstance. In rare instances it is possible for bids on more than one project to be due on the same day and even at or near the same bid times. The estimator is aware of these unusual situations beforehand and must adjust to them.

17.2 PRESSURE

Bid day procedures are part of the estimating process. In Chapter 7 we discussed the aspects of pressure involved in the art of estimating construction work. These aspects were discussed there in a general way in order to give the reader an indication of the effects of pressure on the entire estimating operation. Here we talk specifically about the great increase in pressures on the estimator in the bid day setting. These pressures arise because the estimator realizes that while time is steadily decreasing as bid time approaches, there is a corresponding increase in vital operational functions. Time is ticking away as the subbids grow in number and complexity. Each of the increasing number of subbids must be analyzed in relation to the others. Money figures are increasing in number and size. Possible contradictions, ambiguities, and duplications in the bid documents are now becoming evident. Everything seems to be happening all at once. There are ever-changing low material prices coming in that must be plugged into the estimate sheets. Figures must be added and subtracted constantly; simple arithmetic functions become a chore. Before one set of figures is adjusted in the estimate, a new set of low figures appears. There are constant changes in the estimate, requiring constant care not to forget to replace subtracted figures with added figures and vice versa.

Then, too, subbids are being called in that include work already quantified and priced by the estimator in his estimate of the work of his own forces. If it is suspected that these subbids might be lower or different in the amounts of work they include, these quotes create problems, which increases the pressures. And time keeps passing. In addition, there are alternate and unit price operations to contend with. Both of these operations involve the same problems for the estimator as the base bid, and they add great additional pressures when they are complex and extensive.

There are constant interruptions in concentration (at a time when concentration is vital) by necessary questions from and discussions with the entire construction office organization. Certain questions *must* be directed to and discussed with the estimator. Even the best-organized bid day procedure cannot eliminate all of these pressure-producing situations.

Some of the foregoing observations are examples of things that cause and increase pressures on personnel involved in the estimating operation. To give an idea of the extent of these pressures and their effects on estimating personnel, one need only appreciate the immediate release of tension and the emotional and mental exhaustion that becomes obvious after the bid has been submitted. It will be the rare estimator who will not admit to these feelings. The pressure is always present and will always be in this construction operation. There are things that can alleviate or mitigate it to some extent, but pressure is inherent in the process and must be accepted and mastered by those who wish to engage successfully in estimating operations.

17.3 DEVELOPING THE BID DAY PROCEDURE

The one significant way of easing the bid day portion of the estimating operations is to carefully organize each and every operation to be performed that day into specific segments of time. The ideal organization of operations into various time segments allows just enough time and no more to operations that *must* be performed, leaving any slack time for less important operations. Once organized, these operations must follow the patterns established. If the organized plan of operational procedures is followed, pressure and confusion and thus the risk of error are minimized.

Let us give a concrete example of the organization plan for bid day as a theoretical situation.

Example. A project being bid has an approximated value of $10,500,000 to $11,500,000. It is a fairly detailed and complex office building of moderate size. It has a reinforced concrete frame with a full basement–garage structure. The proposal form asks for seven alternates involving the interior finishes and one involving a substitute facing material for the building's exterior. In addition, a series of unit prices (additions and deductions) are sometimes required. The time for completion is specified. There are three addenda issued to date. Bid time is established as 2 PM EST, June 24, 1979, at the Council Chambers of a city.

There are nine general contractor bidders on this job. The proposal requires that the electrical and mechanical trades be included in the base bid, but the library equipment contract is to be bid directly to the city, and the low successful general bidder is to assume this contract. His fee for this assumption is to be included in his base bid price.

The ABC contracting firm is one of the general contract bidders. ABC is financially able to do this job and is bondable. ABC's office operations are conducted by a total force of 14 people consisting of 4 expeditors, 2 estimators, 4 financial and cost support personnel, and 4 clerical people; in addition, there are 3 executives who function sporadically in and out of expediting, estimating, and financial operations.

ABC has decided to figure concrete forms and placement (excluding resteel), cement finish, masonry and all miscellaneous metals placement, rough carpentry labor, and the conditions of contract as the work of its own forces. The estimator has written up the subcontractor sheets, and they show that there are to be 47 subcontracted labor and material divisions of work; in addition, there are 4 labor only and 9 materials only divisions of work. Invitations to subbid have been sent to 150 various subcontractors and suppliers (thus averaging 3 invitations per subcontract division in the write-up). Prior to bid day (and mostly on the day before) certain estimating work has been completed on the bid.

The total estimate of work by ABC's own forces is as follows:

Forms	$ 565,734
Concrete	377,451
Finish	74,515
Masonry	427,978
Metals	248,954
Other	63,130
Total	$1,757,822
Conditions of contract	315,639
Total	$2,073,461

This total figure is arrived at by the estimator (see Appendix A) for the cost of the work to be performed by ABC's own forces. This part of the estimate has been extended and totaled fully as of the end of the day of June 23, 1979.

The same items of work in the base bid estimate have been figured, extended, and totaled for all work in which ABC's forces are involved, in all of the alternate proposals. A separate individual write-up including all projected subcontractors must be made for each alternate proposal as well. Each addenda might also require its similar write-up.

Individual calls have been made on June 23 to all subcontractors and suppliers who are unique or furnish something special in the bid documents.

Some subcontractors' and material suppliers' written proposals stating the conditions of the proposal but not the price will have been received, with the probability that there will be several more the next morning after the 9:30 AM mail pickup.

The estimating operators have picked up signals that some restlessness and some of the usual prebid maneuvers are going on between subcontractors and the competition, particularly in the mechanical trades.

In furnishing information to the company executive who is responsible for servicing the bid security, the estimator has given the estimated value of the job (on the bonding company's insistence) at $9,650,000. On the basis of this estimate the bid bond has been furnished by the surety. This estimate was based roughly on an 80% and 20% relationship of subcontract amounts to the value of ABC's own work, based on empirical or historical experience with similar work.

Thus the stage has been set for the next day—bid day. The reader should note here that some features of the scenario will not occur this way on every job. For example, it might not be possible to predict the range of the bid until it is closer to bid time, particularly when unusual or highly specialized work is involved. But certain of these steps must always be completed before bid day as a planned operational procedure to minimize bid day problems, which, in any case, will be more than enough for the operators to handle. There is absolutely no excuse to not have a close estimate of the work of one's own forces ready on the eve of bid day. The maximum amount of work that the operator should be left with at this time is to substitute the actual quoted low prices received on bid day for estimated material prices used in work installed by his own forces. There may also be some required adjustments to some estimated ball-park (not taken off) quantities when new and more accurate quantities accompany these material prices. Reinforcing steel tonnage can usually be roughly quantified through the use of a formula based on volumes of concrete quantities. (If 1 ton of resteel is figured for 1 cubic yard of concrete, then 1000 cubic yards requires 100 tons.) The tonnage figure is plugged into the estimate of the work of the contractor's own forces (under the metals write-up). Labor and material unit prices are applied to this tonnage estimate in order to complete the total estimate. On bid day this calls for only a minor adjustment, because all that needs to be done is to substitute the proper tonnage for the supposed tonnage, and its reextending and retotaling it.

All of these prebid day requirements for the estimate are done with the sole purpose of making bid day easier to handle. The more that can be safely and satisfactorily completed or partially completed in the estimate before bid day, the better.

Organizing Forces

From our previous example we see that there are 14 to 15 people available to assist the estimator in his operation on bid day. This is assuming, of course, that their time is exclusively available to the estimator on bid day. Since this is rarely the case, there are certain personnel that are more important to keep available for estimating operations on bid day than others. The estimator must decide on who these people should be and organize them accordingly.

At some point on bid day the estimator must divide the responsibility for the estimate into two specific parts. The first division has the responsibility for any adjustments to the estimate of the work of one's own forces. These adjustments must be made constantly throughout the allowed time on bid day. Ordinarily, the estimator personally takes care of this task, along with any similar adjustments to the alternate or unit prices. The second division of responsibility is the most important part of the estimating operation, and the operational responsibility of this division rests on the person who compiles and tallies the subbids during the course of the bid process on bid day.

Thus there are two main operators who gather and analyze figures on bid day. The rest of the organization is usually assigned the task of taking, noting, and transmitting the telephoned bids for the subbid tally.

The Tally

The operator assigned the task of tallying the subbids has handled the subcontractors since the early phases of the bid period. In Chapter 7 we said that he was also assigned the task of inviting subcontractors to bid and making his own subcontractor write-up for comparison with the estimator's write-up. He has handled subcontractors and their questions throughout the bid period.

On bid day his task becomes preeminent in the construction operation of estimating. Prior to bid day the bid tallier has prepared a convenient tally system for the listing of subcontract divisions on which he records each subbid as it comes in. Each trade or subcontract division has space in the listing for the brief but concise recording of information that is vital to the tallier, eventually allowing him to choose the right low subbid from those submitted. Ordinarily, the vital information that he must know to make his choice is as follows:

1. The name of the subbidding company.
2. The name of the project for which he is submitting his bid.
3. The base bid price.
4. The alternate prices.
5. A statement of tax inclusion or exclusion.
6. A statement by the subcontractor that his bid is based on the plans and specifications.
7. Any specific exclusions or inclusions of work beyond the plans and specifications.
8. A statement on addenda noted and included in the subbid.

Thus a typical entry under the trade division of painting would be as follows:

Painting 8D midtown office complex

XYZ Painting Co.	BB = $75,000 P&S, tax included
	A1 = +$1,000, A6 = −$500, A7 = +$17,000
	Add 1 thru 3 noted and included
CDF Painting Co.	BB = $69,000, tax included
	A1 = +$1,100, A6 = −$1,700, A7 = +$16,000
	Exclusion—no back painting of wood doors included
	Add 1 and 2 noted and included
	No Addendum 3 noted
Smith Bros.	BB = $73,000 P&S, no tax
	A1 = +$1,000, A6 = −$300, A7—no bid
	Add 1 and 2 noted and included

From this and all of the rest of the tallied listings for every subcontract division of the write-up, the bid tallier must come up with the low subbid to be entered on the estimator's own subcontract write-up sheet in the estimate just prior to bid time.

If we return to our prior discussion about soliciting 150 subcontract bids, it is obvious that the recording of all of these bids (if they are all received), plus any and all unsolicited bids, requires a considerable amount of manual work for the bid tallier. Aside from this menial work, the bid tallier is forced to exercise an enormous amount of mental discipline because of the steadily increasing pressures as bid time approaches. The bid tallier *must not* become flustered or confused. He must record the subbids correctly on the tally.

His tally listing should ideally be done in such a way that when the estimator asks for a low subbid figure anytime prior to bid time, the tallier can easily supply the current low figure he has entered at that time. The estimator often requires these current low figures to arrive at a price of the work so that he may give this information to those responsible for markup, enabling them to analyze at least a range of costs of the work on which to base their markup. The time to do this operation is usually as early as when every subcontract division has at least one subbid in the tally listing. The latest practical time to assemble this "cost of the work" price is the time at which it is the easiest for the tallier and estimator to do it together. It is not practical to wait too long to perform this operation.

The bid tally is assembled from the time the first subbid is received until the very last moment allowable before bid time. The last allowable moment calls for a judgment on the part of the estimator as to how much time he needs to summarize the estimate correctly and accurately and for the proposal to be physically filled in and delivered before bid time.

As bid time approaches, the pressures on the bid tallier increase, because usually the number of subbids also increases. Using our previous example, out of the 150 to 250 subbids possible on the subject job, it will not be uncommon for 75% of these to come within the last two hours before bid time. The problems that this late subbidding causes are obvious. As the number of subbids increases, their mere recording takes up more and more time, leaving less and less time for the bid tallier to really analyze and consider each of them. This is unfortunate but still a fact of life in estimating operations.

One can readily see the great amount of technical skill the bid tallier has to develop in his operations. This estimating operation demands a greater amount of skill in performance than any other construction operation relative to the amount of time in which it must be performed and the pressures involved. This operation not only calls for mere technical skills but also for an organized, disciplined mind and the ability to quickly analyze and decide about things involving great amounts of money. These skills are mainly acquired through practice.

For the instant bid there are 50 low responsible prices that must be entered in the estimate. These 50 prices are to be picked out of perhaps 100 to 200 subbids, which are by no means uniform in their content. They can vary widely in inclusions and exclusions, and these variances are not known by the tallier until bid day. There can be great variations in price, even when the contents of the work bid on is the same or similar. Out of this conglomerate of subbids the tallier must be able to distinguish which are low and responsible subbids for each of the 50 categories of work.

To add to his problems, there is always the possibility of the combined subbid. The combined subbid is a bid that includes more than one division of subcontracted work in its price. There is also the possibility of the subbid that varies from the tallier's conception of what the bid should or should not include (when, e.g., the bid is for furnishing labor and material, whereas the tallier's conception of what the bid should include is labor or material only). There is always the possibility of the complimentary subbid, the deliberately misquoted subbid, or simply the bad subbid. The complimentary subbid is a bid that is submitted as a courtesy because it was solicited, but it is not submitted as a serious bid. The problems for the tallier with this type of bid results from the fact that he is not always sure whether it is actually complimentary. Complimentary subbids are usually high and this sometimes causes confusion for the tallier when comparing it to other subbids.

The deliberately misquoted subbid is submitted for various reasons. Many times it is submitted only to obtain information on other subbids in the same trade. These subbids are ordinarily either inordinately high or low and they evoke an immediate response from the tallier, which is just what the misquoting subcontractor wants to happen. The tallier's response tells him something he wants to know. The tallier must handle such subbids very skillfully, because often this behavior by a subcontractor who is unfairly seeking bidding information puts the operator in a "peddling" position, even though he may have had no intentions whatever to deliberately reveal the other bids that he has.

The subbid that is simply wrong raises major problems in the estimating operation on bid day. Assuming that this type of bid is sincere, it is very difficult for the tallier to correct what is obviously a bad situation in the competitive bid process. This problem increases greatly as bid time approaches. If this bad subbid is submitted 10 minutes before the set bid time, then it creates a major dilemma, because a job can be won or lost, depending on how the estimator and tallier handle it.

Example. All bids received to 1:30 P.M. for excavation are in the range of $500,000. At 1:35 P.M. a bid is received from a heretofore reliable subcontractor that quotes $100,000. A wrong bid is indicated and the operator has the choice of ignoring it or informing the subcontractor that his bid is wrong. In this case he should ignore the bid figure and hope that the competition will do the same. A much more difficult situation arises when a wrong bid of $375,000 is submitted at 1:45 P.M. Here the price differential is not so striking and it is possible for this subbid to be either good or bad. If the operator thinks it is wrong, then it should either be adjusted upward by the operator or ignored. But in this case there is a real question as to whether the competition will or can do the same thing.

The tallier thinks that it might be a good bid; what does he do in this case? He takes a risk and uses the bid, but the risk can be minimized if the tallier has closely examined the excavation division of the bid documents and is fully aware of what that division includes and excludes. If he cannot determine the bid's validity by these means, he must rely on the reputation and past performance of this particular subcontractor. The operator must carefully analyze the situation and decide whether or not to use a subbid out of line with the majority of the other subbids on the excavation work.

Another situation that causes real concern to the bid tallier is when a seemingly low bid price might or might not constitute the low responsible bid. A subcontractor may not be the responsible low bidder (even though he has the low price) because he cannot possibly handle the job with the size of his organization, or he may be financially incapable of handling the size of the job, or he might (as a known fact) be in financial trouble on his other work. When these things are known for certain, it makes a very complex problem easier for the operator to handle. When the facts about a subcontractor are not known for certain, however, some very risky problems arise. An unknown, unsolicited subcontractor can create real problems on bid day when he submits a low bid.

There is a realistic rule of caution for the operator to use in the situations that arise out of problems created by uncertain subbids on bid day; it is based on the competitive bid situation. The operator must assume the following:

1. All the subbids have been given out to the competition.
2. It is possible that all of the subbids have been submitted in varying amounts and contents to the competition.
3. There can be no assumption that any two competitors will view all the subbids in the same way relative to their being valid and responsible.

Therefore the construction operator must view these troublesome types of subbids with caution and, after whatever analysis time will allow, must try to even out the competitive consequences of these types of subbids and act accordingly. Whether or not, for example, he decides to broadcast these bids to other subcontractors or even his competition is a matter of competitive judgment based on their potential effect on his estimate. Whether or not he tries to correct the subcontractor (or is even able to) is a matter of competitive judgment. Ordinarily, the toughest problems for the operator arise where he must decide under the pressures of time whether or not to use them, ignore them, or adjust them. Just one of these bids can cause either the loss of the job or result in a serious deficiency in his own bid. The operator must be aware that sometimes his competitive position in the bidding process is hopelessly compromised by these types of bid practices on the part of subcontractors.

As the subbid tally continues toward bid time, the tasks of both the estimator and his tallier increase. As submittals build up, the subbid comparison function required to be made by the tallier increases substantially. All kinds of comparisons must be made. This requirement of comparing the submittals leads to another major problem: the combined subbid that is submitted without a breakdown of its component trade divisions and their respective prices. This makes comparisons extremely difficult, if not impossible. Sometimes such subbids are submitted unintentionally. The tallier must have a breakdown of the combined subbid if he is to compare it to other separate trade division subbids. On contacting the subcontractor who combines its subbid, the tallier can ask for or even insist upon a breakdown. If the subcontractor complies, then the tallier is in a better position to make comparisons. Since this contact with the subbid requires time, which becomes critical as bid time approaches, the combined bid is still a problem. It is when the subcontractor refuses to give a breakdown that serious difficulties result for the tallier. The subcontractor who deliberately does this obviously wants to gain a competitive advantage. He has a competitive advantage because there is no way for the general contractor to compare the price or content of the various segments of the combined bid. One combined subbid can cause a complex situation. The reader should consider how this situation is magnified when a series of combined bids are submitted that do not contain the same divisions of work.

Example. The CDF Company bids acoustic ceilings (including special acoustic treatments), resilient floors, and plastering combined with certain exclusions and inclusions to each division of work for $300,000.

The HJK Company bids acoustic ceilings (not including special acoustic treatments), resilient floors, and toilet partitions for $193,000.

The MNO Company bids only special acoustic ceilings, dry wall, and plaster plans and specs for $395,000.

In addition to these bids, there can also be several other separate bids on each of all the trade divisions included in these same divisions of work.

Consider the dilemma presented to the estimating operator caused by the above situation. What is the true low price for each division of the work or a combination of all of them? Additionally, even if there were a way of logically arriving at the low price for the total of these divisions by adding together the various combinations of low subbids on each separate division of work, the time to do so is unavailable on bid day. Since the bid tallier must arrive at the low subbid figures that are to be included in the estimate, he must concentrate on sorting out figure amounts first and then go on to distinguish among the contents of all of the various bids (combined and separate) in relation to their respective figure amounts.

There are certain things the operator must do in this situation. He must:

1. Come to the closest approximation of the true low price from his analysis of all the subbids.

2. Make judgments on the relative importance (generally based on money amounts) of the combined bids that he has. Certainly, he must concentrate his efforts in the time he has on those bids that most affect his bid price to the owner.

3. Make every effort to get the combining subcontractors to break down their subbids. Very often this can involve threats and coercion to let the erring subcontractors know in no uncertain terms the long-term business effects of their behavior. Failure to convince the subcontractor of the serious effects of his combined bid should result in a very stern reaction by the contractor with this subcontractor, regardless of the outcome of the job at hand.

The tasks of the estimator as well as those of the tallier become more numerous as bid time approaches. Besides having to plug in new figures into the estimate of the work of his own forces, he is faced with two more major tasks that are related to each other but require different operational approaches.

Example. We mentioned that ABC (the general contractor) decided to figure the masonry on the job. The estimator has taken off, priced, and totaled the work at the cost (without markup) of $428,000. At some point during bid day the masonry subbids start to come in. Because of time considerations, the estimator must examine and compare all the subbid figures as they are submitted to his own estimate of the cost of the work. He must do this regardless of whether or not his own price or the first masonry subbid price turns out to be low. He cannot wait for the last possible masonry subbid to be submitted to make the comparison. The point here is that the estimator must act on the assumption that there is a masonry price out there, that it might be low, and that his competition is going to get it. Assuming that this low subbid has been priced plans and specs, then the only choice left to the estimator is to guess at the anticipated overhead and profit the subcontractor has in the price, subtract it from that price, and compare his figure on a cost versus cost basis. The guess, of course, is in the anticipated overhead and profit figure. If the bid is presented with exceptions to the bid documents, then a closer analysis and breakdown of these exceptions becomes an additional task for the estimator.

The second task presented to the estimator during the bid day is more complicated. This arises when a subbid on a major portion of a division of work is submitted in which the exceptions to the work included in the divisions as called for in the bid documents become the rule in the subbid.

Example. Taken as a whole, the specification for the total concrete framing is contained in several divisions of the traditional bidding documents. There is a technical specification on concrete, resteel, miscellaneous metals, and so on. All these work items are involved in and govern the operation of concrete framing. The formwork involved in the framing operation is not usually made a distinctly specified item of work; it being optional in most cases that the contractor be able to price the formwork as work to be done by his own forces or to allow for the taking of subbids for the work. The price of the formwork can thus be a major competitive differential factor in the estimates of the various competitors. The estimator has figured the price of his work according to his projection of the abilities of his company to perform the work in a certain way at a certain cost. This should theoretically allow him to think that he has a competitive advantage over all of the other bidders. Remember that the estimator can make a very detailed estimate of formwork or a general estimate, depending on his company's estimating theories and practices.

Now, on bid day subbids on only portions of the formwork are submitted. The XYZ Company, for example, bids all horizontal form surfaces specifically excluding all vertical surfaces, resteel work, pouring, and miscellaneous metal work installation. This then eliminates from the subbid walls, columns, bulkheads, beam sides, and so forth. In his own estimate the estimator has, by careful detailing, kept these things as separate items. Incidentally, this is an important incentive for the estimator to thoroughly detail the items in the write-up of the estimate of his own work. To do so allows him to easily separate and compare his items against items in subbids that include only portions or segments of a total operation.

But even when he has detailed items, the estimator has a difficult time comparing this type of bid to his own estimate. While this itemization allows him to compare apples with apples in the total price, he must still analyze and compare any supplemental operations that he might have included in his units which reflect on the true comparable price of the subcontractor's work. In which detailed item, for example, is layout included? Is it in some vertical form unit price or in a horizontal unit price? These and many other supplemental operations to the form operation must be considered when comparing the prices. Again, the factors of time and increasing pressures call for the highly trained and organized mind using specific procedures to cope with them.

The reader should keep in mind that these problems extend to the alternate prices as well as to the base bid price when they are required in the proposal. The reader should realize the extension of work required in pricing an alternate that involves great numbers of trade divisions of work, as for example, in the alternate pricing of an entire floor or wing of a building. All of the problems encountered in arriving at the cost of the base bid figure are encountered as well in the large alternate prices when they are required.

Segments of Time

If the office is open for operations at 7 AM the morning of June 24, there are seven hours available to complete all of the operations that take place on bid day. Segmenting this allowable time to operations, however, cannot ever be done exactly on bid day, because the estimating operator has no real control over the course of events during that time. Therefore planning time to operations must be made in a very general way. There are three specific operations, however, where definite segments of time must be rigidly assigned. First of all, a definite time must be established for arriving at figures that indicate the "price range" of the estimate base bid. Secondly, there must be a cutoff time for all outside communications to and from both the estimator and the bid tallier so that they may assemble the cost of the work price for markup. Thirdly, adequate and realistic time limits must be established for the timely delivery of the proposal to the place of its receipt.

The assembly of the "price range" of the total estimate is necessary for several reasons: it allows those who are responsible for marking up the cost of the work to realize the proportions of the job, enabling them to intelligently determine how much markup to add. It eases the estimator's mind in knowing that there is at this point a complete estimate at hand, even though the estimate may go through several corrections before finalizing it for bid; the estimator knows that he has some determined figure, thus eliminating the possibility of an incomplete bid. It also furnishes a broad check for missed or duplicated items. Establishing a "range" is a valuable and necessary step in the estimating bid day process. A specific time must be set aside on bid day to do this operation. This operation requires an uninterrupted session between the estimator and bid tallier to plug in all figures in the subcontract write-up sheets.

Within the closing minutes of bid time, the estimator and tallier must again closet themselves in an uninterrupted session for the purpose of assembling the final cost of the work for markup. During this segment of time they should absolutely not be interrupted. The cost of the work is exactly determined, including all corrections being made to the first assembly of the total estimate (the range). This cost is presented to those responsible for markup and the markup is added. Bond premium costs are calculated and added. These operations result in the base bid figure. (Alternate prices that are major items of work and money are also determined by these same operations.) At this stage the summary sheet of the estimate is finalized (see Appendix A).

The final base bid figure is usually established very close to bid time. Subbids are still being submitted and tallied and immediate corrections in whatever numbers required are made to the final figure. Extreme caution must be exercised at this point so as not to misplace or duplicate figures on the summary sheet. There is every possibility of mistakes occurring at this time because of pressure considerations. *Note:* the time allowed for these last operations *must* include time for the final arithmetic checking of the summary sheet. This must be done on every bid, without exception. This is an area where very serious errors occur in the bid process.

The office copy of the proposal form must then be filled in. Unit prices and other information should have already been filled in well ahead of bid time, except for the major alternates. All supplemental required information should have been taken care of and the main area of concentration at this point is the exact filling in of the base bid price.

We must next consider the time required by the person taking the bid in to properly complete the proposal form and deliver it on time. He should leave the office with the bid package some time prior to bid time, allowing himself time to station himself for telephone contact back with the office. He must determine the time it takes to go from the telephone station he chooses to the place of bid receipt fairly exactly. He has the required copies of the proposal form and other bid documents with him. The proposal form has already been filled in with as much data as was possible to fill in before he left the office, leaving as little to complete the proposal at the telephone station as possible.

He should arrive at his telephone station in good time, after carefully considering the extent of his following operations and his travel to the place of receipt of the bid. Before leaving the office he should have the proper writing tools and telephone change. When we say, "he stations himself," we mean exactly that: he finds a place with a telephone as convenient to the bid location as possible and then calls the office and remains on the telephone until the final figures are ready. In the office that line is kept open at all costs. The stationed operator must tell the office his distance from the place of bid receipt, and the time involved in getting from his station to the place of bid receipt. He does not leave his station.

At this time there are only a very few minutes left to bid time. The operator delivering the bid must fill out the remainder of the proposal form in all copies as required, and he must always read it back to the office for a check. Most base bids are required to be written in both words and numbers and in triplicate. This manual operation takes time and is done under conditions of extreme pressure. The bid, however, must *always* be read back.

At this point the proposal is out of the hands and control of the estimator at the office and is in the control of the person who is to deliver it. It must be delivered on time.

Specific segments of time must be assigned on bid day for operations other than the bid delivery. These periods of time cannot be established in a haphazard, careless manner; they are necessary in all cases and must be considered and planned for very carefully and then strictly adhered to.

Another segment of time that must be considered on bid day is that required by the executive to determine and add the markup. Depending on who is responsible for the markup and how it is arrived at, adequate time must be allowed. Where only one person determines the markup, this presents a different case than when a group makes the determination. With a group there can be a great deal of discussion and even argument involved in this operation. A bid should never be late because of this. The time for operations of markup should have been planned far enough in advance.

Another vital operation—that of calculating contract bond premiums—follows the markup operation, and this operation also requires time.

Corrections

An expedient operation that is utilized to save both time and confusion near bid time is making corrections. It is possible that corrections just before bid time will be necessary on any portion of the content of the estimate. Corrections on newly submitted subbid prices can be made either before or after markup; they can, in fact, be done right up to the time when the person who is delivering the bid writes in the base bid figure on the proposal. The correction operation was developed to minimize the effects of the late submittal of low subbids. There must also be a corresponding entry of each and every correction on the summary sheet. This whole operation calls for extreme care and clear thinking; it is an operational tool that saves valuable time. If it is not done with care and confidence, it can be very dangerous in its application because it automatically changes the amount of the base bid figure. A set procedure is usually established by the esti-mator for the correction process. This involves a system that accurately records the removal of money amounts from the completed estimate, with corresponding additions made. Each and every deduction made on the correction sheet must show a corresponding addition; each and every addition must show a correspond-ing deduction. The reasons why we cannot suggest a specific system for perform-ing this function is because it is essential that the individual company form its own correction procedure which all its operators will understand on bid day. All of its forces must be comfortable with the established system in order to avoid confusion and hesitation at this critical time on bid day. See Appendix A.

The Subcontractor Bid Process

Many subcontractor bidding practices on bid day have been designed to protect their self-interests in the bidding process, and they have been carefully developed by the subcontracting portions of the construction industry. Subbids arrive at all times during bid day, but, generally speaking, most of them arrive very close to bid time; some even arrive after bid time. We must examine these subbidder's practices carefully. The construction operator must understand how they work and, indeed, understand how subcontractors think about their place and impor-tance in the bidding operation in order to properly handle the subcontractor and his bid on bid day.

First of all, the construction operator should realistically evaluate the impor-tance the subcontractor places on his bid under ordinary circumstances. Nor-mally, the subcontractor has extended his best efforts in estimating his work, with the idea that if he is low, he should be entitled to the award of a subcontract from the low general contractor bidder. The general contractor's operator must treat with confidence all of the subbids as a matter of good professional and busi-ness practice.

Secondly, the operator should realize that the experienced subcontractor knows the conditions of time and pressure that exist in the general contractor's office on bid day. He also becomes generally familiar with the procedures that the differ-ent general contractors use in their respective offices. While there may be some small variations in these procedures, they are fundamentally the same, because the nature of the competitive bid process requires it.

The good subcontractor realizes that in order to protect his bid, while at the same time giving practical consideration to the general contractor's problems on bid day, he must set a pattern in his own operations to satisfy and protect his own interests and those of each of the general contractor bidders. He must pro-tect his bid and at the same time make sure that the general contractor can use his bid.

Thirdly, the subcontractor can never be sure which general bidder is going to be the low general bidder, at least in the straight competitive bidding process. Therefore he cannot afford to not bid the complete list of general contractors un-der ordinary circumstances. He may, however, give special considerations to pre-ferred general bidders in his subbidding practices, as well as in his subbid prices. For example, he can give his preferred contractors an earlier quote on bid day, because he knows the value of his early bid to the general contractor's estimating operation. Needless to say, the general contractor who is on the subcontractor's preferred list should endeavor to stay on it. A subcontractor's treatment of preferred general bidders on bid day becomes a distinct operational advantage and this preference should be valued accordingly.

There can be times when a general contractor's treatment of subcontractors on or after bid day seems to be inconsiderate of their true and real importance in the bid process; sometimes it may seem downright dishonest, but this kind of treatment can result purely by accident and without any bad intentions on the part of the general contractor. Ordinarily, it occurs because of the unique time-pressure nature of the bid process operations. Unfortunately, under conditions of pressure or excitement, subbids can be misplaced, mishandled, or even ignored. This leads to the situation where the early cooperating subbidder does not end up with the job. When such a situation does arise, it is mandatory that the operators in the general contractor's office call in the subcontractor and explain the unfortunate circumstances in a clear and forthright manner. While the subcontractor may be dissatisfied with not being awarded the job, he will at least have his trust and confidence in the general contractor restored in part.

The Late Subbid

Every subcontractor has his own thoughts about the bidding process and how it works; he also has an opinion about the individual general contractors with whom he is bidding. Usually, all subcontractors have had bad experiences at one time or another resulting from the bid process. It is probable that all subcontractors at one time or another have been legitimately low on a job and lost it. These and similar kinds of experiences have led to the single most difficult operational problem for general contractors on bid day—the submission of late bids by the subcontractors. The most unfortunate result of this practice is that ordinarily those subbids that are of the greatest significance in the amount of work and that involve the greatest amounts of money on the project are those which come in very late on bid day. The electrical and mechanical trades and excavation are customarily of major singificance on the ordinary construction project. For example, these trades can at times be 60% to 70% of a sewage treatment job proposal both in price and in the amount and complexity of the work. Consider then receiving all of the responsible subbids in these divisions of work in the last half hour before bid time.

Nothing can really be done with this operational practice by the individual general contractor other than having and nuturing good relations with the various subcontractors. The individual contractor has no control over the practice itself, nor does he have real control over the reason for the practice. It results, however, in the main risk that the general contractor takes not only just on bid day but in the entire course of its business venture.

There must be a reason for the late bid. The reason is usually that the subcontractor can best *protect* his bid from peddling by turning it in late. The general contracting segment of the construction industry may take exception to this statement on the grounds that it believes that its participants are professional and ethical people. This is certainly a correct belief in general, but the fact also remains that subbids do arrive late on bid day as a matter of course in the bidding process today, and those general contractors who practice peddling must be held responsible for this phenomenon.

Since the late subbid problem must be faced, a main objective of the construction operator must be to develop operational procedures to cope with the late bid problem.

The procedures he develops to solve the problem on bid day are those that assure the best possible handling of the late bids. This calls for the assignment of personnel in the late stages of bid day to handle the late bids according to their importance in the general contractor's proposal. The operators who are assigned this task must know the relative importance of each late bid received; however, they must also correctly guess which of the technical divisions of the specifications will be bid late. This is necessary because there is simply not time to question, discuss, or compare in any detail among themselves all the contents of the bid documents versus the submitted late subbid. The construction operator will soon learn that the operation of handling the late bids is an area that only practice can make perfect.

The office forces taking subbids must also start to severely limit their conversations with subbidders as bid time approaches. They must develop the means of shutting off conversation with all but the most important bidders. They must realize the limitations of the office's telephone service: only so many calls can be handled at one time. The operator recognizes that there are subbidders who have a figure available but have not been able to make connections or who are deliberately holding off submitting the bid. He needs all the subbids he can get. He cannot afford to let a potential low bidder pass him by. Strict rules, therefore, must be made beforehand for the entire office personnel on the necessity of speed and efficiency in the taking of bids. (There should never be irrelevant conversation allowed at any time on bid day.) Many times field forces and forces connected with other company work must be alerted that the office phones are reserved on bid day exclusively for the bid process.

Taking and Recording Subbids

The construction operator can facilitate the taking of subbids in two significant ways: first, he can have a set-form made for the purpose of minimizing the physical writing of the subbid itself. This form can have headings for all of the specific and commonly required data in printed form, thus leaving only its fill-in to be done by the bid taker (see Figure 22).

Secondly, the estimator must outline and put into a simple list questions that must be asked of specific subbidders when they are taking their bids. These questions must be put in simple language that can be easily transmitted by the bid taker to the subcontractor for answers. This listing of special items of required data should only contain the most important and critical areas of doubt about a specific division of the work. Rarely should this list of questions exceed one page, covering all of the divisions of the work. Remember that clerical forces are taking the subbids. They know nothing about the particulars of the job at hand. In most cases they are capable of taking the straight bid, but they are not capable of asking these important and critical questions unless they are listed. In many cases

ABC INC.

Sub-Bid Information Form

Code of Sub-Bid Taker

Name of Subbidder   **Division(s) Bid** _____

                   Caller   **Telephone**

Base Bid   **Plans and Specs — Yes   No**

                                            Tax Included — Yes   No

Alternate   **Alternate**

Alternate   **Alternate**

Alternate   **Alternate**

Remarks (Inclusions-Exclusions)

Figure 22. A subbid information set-form.

the subcontractor is aware of the reason for the specific question and has its answer. This then calls for a bid taker to make a simple notation on the printed subbid form for the bid tallier's attention. The best that can be done if the subcontractor does not have the answer to a vital question is to notify the estimator or tallier to personally contact the subcontractor. The estimator must judge the importance of the answers to the questions and decide on the need for further conversation relative to time.

All the people involved in taking bids on bid day must be made to understand the critical importance of the accuracy of their fill-ins and notations on the subbid form. Numbers or numerical digits, for example, cannot be forgotten or misplaced, and all of the information required must be completely and accurately recorded on the form. Another seemingly minor detail than can cause trouble in the time-pressure situation is that caused by the sloppy and illegible handwriting of those who take the bids.

The subbids are taken over the telephone; they then go to the bid tallier for their inclusion in his write-up. This involves minor but important operations. The handling of subbid forms must be planned so that they are brought to the tallier in some sort of order, listed by him, and put in a place where they cannot be mixed with other incoming subbids. No subbids should ever be missed because of the mishandling of the bid forms. They must also be kept for confirmation purposes. A simple in and out filing system developed by the tallier for bid day should be sufficient for this purpose. On bid day it is wise that bid takers hold conversations with subbidders to a minimum. The less conversation, the less chance for possible legal disputes, disagreements, or mishandling of subcontractors, or for shopping or peddling practices to take place. Time considerations usually force this restriction on conversation anyway, but conversation about anything other than bid information should never be initiated by the general contractor's forces on bid day.

Sometimes, however, it is essential for the general bidder's information for there to be discussion on specific subcontractor bids. This can result from confusion about inclusions or exclusions, or any point on which the estimator is not clear but that he determines to be of real importance. The estimator must be given time to clarify his uncertainties. In order to avoid confusion and to keep the subbid-taking process going efficiently, very few people must have the authority to have such conversations—one or two at most. The reason for this restriction, primarily, is that with one central authority performing this task confusion, overlap, and a waste of valuable time are avoided. Sometimes the estimator, because of other critical duties at the time, will turn such matters over to an executive, but only under the closest supervision. Many of these conversations revolve around transmitting or receiving official positions of either the general bidder or the subbidder on some major misunderstanding about the nature or extent of subcontractor's bid. Thus the use of an executive for these conversations is most effective in many instances. The combined bid situation would be an example of this sort of instance. The conversation must carry a weight of authority that the contractor will recognize and respond to.

17.4   A SUMMARY OF BID PROCEDURES

We have tried so far to suggest to the reader some practices, procedures, and processes that are commonly used on bid day. In doing so, some of the problems and complexities of bid day were brought to the reader's attention. They were mentioned in order to show the close relationship between the problems encountered on bid day and the need to develop specific operational solutions.

Now we would like to summarize the general contractor's entire bidding process in order to show the complete in-line processing of the estimate to bid time (refer to Figure 23).

The following estimating work is done prior to bid day:

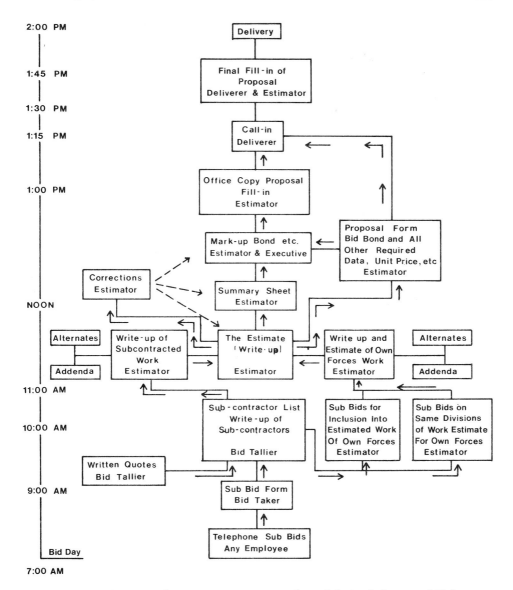

Figure 23. Diagram of estimating operations performed during bid time on bid day.

1. Writing up the estimate.
2. Completely quantifying prices, extending and totaling figures, and checking all estimated work to be done by the contractor's own forces, which includes plugging in estimated figures to items on which the actual low prices will be inserted on bid day.
3. Filling in the summary sheet with the totals of all of the money amounts from the write-up sheets.
4. Checking for the signatures, seals, and any other data required on all cop-

ies of the proposal form and bid security that can be fully determined before bid day. This may or may not include unit prices, percentages of markup for extra work, time of completion, or fee for assignment of separate contractors, depending on how critical they will be as competitive factors.

5. Performing the work on the alternate prices and the addenda.

6. Breaking down the estimated work of one's own forces when it is reasonably anticipated that there will be the need to compare prices with similar or equal subprices on bid day.

7. Listing in the tally all quotations (written or telephoned) received to the end of the day before bid day.

8. Determining assignments for all office personnel to be involved in any way in the bid process the next day.

9. Preparing the office facilities and distributing the working materials and forms to be checked for use the next day.

10. Alerting field personnel that the next day is bid day, in order to avoid their interfering with the bid process.

11. Collecting data and information that might be necessary to use the next day, such as the telephone numbers of permit authorities, information on special, unusual subcontractors and suppliers, a good telephone directory, and so forth.

12. Making sure that the person who delivers the bid knows where the place of delivery is and, if possible, have him determine his call station ahead of time. This operation may require some special arrangement if the office and the place of delivery are far apart.

13. Working up, if possible and practical, a rough range estimate for the purpose of allowing those responsible for markup as much time as possible to think about overhead and profit ahead of bid time.

14. Checking for late addenda.

15. Establishing the central control of estimating operations by making sure that everyone knows where the control of the following day's estimate lies.

16. Checking and rechecking figures, preferably using different computing machines.

17. Suspending discussions of the validity of the estimate of work of one's own forces on bid day. Sometimes it is necessary to do this in the area of general conditions close to markup time, but most, if not all, of these discussions, should be held prior to bid day.

On bid day the following processes and procedures take place up to bid time. Some of them are continuous, some are intermittent, some are extemporaneous, and some result from surprise situations or from unexpected events occurring during bid day. All must be handled quickly, accurately, and efficiently:

1. Everyone in the office takes subbids—as the subbids come in slowly in the morning, the estimator and bid tallier can usually handle the volume. When they

are busy, the calls are automatically given to the other available personnel. As bid time approaches, the volume of the calls increases so that all the office personnel must begin taking and recording bids. Each bid is recorded on a subbid form by taking the subbid accurately and fully in legible handwriting.

2. These subbid forms are given to the bid tallier (including subbids taken by the estimator himself) in some sort of order so that it is convenient for the tallier to easily pick out the information he needs.

3. The tallier lists the pertinent information of the base subbid. Pertinent information consists of the subbid's "plans and specs" verification, its tax inclusion or exclusion, the addenda included, and the alternate prices of the subcontractors. He may or may not list any but the most exceptional inclusions and exclusions. The tally list form is a developed form that is acceptable for use by the entire construction organization. Fundamentally, the form must be designed to meet the need of the tallier to record and to efficiently transmit the information on the low subbids to the estimator. He makes this list form in such a way that it is convenient for him to operate with it, thus giving him confidence in doing the tally operation in the correct way.

4. As certain subbids come in that the tallier knows have vital information for the estimator in his pricing of the work of his own forces, he gives the bid form sheet to the estimator, who records the information he desires and then returns the bid form sheet to the tallier. This type of bid is one that gives the following:

1. The final low material prices that are to be used in the estimate of the work of the contractor's own forces.

2. Equal or similar subbids on the same work or division of work already figured as work of the company's own forces. This also includes subbids that only partially cover the same work figured in the contractor's estimate. This type of bid demands immediate transmittal to the estimator, because the estimator needs all the time he can get to compare these prices to his own. Both base bid and large alternate prices in these subbids must be given to the estimator and, most importantly, any inclusions or exclusions in them must be made known to the estimator.

3. Combined prices that include more or less work than that covered by the estimator in his estimate of the work of his own forces. An example of this type of bid would be where the subcontractor has included all of the masonry construction specified. Where the estimator has figured brick and block masonry only, planning to subcontract the stone masonry, the estimator's price must be compared against the subbid that includes *all* masonry construction.

4. Subbids that include only a portion of the work in a division of work that has been figured totally by the estimator in his estimate of the work of his own forces. Comparisons of this type of bid must be made with various segments of the estimator's prices for the same work. As soon as these types of bids (2, 3, and 4) start to come in, the estimator takes off in his own direction, leaving the tallier and bid takers to do their tasks of recording and tallying the subbids. The estimator begins the task of gathering low prices (materials prices) for inclusion into the estimate of the work of his own forces. He must also start comparing this type of subbid price to his own.

5. The bid taking, recording, listing, and transmittal to the appropriate places and comparisons of the subbids continue until bid time. Patterns are taking shape. As more subprices come in, they are becoming adaptable to comparisons with others. The bid tallier begins the task of comparing the subbids with other subbids on the same work in order to establish the true low responsible bidder. *This comparison of subprices becomes the major operation on bid day.* As the time decreases, the subbids increase. Low figures begin to appear. There are many considerations for each listed item, however, that must be given by the estimator and tallier when the bids begin to show patterns.

Capabilities and financial and organizational capacities of a current low subcontractor start to become significant to the tallier and estimator in spite of the fact that his price is the lowest. These same comparisons must continue from the beginning of the pattern indicating low subbids until the final low subbid figure is determined for its insertion into the estimate itself.

6. The estimator continues to make adjustments to his estimate that are required by the receipt of new low subbids. When the pattern of low figures starts to show, there is ordinarily the need for the estimator to start analyzing and reanalyzing his pricing of the conditions of contract. This analysis of the conditions of the contract continues until markup time. As the final estimate starts emerging, the estimator often changes his mind about the adequacy or inadequacy of his general condition costs.

7. When the listing contains at least one figure for each written-up division of work, the estimator and tallier break off by themselves. Remember that each operator has written up his own sublist. Sitting down together in private, the estimator then calls from his write-up sheet all of the various divisions of work and the tallier gives the low subfigures he has at that time. The estimator inserts this into the subcontractor write-up sheet of the estimate. When the entire list has a number inserted after each item, either the estimator's or the tallier's list is called back. This furnishes a check that all divisions of the work are covered in the estimate.

8. Each of the operators then goes back to continuing his own separate tasks. The estimator totals the subcontract write-up sheet (all the other sheets of the estimate have been totaled and brought over to the summary sheet). He then totals the summary page and checks its extensions. The figures on the summary sheet furnish the estimated and "cost of the work" figure, which is not changed again except by the corrections operation, the addition of overhead and profit, and bond and permit costs. The filled-in summary sheet furnishes a total (but not final) bid price used to arrive at the proposal. The sooner this step (8) can be taken, the better. If the tallier's list has at least one bid for each division of work at 10:00 AM, then this is when this preliminary totaling operation should be done. The preliminary totaling operation furnishes a preliminary base bid figure which from that point on, is only changed by correction figures that are added or deducted from the preliminary base bid figure until the *final* bid price is arrived at. This method of summarizing and then correcting eliminates any necessity for going back and changing the figures in the base estimate sheets.

9. Needless to say, corrections to the estimate must be done very, very carefully and meticulously because this is where serious errors can occur in the purely mechanical aspects of the estimating operation. Corrections can be made as many times as is necessary or required by the change in the subbid prices, but obviously the fewer corrections there are the better. A very carefully thought out method of doing corrections should be set up and strictly followed. Checking and rechecking arithmetic is essential in the correction operation.

10. All during the time described above, alternates are being handled, using generally the same methods of operations. While alternates can be large at times and are thus important competitive factors, the operator must realize the relative importance of the base bid figure versus the alternate prices. Therefore, if time runs out, the concentration of efforts should always be given to the base bid figure.

11. When unit prices required in the proposal form cannot be completed the day before, this becomes another concurrent operation for the estimator to perform along with the operations listed above.

12. At some predetermined time very close to bid time the estimator and tallier must break away again to assemble a very accurate corrected estimate of the "cost of the work" for markup. The corrections up to this time have been listed in order on the summary sheet.

13. The estimate summary sheet is presented to the executive for markup. Markup is made on the latest possible corrected estimate, but corrections can still be made beyond this point. Permit costs and bond premium costs are determined if required. The summary sheet is finalized and it includes the addition of bond, permit, and overhead and profit costs. Arithmetic checks are made again on the summary sheet.

14. The office copy of the proposal is filled in fully.

15. The figures are read slowly and carefully over the telephone to the person delivering the bid, allowing him enough time to write in the figures on all copies of the bid proposal and to read them back.

Any low subbids arriving at this time are ignored. In some rare circumstances they might be analyzed quickly for use, but the estimator must think carefully about making corrections at this time. There is no time left now for checks and therefore any changes involve risks in the misapplication of the new figures. In any case, once the figure is written in on the deliverer's end of the line, the bid process is out of the control of the office.

16. The deliverer takes the bid to the place of delivery and makes sure it gets into the proper hands at the proper place.

17. If the proposals are to be opened and read aloud, then the deliverer stays and records all the information of all general contractor bids at the bid opening. He will note the late arrival of other general bids, and if there are deficiencies or variations in any of the general bids read, then he may or may not protest a nonresponsive or deficient general bid. Protesting competitive bids usually depends on the policy of the company. There is almost always a clause in the bid documents where the owner reserves the right to waive any or all deficiencies or irregularities in the bids. Under certain circumstances, however, this clause is not given effect by the operation of law. One circumstance, for example, that should always be protested in public work is the receipt, opening, or consideration by the owner of a late bid.

17.5   AFTER BID TIME ON BID DAY

When the person who has delivered the bid calls back to the office with the results of the competition, there are things that the construction operator must think about and act on. Usually there is elation if the bid is acceptably low, and there is always disappointment if it is not low.

Let us first examine what must be done if the bid is low. If it is reasonably low, then the operators can be fairly comfortable; but if it is very low, then a very important operation must take place very rapidly. The estimate must be given an immediate and thorough check for the purpose of determining the possibility of error. If the mistake is not found immediately in checking the numbers, then more time must be given to trying to find one elsewhere in the estimate. An error of major proportion usually explain the lowness of the bid and the contractor must make a difficult choice in fairly good time. When the amount of the error exceeds the bid security, this situation might call for one choice, whereas an error within the amount of the bid security might call for another. Then the entire job must be analyzed for the possibility of profit or loss in relation to the amount of the bid security should there be a default.

Mistake is still a valid excuse in contract law under certain circumstances. The construction contract proposal, however, demands security for the validity of the proposal itself. This security guarantees to the offeree that the offeror will come to contract within a time stated in the proposal. This makes claiming mistakes within the bid process more difficult. Recently, owners have become very serious about bid security defaults. Customarily, they have held on to the bid security of the three lowest bidders until such time as they can come to contract with one of them. Sometimes this security is held by the owner on the theory that the economics of the situation call for coming to contract with one of the bidders rather than calling a default on the security, with subsequent need to rebid the project. It is, however, becoming more common for owners to call a default on the bid security.

What chance the low bidder has on successfully claiming a mistake depends to a very large extent on the quickness of the communication of his mistake to the owner, with substantial and convincing proof of the mistake. The owner may consider, for example, at a time close to the occurrence of the mistaken bid, that it is not worthwhile to have a contractor on the job who is too low. He may suspect that if he calls default on the bidder, the bidder will come to contract but will do so in a very unhappy frame of mind to say the least. It then becomes a matter of business and economic judgment for the owner to decide whether or not he should exercise his legal rights under the bid security. But in any case, the faster the operator can determine his error, make a choice as to whether or not to claim a mistake, and communicate his decision, the better his position will be if he is faced with claiming a mistake.

The Unsuccessful Bid

There is the natural tendency on the part of operators to worry about even the good low successful bid after the elation of being low has subsided. There is little need for this worry, because from this point on it becomes the responsibility of other parts of the operational organizations to construct the job and make money.

It is when the bid is high that the construction operator should be concerned about his abilities. Different estimators feel different ways about being second bidder as opposed to being the highest bidder. Some will even express the pride in being second on the theory that their estimate was the closest to being low. Others despise being second, believing that luck was the prime reason for failure, and not their estimating ability. But in any case it is wrong and counterproductive to dwell on a lost bid. It is a waste of time and effort to go over the "ifs" and "buts" of a particular unsuccessful estimate. The experienced estimator will quickly forget the lost job and move on to the next one, without dwelling on the past. It is only when a pattern of high bids occurs that time must be spent on analyzing the bids. There is an exception to this way of thinking, however.

It is almost always beneficial for an unsuccessful bidder to examine the subcontractor's bidding practices on or after bid day. If there are, for example, a great many late bids that occur on a bid day, which the estimator suspects caused his high price, then he must determine if this is coincidence or if there is a reason for this situation.

The Successful Low Bid

It is best, on being the successful bidder, to start gathering information for future use on the job for the remainder of bid day. Little conversation should be held for the rest of the day with subcontractors, who, by this time, know who the low bidder is. The initial processes of getting confirmations on subbids should start at this time and proceed until the subcontracts are negotiated and set. Any uncertainties remaining about the true low subbidders should be itemized for future subcontracting operations. The process of demanding written confirmations of subbids can start the morning after bid day.

Attitudes

Win or lose, realistic attitudes, affirmative thinking, and looking forward to the next estimate should prevail among the estimator's attitudes on bid day. His colleagues in the entire operational organization should offer him understanding and encouragement. Adverse comments or snide remarks by others have no place here. One can only recall the response given by the old, experienced, tried and true estimator who, on being unsuccessful on bid day, would simply say to those who intimidated him, "Who do you think I am—God?"

The successful estimator should always try to make sure that bid day is handled with skill and good organizational systems. There can be little doubt that the correct exercise of this skill and the proper use of organizational systems contributes to the winning of bids. There can be no doubt that operational incompetence and sloppy use of organizational systems on bid day are the main causes for either a general contractor being a high bidder or being too low in its price—in either case it is a loser.

18

The Job Meeting

18.1 AN EXPEDITING TOOL

The job meeting is a part of the construction operational process of expediting the work of the project. It is essentially a planned gathering of interested job participants who possess information and knowledge about the work at one place at a regularly scheduled time during the progress of the job. The job meeting is an important construction operation that requires skill and expertise on the part of the construction operator in both its handling and participation. There are certain skills that the construction operator must learn that are necessary for him to effectively conduct or participate in the job meeting. These skills require developing expertise in the following:

1. Calling meetings.
2. Handling meetings and participating in them.
3. Utilizing the information provided in the meeting as a current and continuing operational tool and also keeping this information as a matter of record.

Customarily the procedures of the job meeting operations are set indirectly by the contractual obligations of the individual construction project. Although the job meeting may not be specifically called for in the contract, the practical value of such an operation is immediately evident in satisfying other definitive contractual obligations. An example of this value occurs when the contract specifies the time for completion. The job meeting, being a time-saving device, serves as a means of forcing considerable movement by the contracting parties toward satisfying the obligation of time for completion.

The construction operator should carefully study the proper means and methods of operating with the job meeting. The meeting is an extremely useful tool for recording facts at a particular time. Even more importantly, it can produce information based not on fact but on individual opinions or viewpoints at a particular time in the progress of the work.

The Reason for Holding Job Meetings

Depending upon who calls and chairs the job meeting, the reasons for holding job meetings can be substantially varied. If the owner chairs, he might have his own ways and means of holding the meeting that serves some particular purpose. Similarly, if the general contractor or the architect/engineer chairs, they also might have specific means and methods of holding the meeting to gain some special objective. But no matter who chairs the meeting, the overall reason for the job meeting must always include the objective of expediting job progress. This should be held firmly in the construction operator's mind, regardless of the nature of his participation in the operation.

The reader might ask here why it is necessary to hold job meetings when, for example, correspondence might solve problems of job progress just as well. He might ask why a job meeting should be necessary where good field supervision is evident from the status of the project and might question the sense and practicality of having a person-to-person, face-to-face talk on matters involving the contract. The answers to these questions are that the job meeting serves the specific purpose of exchanging information informally on the job and it is the informal nature of the proceeding that furthers job progress. The job meeting is a deliberate attempt to discuss, analyze, and solve job problems without the restrictions imposed by formality. It is held and conducted through oral communication to enhance feelings of cooperation and openness and to promote a real understanding among the proper participants.

18.2 LOCATION

The job meeting can be held at almost any location, but the best place is on the site of the work. For the purposes of this discussion the place of the job meeting is examined in this context. The main advantages of holding the meeting immediately on the site is that it facilitates the purpose of the job meeting through the ability of the participants to observe the physical state of the work in place. The work is present for all to examine. Its progress and many of its problems, if they are matters of dispute or confusion are easily ascertainable by inspection. Besides the physical status of the building, physical facts relating to difficulties and problems are also there to see firsthand.

When job meetings are not held at the site, it is usually necessary that the participants visit the site anyway in order to be fully aware of the state of the job. It is often necessary, for example, at a job meeting to be able to determine and compare physical facts that are involved in disputes arising over the interpretation of the drawings and specifications. By simply looking out the window or walking out on the site to establish these facts is a practical and economic use of valuable time.

The construction operator should realize that at the very least a minimum degree of comfort should be provided for the job meeting. A space that is too crowded, for example, does not lend itself to accomplishing the meeting's practical purpose of open and frank discussion. It is necessary to provide adequate facilities for the meeting at the job site.

18.3 THE COMPOSITION OF A JOB MEETING

The Principals

We have already said that different parties can chair and control a job meeting: sometimes it will be the contractor, sometimes the owner or the architect. In certain contractual systems (e.g., the separate contract system or a system using the construction management concept) the owner or his representative must usually chair the job meeting in order for it to be effective. Many private owners (particularly in the manufacturing industries) prefer to chair the meeting, because of serious and complex problems relating to the maintenance of their own production schedules. Architects will generally want to conduct the meetings where they have had a continuing relationship with an owner. Many public bodies and authorities who maintain construction inspection staffs also prefer to run the job meeting on their projects. It is often mandatory that they do this when there are interrelated multiple general contract projects on a site. Both architects/engineers, owners, and staff personnel usually handle job meetings with considerable skill and expertise. The construction operator should be aware of this fact and learn their means and methods to his advantage.

Generally speaking, there are three principal parties involved in the job meeting process: the architect/engineer, the owner, and the general contractor (or main separate primes). But there are also other important principals usually required for the individual job meeting. Subcontractors and material suppliers who are critical about ongoing or subsequent operations are often the most important people to have at a particular meeting. Many times they have exclusive information that is important to transmit to all the participants on the project. Problems such as deliveries, strike potential, anticipated shortages, and similar important information surface at the job meeting. Bringing these problems out into the open and frankly discussing them expedites the job and eliminates or at least alleviates unpleasant surprises and their resultant costs.

The need for the participation of the various types of principals in the job meeting depends on the skill of the operator in identifying the real value of having them there. Their importance relative to current job progress or delay is something that is not always apparent, and their importance down the line in job production can be even more obscure. The new operator must learn to analyze and distinguish the importance of these participants as the circumstances at any one time dictate.

The Chairman

No matter which principal is given the task of chairing the job meeting process, there must be someone in control of its proceedings. The chair is required to do the following:

1. Give at least initial notice to all parties on the project of the fact that job meetings will be held on the project and that their attendance will be expected when they are summoned.

2. Select the participants for the individual job meeting and give them notice of its time and place.

3. Name the place of the meeting and see that facilities for the meeting are provided.

4. Provide an agenda to the participants (and all affected nonparticipating parties on the work so they may attend if they think it necessary). Usually it saves time to provide the agenda ahead of time to the selected participants.

5. Provide for the taking, transcribing and editing of the minutes of the meeting, which must be transmitted to all affected parties. It is important here for the construction operator to realize that nonparticipating parties may have vital interests in the subjects discussed in these job meetings. Minutes provide an easy and efficient method of broadcasting pertinent information, provided they are communicated properly.

6. Hold the meeting using set procedures that will allow for informality in order to maximize the free and unhibited exchange of information.

7. Control the agenda and guide the discussions on its items in such a way as to gain the maximum amount of information on the most important and pertinent problems.

The chairman must direct the meeting and keep it under control. Because of the informality of the process and the fact that many different personalities are directly involved, this sometimes becomes a very difficult and demanding task.

It is best for the chairman to set and keep a schedule for the meeting firmly in his mind. Job meetings, unless they are closely controlled, have a tendency to stray from the agenda items if the chair allows them to.

The Participants

Participating in a job meeting is not the same operation as chairing it. While the chair directs and controls the entire meeting, the participant is usually confined and restricted by time and the agenda to playing a lesser role. Therefore what a participant says and does in a job meeting has to be carefully planned so that his individual goals in the meeting can be attained.

The ability to listen carefully is a decided operational advantage to the construction operator when he is a job meeting participant; this includes the ability to not only listen to the proceedings of the meeting but to also become sensitive to its tones and undertones. Many times such things as the spirit of cooperation on the job or attitudes expressing dissatisfaction can be picked up from the conversations in the job meeting. Such things are not always evident in the written minutes of the meeting, yet they can be extremely important to the participants in their evaluations of the job's progress and problems on the job.

18.4 THE AGENDA

What takes place in a job meeting? What should be discussed? What should be accomplished? These and similar questions can be answered by examining a typical agenda. The agenda items listed here contain the items ordinarily discussed at construction job meetings.

1. The reading of or reference to prior minutes.
2. Safety.
3. Cleanup.
4. Progress.
5. Performance problems.
6. Impossibility of performance problems.
7. Job sequencing problems.
8. Area of work problems.

9. Changes and changes in the work problems.
10. Problems arising out of the interpretation of the technical contract documents or from discrepancies therein.
11. Access problems.
12. Storage problems.

The agenda must allow the job meeting to be held in some semblance of order and to be constructive in nature; it ordinarily reflects the thinking and biases of its maker. Thinking in a proper way about the design of a good agenda results in a system being established for a particular job that becomes familiar to the participants as the job progresses. This saves time. Some construction operators, for example, will find that the best results are obtained by first cleaning up items of the agenda that are only reminders of contract obligations (such as safety and cleanup) and leaving the more difficult problem areas for the end. Some chairmen choose to write the agenda the other way round.

It is probable that some items on the agenda cannot be solved within the meeting itself. Some problem areas either must be referenced to other parties who are not present to give information or reply to questions presented. Not every question must be answered on the spot, nor must every problem be solved then and there. It is often sufficient just to have the problem mentioned in the meeting to have a movement toward solution. The chairman must develop a skill in determining which problems must be solved on the spot and which can or should wait for further and deeper analysis and participation. The problem with this last statement is that a point which may seem of lesser immediacy for discussion to the chairman may be of great importance to one or more of the participants at the time.

There is a danger area for the construction operator to be aware of in handling the agenda. When discussions diverge extensively from the planned agenda items (particularly in the area of dealing with problem items), the chairman must exercise considerable discretion and determination to bring the meeting's discussion back on course. This is necessary, because as such divergences increase, more problems surface, raising greater complexities that involve more parties (that may or may not be present). Ordinarily, it is not very wise for the chairman or the participants to assume that such a problem has been solved when this happens. It is better to refer the problems to further study by all of the parties involved. Depending upon the period between job meetings and the necessity for a quick solution, it is better to defer any proposed solution until all parties concerned can have their input. It is important that the participants do not gain the impression that solutions have been found in job meeting discussions when in fact they have not. It is the chair's responsibility to clearly identify the true status and resolutions of the problems.

While the agenda is the statement of the chair's view of important items of discussion, other pertinent items should be accepted and brought into the meeting by the chair or scheduled in the agendas of future meetings. The job meeting agenda should not consist of only those items the chair has determined to be the subjects of discussion. Again, discretion and skill are needed by the construction operator to include all legitimate items raised for discussion (provided they can be handled within a reasonable time frame) in the job meeting. This may force the postponement of some items and the addition of more important ones. The chairman must call on a rather thorough knowledge of construction operations to do this properly. He already has enough problems in controlling a job meeting without the spontaneous addition of any but the most important items to the agenda, yet the process will not work if rigid and autocratic attitudes are taken.

The length of any meeting of this sort is of critical importance. Again, the operator must use good judgment as to how long the agenda should be. If the meeting is too long, much time can be wasted. Sometimes inclusion of participants who are not directly involved in a specific agenda item increases the amount of time spent on an item, or they may ask to be excused to save time. To excuse them from parts of the meeting however in many cases defeats the overall purpose of disseminating information. Even though the minutes of the meeting should pick up at least the general points of discussions, it is usually much more beneficial if the participants are present at all discussions.

Very often participants, when called to a meeting, rightfully claim that the meetings waste time and money, and they do not attend. The way for the chair to change this attitude is to give them notice that the risks of nonattendance are theirs. Unreasonably lengthy meetings should not be necessary if the chairman has a well-planned, reasonably short agenda. Operating under this type of agenda, the participants at a job meeting should, using good common sense, be able to accomplish a tremendous amount of work.

Another item that may appear on the agenda has to do with affording consulting experts opportunity to give their reports on problems. This is frequently necessary on jobs that have special technical problems on the job site. Ordinarily, consultants are called in to give impartial opinions on matters that affect the project in one way or another. It is excellent practice to require the consultant to give plain English reports of their findings, in person, for the benefit of interested participants; a question and answer period should also be allowed.

There is one exception to this suggestion: it is rarely beneficial to include legal consultants in a regular job meeting. Usually, this type of assistance turns out to be counterproductive to the spirit and objectives of the meeting and it often introduces irrelevancies and wastes valuable time. It is better to schedule special and much more exclusive meetings to handle legal consulting problems.

The construction operator must always be aware of the consequences of deliberately including items on the agenda that reflect adverse situations on the project. This is necessary for two reasons: first, the meeting is recorded, and the parties of a dispute are at a disadvantage when unguarded, intemperate, or informal statements are recorded. The second reason is more important: if the participants are aware of any constraint on their statements of views and attitudes, the whole job meeting is affected. One rarely carefully composes a statement made in a job meeting. It is very unlikely, for example, that a quick statement, made in anger, on an adverse situation will have been carefully thought out; yet it becomes a matter of record. Statements and discussions on adverse situations are best left to correspondence. Real disputes are not good subject matter for job meetings.

18.5 THE MINUTES

The minutes of the meeting are its written record and the chair is responsible for taking them. In some cases a stenographic record is made; however, on most construction jobs the recording of the minutes is done in a very general and informal way. There are different reasons, both good and bad, for the chair's choice of the recording method. Stenographic recording is obviously more accurate; however, less accurate methods of recording discussions encourage freeness and frankness in discussions. With proper editing and communication techniques, it is probably preferable to be as informal as possible in the recording of the minutes in the initial meetings.

We should interject here with an obvious question the reader might have: why not use a tape recorder? Usually the tape-recording method will be rejected by most parties familiar with the construction industry. Many times this is based on a personal bias against being exactly quoted. In any case, taping destroys the informality of a discussion and puts a damper on free and open expressions of views and opinions.

It is necessary for the chair and the participants to completely understand the whole process of the job meeting minutes operation. The minutes are definitely a move toward formalizing the job meeting process: they constitute more than just a reminder of what took place, what was discussed, or what was agreed upon at the meeting; they become the permanent record of important matters concerning the job and its problems. This means that the minutes can have substantial legal significance in the future. The construction operator must recognize this fact and remember it constantly in processing this operation. They are a record of opinions at a particular point in time. For example in an adverse situation when determining what someones attitude was at a particular time in the past is critical, it might be determined from the minutes. As the job progresses and the minutes of the meetings increase in number, it is essential that the operator realize that they constitute an evidential record.

The Minutes Process

The chair takes the minutes, as well as transcribes, edits, publishes, and transmits them. While the minutes of a job meeting are not usually taken word for word, they must be taken in such a way as to truly reflect the essence of the particular point discussed. The transcription of the minute notes serve as a reminder of the pertinent and relevant points discussed. Editing is done either at the time of transcription or afterward as a separate operation. The construction operator must develop his own system for processing the operations of transcribing and editing. After these are completed, the minutes are published and distributed to all interested parties. We repeat that they must be distributed; they are of little value to anyone if they are not.

In the process of taking, transcribing, and editing the minutes, it is important that the construction operator make a true statement. The minutes should be composed as impersonally and objectively as possible. There is little value in relating, for example, the personal behavior of the participants other than their negative or affirmative responses on particular points. There will be little to gain from an overly subjective presentation of what occurred at the meeting.

When the operator transmits the published minutes, he should always, either by a cover letter or on the document itself, make a statement to the following effect:

> These minutes have been made according to the best recollection of the publisher and are his impression of the content of the job meeting of _____.
> Should the reader object to the contents of these minutes or discover mistakes, discrepancies, or deficiencies therein, it is mandatory that he immediately respond to them; otherwise the publisher shall consider them accepted.

The most important part of this statement is the demand for an immediate response when there are objections to its content. Immediate responses allow the minute taker to correct the minutes, if necessary, before the next job meeting. Sometimes reediting and correcting the minutes is a difficult task; but it is an essential one, for if the minutes of the instant meeting are not accepted because of uncorrected discrepancies, at the next meeting this will only cause confusion, trouble, and delays.

The Use of the Minutes

When the minutes have been distributed, there is one simple rule for the recipients to follow: read them. Too often these important documents are filed away and go unnoticed by the construction operator. This is not a responsible thing to do: first of all, they have a practical value as a reminder of the contents of the job meeting, and secondly, there is always the possibility of the minutes being incorrect. If they are incorrect, it is absolutely necessary to have them corrected. From an operational standpoint published minutes that are not thought to be correct can cause a great deal of trouble. Many various parties usually depend upon what they say. When parties that were not participants receive the minutes, the consequences can be especially serious when items affecting their operations were discussed at the meeting.

There are times when there can be no resolution of the differences in recollections and impressions of the job meeting itself. If these things cannot be worked out to the complete satisfaction of all the parties, at least these differences must be noted (usually by a nonacceptance of the minutes at the next meeting). This may call for a reanalysis and more discussions in subsequent job meetings. In any case, mistakes in the record must be corrected.

If there is one valuable and handy tool for the construction operator to use for keeping up-to-date and fully informed on the project and its progress and problems, it is the minutes of the job meeting. These minutes, along with job logs and reports, constitute a part of the job's written history.

18.6 HANDLING THE PROCESS

As well as describing the formalized processes and procedures that are used in this construction operation, we must, in addition, discuss the techniques required to handle its function. In this area it is necessary for the construction operator to develop and use his skills in handling people. The job meeting requires some management capabilities, but basically it is simply a matter of getting as much information from the participants as is necessary to accomplish its goals. The operator does not necessarily tell people what to do in a job meeting; he is not always required to demand action on a certain point. It is important that he make the parties on the job aware of problems by opening up avenues of discussion. When he has done this, he is on the way to the solutions of these problems through the cooperative efforts of the participants. He does not solve the problems himself nor impose solutions, except as a last resort; the participants, through mutual understanding, will solve the problems themselves.

In order to do this the operator must give the participants incentives for cooperation and show them the benefits of working together. He must stress these points in the first meeting and continue to do so throughout the job. It is rather simple, for example, to explain to the participants that more money is to be made through a mutual give and take than otherwise, and this must be made evident in all the job meetings.

In handling a job meeting, the operator rapidly learns that he is dealing with all different types of people. The different temperaments and attitudes of the participants demand that he (not they) adjust to these things; it is not as if he had the exclusive choice of who the participants should be—he takes them as they come. As a participant, he ordinarily has no say at all as to whom he is going to be dealing with.

It is true that the chair should be able to control attitudes of outright resistance or unreasonableness, but as a rule the meeting includes all kinds of personalities that have the strengths and weaknesses of human nature. Ordinarily, reasonable participants in the job meeting operation will cooperate to meet common goals. We admit that each participant, including the chairman, is there basically to protect or push items of self-interest, but this self-interest should not be allowed to undermine the goals of this construction operation—it must be constructively directed. The use of good, solid logic and reason should make the job meeting a successful operation. At those times when logic and reason do not prevail, it is necessary that all the participants quickly recognize what is happening. The job meeting, as a helpful tool, is invaluable when used properly; but when the participants in a job meeting process become unmanageable or interested solely in some exclusive, selfish interest, this can be a clear detriment to the job. The chair should then exercise special care to turn or shift away from these nonproductive influences.

After a job meeting procedure has been established for a particular job and a few meetings have been held, it becomes easier for the participants to recognize each other as personalities. Normally, they adjust to each other. But not all businessmen are cooperative or even reasonable all of the time. The chair must be fair when dealing with such parties. Attitudes of mutual respect will allow the various participants to see clearly how productive the job meeting process can be if it is managed properly.

18.7 DEVELOPING METHODS

What does this particular operation demand from the construction operator? Does it require extensive experience? The answer to the former question is that it demands good, solid, imaginative construction thinking, as well as quick or extemporaneous thinking because of the need for the spontaneous nature of its discussions. There is little likelihood that a question or its response made in a job meeting will have ever been given the same forethought and attention to detail as, for example, that of a composed letter; yet the job meeting is as excellent a medium for obtaining valid information as any other formal communication process.

As informal as a job meeting may be, the construction operator should not consider it as a haphazard or unplanned operation. Quite to the contrary, he must consider it extremely important to carefully plan for the meeting either as a chairman or as a participant. Through skillful planning, the development of solid practical methods will follow. The construction operator should realize that he must learn how to wisely use the widest range of his personal behavioral skills and traits, because he must utilize them in dealing with the wide variety of the other participants in a job meeting.

As to the latter question, we can only recommend again that the construction operator (particularly the young one) learn to listen carefully and become sensitive to all that he hears. Perhaps at the beginning of his career he will listen more than he will actively participate in job meetings. Even when he grows in experience, the ability to listen carefully to others is clearly the most important attribute to develop in the job meeting operation.

No amount of instruction in this book could even partially deal with the guidelines for proper personal behavior or conduct when operating with others. It is enough to say that these things are best developed through experience. It is fortunate that we can end this text on this point, for it serves to reemphasize a central theme in conducting successful construction operations that we hope the readers will remember and remind themselves of from time to time.

Appendix A

A Sample Estimate

ABC AND ASSOCIATES, INC.

TITLE	*MIDTOWN OFFICE COMPLEX*	SHEET NO. 1
OWNER	*C.D.E DEVELOPMENT CORP*	JOB NO. *8103*
LOCATION	*COLUMBUS, OHIO*	DATE *1/16/81*
ARCHITECT	*FGH Associates A/E*	ESTIMATOR *R.P.M*

PG	GENERAL CONDITIONS	QUANTITY	UNIT	LABOR	UNIT	MAT'L	TOTAL
	PERSONNEL						
GC1	Superintendent			30000		—	30000
	Assistant Superintendents			7300		—	7300
	Layout Engineers			2800		—	2800
	Engineers' Helpers			1000		—	1000
	Timekeepers			8000		—	8000
	Clerks						
	Stenographers						
	Tool Room Men						
	Waterboys & Supplies			2800		150	2950
SGC1B	Traffic Men			—		150	150
SGC12	Watchmen			12000	Sub?	—	12000
GC14	Fire Patrol			800		200	1000
	FIELD EXPENSES						
GC	Temporary Field Offices			1600		2300	3900
GC	" Labor Shanties			500		500	1000
GC	" Toilets			100	R.	1250	1350
	Heat & Light Temporary Buildings			100	S	400	500
	Storage Yard Rental			—	R	2700	2700
	Parking			—		1200	1200
	Office Supplies			—		100	100
	Office Equipment & Furniture			—		150	150
	Temporary Telephones			—		3800	3800
SGC6	Signs	1 pc.		100		650	750
SGC5	Progress Photographs			—		1200	1200
	Additional Plans & Specifications			—		500	500
	Mileage			—		—	—
	Subsistance			—		—	—
				—		—	—
	Raise In Wages			—		—	—
	Payroll Insurance			—		See S.S.	—
	Pensions			—		See S.S.	—
				—		—	—
				67100		15250	82350

TITLE						SHEET NO. 2			
OWNER						JOB NO. *8103*			
LOCATION						DATE			
ARCHITECT						ESTIMATOR			

PG.	GENERAL CONDITIONS	QUANTITY	UNIT	LABOR	UNIT	MAT'L	TOTAL
	FIELD EXPENSES [Cont.]						
	Travel Time for Trades						
GC.	Building Permits & Fees					See Summary	
GC.	Tests					See Concrete	
SGC.3	Builders Risk Insurance					Owner	
	JOB REQUIREMENTS						
SC18	Temporary Light & Power Hook-up			500		See Elect S.B.	500
SC18	Electric Current					12000	12000
SC17	Temporary Water Hook-up & Wells			200			200
SC17	Payment for Water					700	700
GC	Temporary Stairs & Ladders			300		200	500
GC.	" Runways			300		200	500
GC	" Roads			1300		200	1500
GC.	Maintain Permanent Roads			1300		1000	2300
GCB	Cutting and Patching			850		200	1050
GC.	Barricades & Warning Lights			200		100	300
	Fences				S.C.	5400	5400
	Sidewalk Bridges & Fans			300		200	500
	Temporary Partitions			4200		700	4900
	" Closures						
	" Roof Protection			1000		500	1500
	Miscellaneous Safety Work			300		100	400
GCn	Fire Protection			300		300	600
GCn	Fire Extinguishers					300	300
GC21	Glass Cleaning					See Glaz Sub	
GC19	Glass Breakage					1800	1800
GC21	Aluminum Cleaning					See Met Sub	
GC19	" Protection					500	500
				11050		24400	35450

PG.	GENERAL CONDITIONS	QUANTITY	UNIT	LABOR	UNIT	MAT'L	TOTAL
	JOB REQUIREMENTS (cont.)						
GC I	Weekly Cleanup			1000		200	1200
	Final Cleanup	100000 □	.03	3000	.005	500	3500
	Rubbish Chutes			75		75	150
	Stakes & Batterboards			300		150	450
	Pumping after Excavation			1200		1200	2400
	Trucking from Yard and Freight			800		3200	4000
	Jobsite Trucking			800		350	1150
	Punch List Work			1000			1000
	Spot Overtime			7500			7500
	EQUIPMENT SET-UP						
	Hoisting Equipment for Concrete			20000		37000	57000
	" " Masonry			5000		13500	18500
	Misc. Equipment for Concrete			3000		8300	11300
	" " Masonry			1750		1700	3450
	Small Tools & Surveying Instr.					1500	1500
	Tarpaulins - General Use					300	300
	Elevator			6300			6300
	Hoist for Subs			2700			2700
	Air Compressor					3600	3600
	WINTER REQUIREMENTS						
GC. SC.	Cold Weather Protection-Concrete		2 M.	7000		3500	1
	" " " Masonry		3 M	6000		6000	12000
SC.6	Space Heat	100000 □	.033	3300	.15	15000	18300
	Snow Removal			500			500
	Frost Removal					900	900
	SPECIAL REQUIREMENTS						
	Fringes & Insurance	87175 ᵒᵒ			.34	29639	29639
				71225		126614	197839
				149315		166264	315639

TITLE								SHEET NO. **4**	
OWNER								JOB NO. *8103*	
LOCATION								DATE	
ARCHITECT								ESTIMATOR	

Formwork I	QUANTITY		UNIT	LABOR	UNIT	MAT'L	TOTAL
Spread Footing Forms	3683 ⌀		1¹²	4125	.18	663	4787
Wall Footing Forms	2380 ⌀		1¹⁹	2832	.20	476	3308
Grade Beam Forms	3010 ⌀		1³²	3973	.20	602	4575
Pier Forms	805 ⌀		2²⁴	1803	.33	266	2069
Wall Forms 0-8	38500 ⌀		1⁸⁶	71610	.10	3850	75460
0-16	12380 ⌀		.97	12008	.14	1733	13741
0-24	1900 ⌀		1³⁶	2584	.22	418	3002
Corners	7050 L.F.		.75	5287	.93	6557	11844
Scaff	14280 ⌀		.10	1428	.10	1428	2856
Curved Wall Form 0-8	3016 ⌀		3⁰⁰	9048	1¹²	3377	12425
0-16	2003 ⌀		.78	1563	1⁰²	2043	3606
Corners	1112 L.F.		.75	834	.93	1033	1867
Scaff	2003 ⌀		.10	200	.10	200	400
Supported Slab Forms (Flat)	39814 ⌀		.93	37027	.22	8759	45786
" " " (Dome)	62816 ⌀		.98	61559	.28	17588	79147
" " " (Small)	1815 ⌀		2¹⁶	3920	.17	308	4228
" " " (Make up)	102630 ⌀		.10	10260	.05	5131	15394
Beam Forms (Interior)	9312 ⌀		1⁴⁶	13595	.27	2514	16109
" " (Spandrel)	4315 ⌀		1⁵²	6558	.28	1208	7766
Column Forms □	2316 ⌀		2⁸⁶	6623	.28	648	7271
" " ○	318 L.F.		6⁰⁰	1908	11⁰⁰	3498	5406
Stair Forms (Reg)	1015 L.F.		3⁸⁵	3917	1⁰⁰	1015	4932
" " (Special)	1618 ⌀		4⁵⁵	7378	1²²	1973	9351
Curb Forms	363 L.F.		3⁰⁰	1089	.15	54	1143
Base Forms	1411 ⌀		1⁰²	1440	.10	141	1581
Constr. Jts (Ex)	2380 L.F.		.73	1737	.12	286	2023
" " (Cont)	821 L.F.		.60	492	.10	82	574
" " (Water Stop)	614 L.F.		1¹²	687	3⁰⁰	1842	2529
Insurance, Pensions, etc.							

TITLE						**SHEET NO.** 5		
OWNER						**JOB NO.** 8103		
LOCATION						**DATE**		
ARCHITECT						**ESTIMATOR**		

	QUANTITY	UNIT	LABOR	UNIT	MAT'L	TOTAL
Formwork II						
Keyway.	4815 L.F.	.20	963	.10	481	1444
Cant Strip.	3018 L.F.	.22	663	.16	483	1146
Screeds & Bulkhds (Slab on Gr.)	14803 ⊡	.075	1110	.18	2664	3774
" " " (Supp. Slab)	102830 ⊡	.09	9254	.18	18509	27763
" " " (Small)	1814 ⊡	.095	172	.18	326	498
" " " (Roof Fill)	9316 ⊡	.075	680	.10	931	1611
Exterior						
Retaining Wall forms 0-8	2117 ⊡	1.25	2667	.10	211	2878
0-16	286 ⊡	.74	211	.15	43	254
Corners	483 L.F.	.75	362	.93	449	811
Grade Beam. Short Wall forms	933 ⊡	1.20	1120	.22	205	1325
Curb form. (Reg)	418 L.F.	2.38	994	.18	75	1069
" " (Curb & Gutter)	803 L.F.	2.00	1606	.26	208	1814
Screeds & Bulkhds (Sidewalk)	1001 ⊡	.095	95	.18	180	275
" " " (Pavement.)	1090 ⊡	.10	109	.18	196	305
Haul Lumber	184137 ⊡	.05	9206			9206
Handle & Clean Lumber	184137 ⊡	.03	5524			5524
Oil forms.	156326 ⊡	.02	3126	.025	3408	7034
Form Accessories (Wire & Nails)	184137 ⊡			.054	9942	9942
(Wall Ties & Acc)	44616 ⊡			.32	14277	14277
Insurance, Pensions, etc.	313347 ∞			.42	131606	131606
			313347		252387	565734

TITLE				SHEET NO. 6		
OWNER				JOB NO. 8103		
LOCATION				DATE		
ARCHITECT				ESTIMATOR		

CONCRETE I	QUANTITY	UNIT	LABOR	UNIT	MAT'L	TOTAL
CONCRETE 6 SAX	3696 C.Y.			44⁵⁰	164383	164383
CONCRETE 5.5 SAX	2286 C.Y			43⁰⁰	98298	98298
CONCRETE 5 SAX.	431 C.Y			41⁵⁰	17866	17866
CONCRETE LEAN.	25 C.Y			36⁰⁰	900	900
CONCRETE FILL.	75 C.Y			52⁰⁰	3900	3900
CONCRETE TES.	6411 C.Y			.30	1923	1923
CONCRETE ADMIX	5223 C.Y			.80	4178	4178
WASTE @ .02	128 C.Y	7⁰⁰	896	42⁰⁰	5376	6272
SPREAD FOOTINGS	913 C.Y	5⁵⁰	5021			5021
WALL FOOTINGS	354 C.Y.	5⁷⁵	2035			2035
GRADE BEAMS	58 C.Y	6⁰⁰	348			348
PIERS	23 C.Y	18⁰⁰	414			414
WALLS (HIGH)	436 C.Y	7⁸⁰	3400			3400
WALLS (OTHER)	336 C.Y	7⁰⁰	2352			2352
WALLS (CURVED)	143 C.Y	7⁵⁰	1073			1073
SLAB ON GRADE	260 C.Y	5⁰⁰	1300			1300
SUPP SLABS (REG)	720 C.Y	8⁰⁰	5760			5760
" " (DOME)	2226 C.Y	6²⁰	13801			13801
" " (SMALL)	28 C.Y	22⁰⁰	616			616
BEAMS	602 C.Y	6²⁰	3732			3732
COLUMNS	128 C.Y	18⁰⁰	2304			2304
STAIRS	37 C.Y	30⁰⁰	1110			1110
STAIR FILL	763 Φ	2¹⁰	1602			1602
CURBS	18 C.Y	32⁰⁰	576			576
CURBS						
BASES.	14 C.Y	30⁰⁰	420			42
ROOF FILL.	75 C.Y	5⁵⁰	412			412
SET UPS S.O.G.	14803 Φ	.04	592		148	740
" " S.S REG	39814 Φ	.05	1990		797	2786
Insurance, Pensions, etc.						

ABC AND ASSOCIATES, INC.

TITLE							SHEET NO. 7		
OWNER							JOB NO. 8103		
LOCATION							DATE		
ARCHITECT							ESTIMATOR		

CONCRETE II	QUANTITY	UNIT	LABOR	UNIT	MAT'L	TOTAL
Set Up (Dome)	62816 #	.06	3768	.02	1256	5024
" " (small)	11 pcs	6⁰⁰	737	1¹⁰	11	748
EXTERIOR						
Walls (retaining) (incl ftgs)	138 cy	7³⁰	1007			1007
Grade Beams & Short Wall	12 cy.	5⁷⁵	69			69
Sidewalk (reg)	15 cy	12⁰⁰	180			180
3-18 Sidewalk (A. Surf)	8 cy	16⁰⁰	128	↑	↑	160
Curbs	8 cy	40⁰⁰	320	4⁰⁰	32	320
Curb & Gutter.	19 cy	14⁰⁰	266			266
Pavements	43 cy	18⁰⁰	774			774
Insurance. Pensions, etc.	56803⁰⁰			-.38	21585	21585
			56803		320648	377451

378

TITLE					SHEET NO.	8			
OWNER					JOB NO.	8103			
LOCATION					DATE				
ARCHITECT					ESTIMATOR				

Cement Finish	QUANTITY		UNIT	LABOR	UNIT	MAT'L	TOTAL
Monolithic Finish	109247	☐	.23	25127		—	25127
Float Finish	10816	☐	.16	1730		—	1730
Sidewalk & Pump Finish	2091	☐	.42	878		—	878
Cure Concrete Slabs	122144	☐	.035	4275	.02	2442	6717
" " Walls	19316	☐	.04	772			772
Finish Curbs	363	LF	1.75	635			635
" " (ext)	418	LF	1.40	582			582
Finish Curb & Gutter	803	LF	1.44	1156			1156
Finish Stairs	3781	☐	.90	3403			3403
Finish Stair Pans (T&r)	1051	LF	1.05	1104			1104
Finish Bases	1412	☐	1.05	1482			1482
Finish Roof Fill	9316	☐	.18	1677			1677
				—			—
Set Column Base Plates	36	pcs	11.00	396		—	396
Grout Bases	7	pcs	73.00	511		920	1431
Grout Misc Iron	73	pcs	12.00	876			876
Rub Concrete	17419	☐	.04	697			697
Rub & Bag Conc Surfaces	2318	☐	.05	116		116	232
Grinding	1000	☐	.10	100		100	200
Patch Wall Tks.	44616	☐	.02	892		—	892
Scaffolds	144616	☐	.001	145		145	290
				—			—
Aggregate Surface Finish	324		1.00	324	1.50	486	810
Insurance, Pensions, etc.	46878	00			.41	23428	23428
				46878		27637	74515

TITLE					SHEET NO. 9	
OWNER					JOB NO. 8103	
LOCATION					DATE	
ARCHITECT					ESTIMATOR	

MASONRY I	QUANTITY	UNIT	LABOR	UNIT	MAT'L	TOTAL
EXTERIOR FACE BRICK (MAT)	203 M	—		156⁰⁰	31668	31668
LABOR	203 M	302⁰⁰	61306	—		61306
MORTAR	203 M	—		32⁰⁰	6496	6496
INTERIOR FACE BRICK (MAT)	13 M		-	156⁰⁰	2028	2028
LABOR	13 M	346⁰⁰	4498	—		4498
MORTAR	13 M	—		32⁰⁰	416	416
BACKUP 12"	3218 pcs	.88	2833	.80	2574	5407
8"	26344 pcs	.79	20811	.73	19231	40042
6"	2222 pcs	.68	1510	.70	1555	3065
4"	4165 pcs	.65	2705	.65	2705	5410
Spec						
MORTAR	23363 C.F.	—		.32	7476	7476
PARGING	34623 Φ	.10	3462	.10	3462	6924
CAVITY WALL (INSUL)	7828 Φ	.20	1565	.42	3287	4852
REINFORCING	34623 Φ	.10	3462	.20	6924	10386
INTERIOR BLOCK PARTITIONS 12"	1116 pcs	1¹²	1249	.80	892	2141
8"	16211 pcs	1⁰⁰	16211	.73	11833	28044
6"	6211 pcs	.85	5279	.70	4347	9626
4"	8994 pcs	.82	7375	.65	5846	13221
Spec	713 pcs	.90	641	1¹⁰	784	1425
MORTAR	18080 C.F.		—	.32	5779	5779
STRUCT FACING TILE 8" II	1120 pcs	.80	892	2²²	2486	3378
8" I	384 pcs	.80	307	168	645	952
4" I	1814 pcs	.80	1451	165	2993	4444
T I	200 pcs	1⁰⁰	200	3⁵⁰	700	900
T II	123 pcs	1⁰⁰	123	4¹²	504	627
T III	125 pcs	1⁰⁰	125	5⁶⁰	700	825
T IV	73 pcs	2⁰⁰	146	7⁰⁰	511	657
MORTAR	1760 CF	—		.80	1408	1408
LINTELS 12 T 5'-0	239 LF	2⁰⁰	478	4⁰⁰	948	1426
8" "	417 LF	1⁰⁰	417	3⁰⁰	1251	1668
6" "	129 LF	1⁰⁰	129	2⁵⁰	312	441
REINFORCING (INT)	41837 Φ	.10	4184	.20	8367	8785
Insurance, Pensions, etc.						

TITLE			SHEET NO. 10
OWNER			JOB NO. 8103
LOCATION			DATE
ARCHITECT			ESTIMATOR

Masonry II	QUANTITY	UNIT	LABOR	UNIT	MAT'L	TOTAL
Cut Stone Set	413 pcs	28 00	13244			13244
Mot.	38			→	9316	9316
Mortar	2914 C.F			.90	2623	2623
Sample	1 pc	175 00	175			175
Clean Brick	30080 ☐	.01	300	5.05	1504	1804
Clean Block	54283 ☐	.01	543	5.02	1086	1629
Clean Tile Sft	3000 ☐	.05	150		300	450
Clean Stone	1419 ☐			5.20	283	283
Scaffolds (Exterior)	33200 ☐	.15	4980	.20	6640	11620
Scaffolds - (Interior)	76460 ☐	.18	13762	.20	15292	29054
Masonry Access. M.S.	2 Mon	450 00			900	900
D.B	1 pcs	P.			416	416
M.M.	2 Mon	1100 00			2200	2200
Insurance, Pensions, etc.	170747 00			.48	18543	18543
			170747		257231	427978

TITLE						
OWNER						
LOCATION						
ARCHITECT						

SHEET NO. 11
JOB NO. 8103
DATE
ESTIMATOR

METALS	QUANTITY	UNIT	LABOR	UNIT	MAT'L	TOTAL
Re Steel Materials	486 T				See Sub Sht.	
Set Resteel	486 T			5 256°°	124416	124416
(Light)	66 T			5 100°°	6600	6600
Access.	486 T			5 4°°	1900	1900
C. Welds	160 pcs	25°°	4000		See Sub Sht	4000
Wire Mesh 6x6	16283	.18	2930	.25	4071	7001
" " 6x6	69005	.25	17276	.28	19346	36622
" " 6'00	1199	.30	360	.22	264	624
Handle & Set Misc Iron (Loose)	58 T	170°°	9860		See Sub Sht	9860
Set Column Anchor Bolts	36 sets	30°°	1080		"	1080
Set Misc Bolts	383 pcs	3°°	1149	3°°	1149	2298
Set Machine Anchor Bolts	7 sets		406	12°°	84	890
Set Shelf LS	712 LF	5°°	3560			3560
Set Facia (P.C.) Anchors	739 LF	4³°	3177			3177
Set Dovetail Slot.	14200 LF	.15	2130	.40	5680	7810
Set Pre Cast Anchors	1100 pcs	2°°	2750			2750
Templates	93 pcs	10°°	930			930
Insurance, Pensions, etc.	49603°°			.40	19841	19841
			49603		183351	248954

382

TITLE		SHEET NO. 12
OWNER		JOB NO. 8103
LOCATION		DATE
ARCHITECT		ESTIMATOR

Misc. Carpentry & Specialties.	QUANTITY	UNIT	LABOR	UNIT	MAT'L	TOTAL
Set H.M. Frames	339 pcs	38 00	12882		See Sub Sht	12882
Set H.M. Doors	128 pcs	70 00	8960		See Sub Sht	8960
Set Misc. Rough Nailer ⎤	1428 L.F	.58	828			828
Sleep. ⎬ See F.C for Balance	938 L.F	.72	675			675
Furring ⎦	932 ⊕	.72	670			670
Set Toilet Assess. P.H	72 pcs	18 00	1296			1296
S.N.D	18 pcs	25 00	450			450
T.D.	36 pcs	25 00	900		8285	9185
M's	72 pcs	22 00	1694			1694
M.C	18 pcs	28 00	500			500
Set Folding Partitions	3 pcs	126 00	378		See Sub Sht	378
" " "	3 pcs	190 00	570		See Sub Sht	570
Lumber Materials 2×4 ⎤	1010 L.F			.48	484	484
2×6 ⎟	294 L.F			.57	168	168
2×8 ⎬ See Comp.	1093 L.F			.66	720	720
2× Sp. ⎟	899 L.F			.59	530	530
4×4 Tr ⎟	400 L.F	.25	100	.40	160	260
Asses. ⎦	146 L.F	.50	73	.40	58	131
Set Flagpole	1 pc	150 #	150		1100	1250
Base	1 pc	150 00	150		385	535
Set Incinerator	1 pc	675 00	675		7350	8025
Insurance, Pensions, etc.	30951 00			.42	12999	12999
			30951		32239	63190 ✓ ×10

383

ABC AND ASSOCIATES, INC.

TITLE						SHEET NO. 13			
OWNER						JOB NO. 8103			
LOCATION						DATE			
ARCHITECT						ESTIMATOR			

Demolition	QUANTITY	UNIT	LABOR	UNIT	MAT'L	TOTAL
Remove Existing Brick Struct	5346 C.F	.60	3208			3208
Remove Rubbish	5346 C.F	.30	1603			1603
Haul Rubbish	10 Lds		—	55 00	550	550
Demo Permit	1 pc				700	700
Disconnects	3 pcs				1900	1900
					(Use Sub-Bid)	
Insurance, Pensions, etc.	4811 00			.39	1876	1876
			4811		5026	9837

ABC AND ASSOCIATES, INC.

TITLE		SHEET NO. 14
OWNER		JOB NO.
LOCATION		DATE
ARCHITECT		ESTIMATOR

SUBCONTRACTS I	QUANTITY	UNIT	LABOR	UNIT	MAT'L	TOTAL
DEMOLITION					—	6000
EXCAVATION, BACKFILL, GRADING						527000
LANDSCAPING						48300
ASPHALT PAVING						41000
CEMENT FINISH						SEE EST.
RE STEEL (MAT)						198000
RE STEEL (LABOR)						SEE EST.
MASONRY						SEE EST.
CUT STONE (MAT)						SEE EST.
STRUCTURAL PRE CAST						19365
PRE CAST FACINGS						189200
GRANITE & MARBLE F&I					F&I	59300
STRUCTURAL STEEL						134000
MISC IRON						
METAL SIDING						
METAL DECK (STR)						
METAL DECK (ROOF)						9600
STEEL STAIRS & H.R.						43850
LONG SPAN STEEL JTS.						SEE S.S.
ROUGH CARPENTRY						SEE F.C.
FINISH CARPENTRY						498300
MILLWORK						
HOLLOW METAL DOORS & FRAMES.						53000
HARDWARE					ALLOW	100000
WATERPROOFING & DAMPROOFING						21300
ROOFING & SHEET METAL						53200
SPECIAL PROTECTIONS						8000
METAL SASH & DOOR						37500
WINDOW WALL						437600
METAL DOORS (S.S.)						15350
SPECIAL (REVOLVING)						30200
GLASS & GLAZING					SEE W.W ALSO	73200
ACCOUSTIC CEILINGS						119700
Insurance, Pensions, etc.						

385

TITLE		SHEET NO. 15
OWNER		JOB NO.
LOCATION		DATE
ARCHITECT		ESTIMATOR

SUBCONTRACTS II	QUANTITY	UNIT	LABOR	UNIT	MAT'L	TOTAL
Special Ceilings						4300
Lath & Plaster						402000
Drywall						1
Ceramic Tile						69000
Quarry Tile						1
Resilient Flooring						93700
Special Floorings						26700
Terrazzo Floors						16370
Carpeting						98300
Painting & Finishing						141300
Special Wall Coverings						1
Vinyl Wall Coverings						26200
Metal Toilet Partitions						22000
Folding Doors						10000
Folding Partitions						73600
Site Improvements & Equipment						7300
Kitchen Equipment						244000
Special Equipment Mov. Part.						169300
Comp. Rm. Floor.						42700
Special Sheeting & Shoring						119000
Elevators						278000
Escalators						
Mechanical						2163000
Electrical						1814000
Insurance, Pensions, etc.						
						8543735

			LABOR	TOTAL	

TITLE _____ **SHEET NO.** _16_ **SUMMARY**
OWNER _____ **JOB NO.** _8103_
LOCATION _____ **DATE** _____
ARCHITECT _____ **ESTIMATOR** _____

SHT.		SIGNIFICANT QUANTITY		LABOR	TOTAL	
	ABC AND ASSOCIATES WORK					
4	Form Work			313347	565734	✓
5	Concrete Work			56803	371451	✓
6	Cement Work			46878	14515	✓
7	Masonry			170747	471918	✓
8	Resteel & Metals			49603	248954	✓
	Specialties			30951	63190	✓
	Demolition & Connection			SEE SUB	SEE SUB	SHT
	Carpentry					
	ABC AND ASSOCIATES WORK		Sub-Total ✓	668329	✓ 1757822	✓
1-3	General Conditions			149375	315639	✓
			Sub-Total ✓	817704	2073461	✓
	Subcontracts				8543735	
	Sales Tax	890000.06			35600	
	Allowances				SEE SUB SHEET	INCL.
	Bldg. Permit				3400	
	Mech., Elect., Elev.					
	Sub Contracts to be Assumed					
			Sub-Total ✓		10656196	
			Subcontractor Bonds			
			Correction #1		DED 29700	
			Total		10626496	
			Correction #2		ADD 37000	
			Total		10663496	
			Correction			
	Due Date 2/3/81 2ºº PM EST.		Total		10663496	
	Due CLEVELAND, OHIO	→	Bond ✳		72000	
	Bid Security - Check	→	OH & P ✳		530000	
	Bid Security - Bond YES !! .65/M		Total		11265496	
	Liquidated Damages YES @ 500ºº C.O. 730 D.		BID		11264900	
	Cube	Cu. Ft.				
	Area 112 M F. 38 - $102ºº ↔	Sq. Ft.				
	Addendums 5					

CORRECTION #3 (LATE)   + 20000
        BID   11,284,900

TITLE						SHEET NO. *Correction*		
OWNER						JOB NO.		
LOCATION						DATE		
ARCHITECT						ESTIMATOR		

1st Corrections 145	QUANTITY	UNIT	LABOR	UNIT	MAT'L		TOTAL	
Excavation				*in est*	527000	*New*	518000	
Mech.					2163000		2650000	
Elevators					278000		275000	
Accous.					119700		115000	
					3087700		3058000	
					3058000			
				Deduct 29700				

~~~

2nd Correction 130								
Excavation				*in est*	518000	*New*	508000	
Window Wall					437000		430000	
Electr.				*	1814000		1868000	
					2769000		2806000	
							2769000	
						Add	37000	

Correction #3								
ABC - Supervision						*Add*	20000	

| Insurance, Pensions, etc. | | | | | | | | |

Appendix **B**

A Sample Subcontract
and Subcontract Change Order

To: XYZ Excavation Co. 16 Apple Street Cleveland, Ohio Subcontracor	**A B C Inc.** **General Contractors** **100 Main St.** Columbus , Ohio	Date 6/17/81	No. 8102-2
		Delivery On Site	
Attn: Mr. John Doe		As per Job Superintendent	
	A Subcontract		
Phone: (614) 878-6398		Page 1 of 2	

This is your authority to furnish all labor, materials, equipment, taxes and insurance necessary to do.

THE EXCAVATION WORK

all in accordance with the entire contract documents for the Midtown Office Complex, Cleveland, Ohio; C. D. E. Development Co., Owner; F. G. H. Associates, Architects/Engineers, Inc. their job No. 7816, and, in particular; the Advertisement for Bids dated 12/18/80; The Proposal Form dated 2/24/81; the Agreement dated 4/23/81; the General Conditions dated 6/7/80; the Supplemental Conditions dated 6/17/80; The Special Conditions dated 6/17/80; the Technical Specifications dated 6/17/80; and the Drawings A thru A23 dated 6/17/80; A24 and A25 dated 8/2/80; and the Drawings S1 thru S12 dated 6/17/80; Drawings M thru M16 dated 6/17/80; Drawings E thru E8 dated 6/17/80; Site Drawings 1 and 2 both dated 6/17/80; and Addendum #1 dated 12/24/80; Addendum #2 dated 12/28/80 and Addendum #3 dated 1/20/81 and especially as called for in Section 2 of the technical specifications.

for the Lump Sum Price of $508,000.00

1). State Sales and Use Tax included.
2). All shop drawings will be submitted for approval in one (1) sepia and two (2) prints.
3). All brochures will be submitted in nine (9) copies.
4). Samples shall be submitted in triplicate (3) properly marked and identified.
5). The subcontractor billings and invoices accompanied by its partial waiver of lien shall be received in the office of ABC, Inc. no later than the 20th. of the month.
6). The subcontractor shall not, repeat shall not, commence work on the projects site without first having furnished proofs to ABC, Inc. of coverage as required in the Contract Documents. These proofs must be in the physical possession of ABC. Inc.

Accepted by:

_____ _____   _____   _____
Contractor   Date   Date

To: XYZ Excavation Co. 16 Apple Street Cleveland, Ohio Subcontractor Attn:	ABC Inc. General Contractors 100 Main St. Columbus, Ohio	Date 6/17/81	No. 8102-2
Phone:	A Subcontract	Delivery As per Page __2__ of __2__	

7). This is an Equal Opportunity Employment Project and the subcontractor shall be and remain in compliance with the terms of all applicable legislation and regulations on equal employment.

8). ABC, Inc. will furnish the proper form of payroll reporting under Davis Bacon requirements. The subcontractor shall report their payrolls as required by the provisions and regulations on its Act.

Exclusions: The subcontractor shall not be required to install the required sheeting and shoring. This work shall be done by another contractor. The subcontractor agrees that it will cooperate with the sheeting and shoring contractor in coordinating their work.

Inclusions: The subcontractor shall furnish and install all crushed stone underfill under all slabs on grade specified in Division - 3 - 3. 13 of the technical specifications.

Accepted by:

_____   _____   _____   _____
Contractor   Date   Date

To: XYZ Excavation Co. 16 Apple Street Cleveland, Ohio	**ABC Inc.** **General Contractors** **100 Main St.** **Columbus, Ohio**	**Date** 8/17/81	**No.** 8103-1
Attn: Mr. John Doe		**Delivery** On Site	
Phone: (614) 878-6398	**A Subcontract**	**As per** Job Superintendent	
		Page _____ of _____	

CHANGE ORDER

For work done in accordance with Bulletin #3 dated July 16, 1981.

Add to Subcontract Price $10,000.00

Total Subcontract Price
Including Change $518,000.00

Accepted by:

_____   _____   _____   _____

Contractor   **Date**   **Date**

Index